Jan Mayen Island in Scientific Focus

NATO Science Series

A Series presenting the results of scientific meetings supported under the NATO Science Programme.

The Series is published by IOS Press, Amsterdam, and Kluwer Academic Publishers in conjunction with the NATO Scientific Affairs Division

Sub-Series

I. Life and Behavioural Sciences	IOS Press
II. Mathematics, Physics and Chemistry	Kluwer Academic Publishers
III. Computer and Systems Science	IOS Press
IV. Earth and Environmental Sciences	Kluwer Academic Publishers
V. Science and Technology Policy	IOS Press

The NATO Science Series continues the series of books published formerly as the NATO ASI Series.

The NATO Science Programme offers support for collaboration in civil science between scientists of countries of the Euro-Atlantic Partnership Council. The types of scientific meeting generally supported are "Advanced Study Institutes" and "Advanced Research Workshops", although other types of meeting are supported from time to time. The NATO Science Series collects together the results of these meetings. The meetings are co-organized bij scientists from NATO countries and scientists from NATO's Partner countries – countries of the CIS and Central and Eastern Europe.

Advanced Study Institutes are high-level tutorial courses offering in-depth study of latest advances in a field.
Advanced Research Workshops are expert meetings aimed at critical assessment of a field, and identification of directions for future action.

As a consequence of the restructuring of the NATO Science Programme in 1999, the NATO Science Series has been re-organised and there are currently five sub-series as noted above. Please consult the following web sites for information on previous volumes published in the Series, as well as details of earlier sub-series.

http://www.nato.int/science
http://www.wkap.nl
http://www.iospress.nl
http://www.wtv-books.de/nato-pco.htm

Series IV: Earth and Environmental Sciences – Vol. 45

Jan Mayen Island in Scientific Focus

edited by

Stig Skreslet
Faculty of Fisheries and Natural Sciences,
Bodø Regional University,
Norway

Kluwer Academic Publishers

Dordrecht / Boston / London

Published in cooperation with NATO Scientific Affairs Division

Proceedings of the NATO Advanced Research Workshop on
Jan Mayen Island in Scientific Focus
Oslo, Norway
11–15 Novenber 2003

A C.I.P. Catalogue record for this book is available from the Library of Congress.

ISBN 1-4020-2956-X (PB)
ISBN 1-4020-2955-1 (HB)
ISBN 1-4020-2957-8 (e-book)

Published by Kluwer Academic Publishers,
P.O. Box 17, 3300 AA Dordrecht, The Netherlands.

Sold and distributed in North, Central and South America
by Kluwer Academic Publishers,
101 Philip Drive, Norwell, MA 02061, U.S.A.

In all other countries, sold and distributed
by Kluwer Academic Publishers,
P.O. Box 322, 3300 AH Dordrecht, The Netherlands.

Printed on acid-free paper

Printed in the Netherlands.

Table of Contents

NATO ARW Participants

Names and Addresses

Susan Barr, Norwegian Directorate for Cultural Heritage, PO Box 8196 Dep, 0034 Oslo, Norway.

Krzysztof Birkenmajer, Institute of Geological Sciences, Polish Academy of Sciences, Senacka 1, 31-002 Krakow, Poland.

Vladimir I Byshev, PP Shirshov Institute of Oceanology, Russian Academy of Sciences, Nakhimovski pr. 36, Moscow 117851, Russia.

Geir Wing Gabrielsen, Norwegian Polar Institute, Polarmiljøsenteret, 9296 Tromsø, Norway.

Piotr Glowacki, Polar and Marine Research Department, Institute of Geophysics, Polish Academy of Sciences, ul. Ksiecia Janusza 64, 01-452 Warszawa, Poland.

Bjørn Gulliksen, Norwegian College of Fisheries, University of Tromsø, Breivika, 9037 Tromsø, Norway.

Louwrens Hacquebord, Arctic Centre, Groningen Institute of Archaeology, University of Groningen, PO Box 716, 9700 AS Groningen, The Netherlands.

Jon O Methlie Hagen, Institute of Geography, University of Oslo, PO Box 1042 Blindern, N-0316 Oslo, Norway.

Hans-Jûrgen Hirche, Alfred-Wegener-Institut für Polar- und Meeres-forschung, Colombusstr., 27515 Bremerhaven, Germany.

Keith A Hobson, Prairie and Northern Wildlife Research Centre, Canadian Wildlife Service, 115 Perimeter Road, Saskatoon, SK S7N 0X4 Canada.

Leonid B Klyashtorin, Federal Institute of Fisheries and Oceanography (VNIRO), Russian Academy of Sciences, Krasnoselskaya 17, Moscow 107140, Russia.

Volodymyr G Kyyak, Institute of the Carpathians Ecology, National Academy of Sciences of Ukraine, Kozelnytska Street 4, 79026 L'viv, Ukraine.

Igor A Melnikov, PP Shirshov Institute of Oceanology, Russian Academy of Sciences, Nakhimovski st. 36, Moscow 127324, Russia.

Jan Piechura, Institute of Oceanology, Polish Academy of Sciences, ul. Powstancow, Warszawy 55, 81-712 Sopot, Poland.

Vladimir F Romanov, Arctic and Antarctic Research Institute, Russian State Science Center, Bering st. 38, 199397 St.Petersburg, Russia.

Kirsti Rouvinen-Watt, Department of Plant and Animal Sciences, P.O. Box 550, Truro, Nova Scotia, B2N 5E3 Canada.

Graham B Shimmield, Dunstaffnage Marine Research Laboratory, P.O. Box 3, Oban, Scotland, PA3 4AD, United Kingdom.

Gintautas Stankunavicius, Department of Hydrology and Climatology, Vilnius University, Ciurlionio St. 21, Vilnius 2000, Lithuania.

Stig Skreslet, Faculty of Fisheries and Natural Sciences, Bodø Regional University, Mørkvedbukta, N-8049 Bodø, Norway.

Marko Vana, Institute of Environmental Physics, University of Tartu, Ülikooli Street 18, 50090 Tartu, Estonia.

Anthony W Walne, Sir Alister Hardy Foundation for Ocean Science, 1 Walker Terrace, The Hoe, Plymouth, England PL1 3BN, United Kingdom.

Jan M Weslawski, Institute of Oceanology, Polish Academy of Sciences, ul. Powstancow, Warszawy 55, 81-712 Sopot, Poland.

Foreword

The reader of this book will experience its scientific pluri-disciplinarity which reflects the wide range of expertise that convened for the NATO Advanced Research Workshop (ARW) in Oslo, Norway 11-15 November 2003, addressing a Joint International Scientific Observation Facility on Jan Mayen Island. According to NATO requirements the participants consisted of representatives from nations that were once separated by the Iron Curtain during the Cold War era. During that era, scientific disciplines were allowed to evolve in different directions within each block of nations, creating differences in the conception of problems, methodological approach, terms of expression, and ways of communication. It has been a challenge, but also a privilege to experience such differences, both during the NATO ARW itself, and in the editing of the contributions.

The two Co-Directors were faced with making scientists from different disciplines, research traditions, and nations understand each other. It was recognised that each of the participants were close to lay-men outside their own field of science. This was over-come by stimulating the use of simple words or scientific terms that are widely used in common English, rather than use specialised scientific terms and slang typical of each particular discipline. This has also been the ambition in the editing of this book, to make the information availble to not only members of the scientific community, but also to people working in public administration and politics. Hopefully, I have succeeded to some extent, but recognise that some parts may not be easily understood by other than the informed reader.

Acronyms are widely used in many scientific communities. To include other readers as well, the acronyms have been explained in the text, and they have also been inserted and defined in the index table at the end of the book, with reference to pages where they occur. I sincerely hope that the number of words that have been indexed is sufficient to make it easy to navigate among the wealth of information that is contained in this book.

Preface

This book is the product of a long process, starting during my flight back to Norway after the IGBP/GLOBEC Open Science Meeting in Paris 17-20 March 1998. I had observed plans presented by the World Climate Research Programme to implement a study of Climate Variability and Predictability (CLIVAR). My attention was drawn to the emphasis on Atlantic Thermo-Haline Circulation and the role of Greenland Sea Deep-Water formation. From my days as a very young research student spending two summers on Jan Mayen Island, facing the Greenland Sea when fog permitted, I realized that this particular CLIVAR task would not be an easy one. The waters are inhospitable even in summer and clearly worse during winter. The operation of ships would be costly and therefore infrequent and maybe very inadequate in relation to scientific wishes. It came to my mind that Jan Mayen itself was a research platform solidly moored on the Mid-Atlantic Ridge, and with permanent access to waters in question. It was tempting to explore the possibility of establishing a research station that could face science ranging from unique local studies to Global problems.

In the year that followed, I discussed the idea with international colleagues who I met on different occasions. I understood that many became intrigued by the possibility of running a research station on Jan Mayen Island. As member of a group advising the County Governor of Nordland about environmental management, monitoring and research on Jan Mayen I understood that a research station on the island should be multidisciplinary and address all fields of science, including anthropology and history. The infrastructure built to serve the Loran-C navigational system based on the island had the capacity to accommodate more persons than the crew operating it. The question was rather to line up convincing arguments for doing scientific research on the island and in its surroundings.

The NATO Science Programme "Bringing Scientists Together for Progress" offered a feasible framework for an Advanced Research Workshop that could identify potential scientific ambitions. It required that a

NATO ARW would identify new important topics and directions for future research. It was mandatory that it promoted working relationships between scientists from NATO countries and Partner Countries, i.e. previous Soviet republics and satellite nations.

I was very fortunate to have served as field assistant for Krzysztof Birkenmajer, an extremely experienced Polish geologist doing contract work for the Norwegian Polar Institute on Svalbard in 1962. In December 1999, I renewed my contact with him, now a Professor Emeritus in Krakow, Poland. He was delighted and consented to join as my Partner Country Co-Director. We enlisted an Organizing Committee consisting of Dr. Vladimir Byshev at the PP Shirshov Institute of Oceanology in Moscow, Russia, Professor Jens Meincke at the Insitut für Meereskunde, University of Hamburg, Germany, and Dr. Peter Wiebe at the Woods Hole Oceanographic Institute, USA. Sadly, only weeks before the NATO ARW came to reality the two latter members became victims of "force majeur" and could not enjoy the fruits of their support.

The Organising Committee carefully selected a limited number of scientists, securing a balanced representation of NATO and Partner countries, and a wide range of scientific training and experience. Some participants were seasoned polar field-workers while others were selected for their expertise in a particular field of science relevant to Jan Mayen research development. The final result was also thanks to Dr. Marek Lewandowski, a NATO representative who provided much appreciated assistance in the application process.

Our NATO ARW addressing a Joint International Scientific Observation Facility on Jan Mayen Island took place in Oslo, Norway 11-15 November 2003. Twenty two scientists offered key speeches and participated in group work where one session was devoted to information from six Norwegian public servants. The Co-Directors got time to contribute actively in the scientific work thanks to the reliable and pleasant support from our NATO ARW Secretary, Dr. Eric Mathey. After the meeting, I have been assisted by an editorial committee that helped draw together reports and conclusions. They are Krzysztof Birkenmajer, Louwrens Hacquebord, Keith A Hobson, and Igor A Melnikov.

I thank all involved persons for their contributions before, during and after the NATO ARW in Oslo. It will be to their credit if a research station on Jan Mayen materialises. I would in particular draw attention to my Co-Director "Kris" Birkenmajer, but not only for his transfer of scientific achievements in polar research to our NATO ARW. He introduced me to the Arctic and proved personal qualifications that I have tried to follow, both in science and in life. I am very happy for the opportunity to renew our friendship over a matter of importance and acknowledge that without this man our venture would never have occurred.

Acknowledgements

This book is the proceedings volume from the NATO Advanced Research Workshop on a Joint International Scientific Observation Facility on Jan Mayen Island, organised in Oslo, Norway 11-15 November 2003. The work was funded by the NATO Public Diplomacy Division Science Programme by awarding grant No. EAST.ARW.979238.

The contributions to the book have been subject to peer review by RT Barrett, WK Dallmann, K Eiane, A Foldvik, E Fuglei, S Grønås, H Hop, M Haakstad, KJ Holmén, E Isaksson, TS Larsen, OJ Lønne, SO Lundestad, O Nakken, MG Nettelbladt, PK Reymert, JA Sneli, and D Vongraven.

Chapter 1 NATO Advanced Research Workshop

Research Operations on Jan Mayen Island

Feasibility, Scope and Legitimacy

Stig Skreslet
Faculty of Fisheries and Natural Sciences, Bodø Regional University, Norway
E-mail: stig.skreslet@hibo.no

Abstract: The remoteness of Jan Mayen Island makes it rather inaccessible for scientists and the ruling arctic conditions require logistics that secure conditions for field research. The scope of research interests ranges from Global climate and hemispheric meteorology, seabed tectonics and volcanology, and geophysics of the World Ocean and the proximal Nordic Seas, to ecological relationships on the island itself and in the marine systems that surround it, including anthropogenic influence and the cultural heritage. Closing of the Loran-C navigational station on the island may allow use of existing infrastructure for various scientific purposes. A NATO Advanced Research Workshop was summoned to discuss options and recommend initiatives.

Key words: Greenland Sea, Norwegian Sea, geosciences, marine resources, environmental protection, management, research station

1. INTRODUCTION

Jan Mayen Island has a remote position at about 71°N 8°W, being a volcanic pillar on the northern end of the Mid-Atlantic Ridge. It is surrounded by the deep basins of the Nordic Seas that is composed of the Greenland, Norwegian and Icelandic Seas (Fig. 1).

The discovery of the island is attributed to Jan Jacobszon May, a whaling captain who gave rise to its Dutch name Jan Mayen Eylant that simply means the island of Jan May. In daily terms the island is usually referred to as Jan Mayen.

S. Skreslet (ed.), Jan Mayen Island in Scientific Focus, 1-11

To most Norwegians Jan Mayen is a place positioned in the middle of nowhere, although the daily weather forecasts frequently refers to it as a place where stormy weather is generated, which affects their everyday life.

To the international community of scientists the island has a unique position in relation to their fields of science, irrespective of their speciality and whether their work addresses Global, regional or local problems. However, few have had the opportunity to work there or obtain data from the area, due to its inaccessibility and the lack of access to logistics.

There is a meteorological station on Jan Mayen Island, and a navigation facility run by the Norwegian government. The infrastructure contains integrated living quarters and subsistence facilities that are spacious and hold high standards. However, in modern navigation satellite positioning systems have outmoded the Loran-C navigation system that will probably be phased out in 2005. This raises the opportunity to take over the facility and run it as an international scientific research base.

The present contribution raises some points that legitimise the major efforts required to facilitate the adaptation of logistics and infrastructures into a useful scientific work-bench. They range from entirely global perspectives to the particular problems that can only be solved in a local and unique site on the island itself.

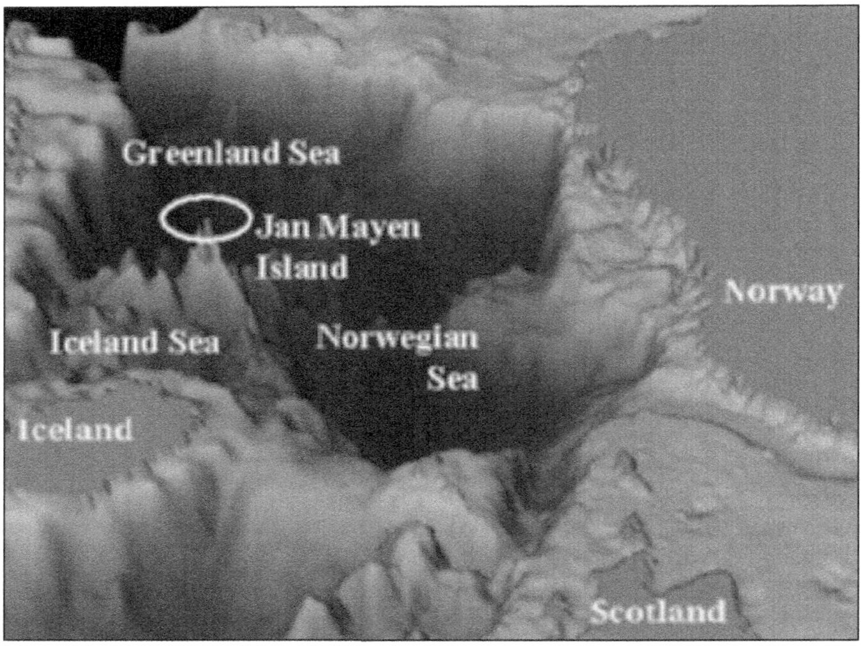

Figure 1. The geographical position of Jan Mayen Island in the Nordic Seas.

2. JAN MAYEN IN SCIENTIFIC PERSPECTIVE

2.1 The Global Scope

The Jan Mayen area is in focus of several international programs. One that addresses Climate Variability and Predictability (CLIVAR 1998) is initiated by WMO, ICSU and UNESCO. A particular field of interest in this program is the World Ocean's Thermo-Haline Circulation (THC). It plays important roles in the distribution of heat by oceanic currents and exchange of water with the atmosphere. The THC is driven by the sinking of surface water in two regions of the North Atlantic. One is the Labrador Sea and the other is the Greenland Sea (Fig. 2). Radiation of heat from the sea surface is the driving force that causes the surface water to increase its gravity.

The Greenland Sea water sinks to an intermediate depth and flows southwards in the Norwegian Sea to the Greenland-Scotland Ridge where it leaves the Norwegian Sea through the Faeroe Bank Channel. It cascades to the bottom of the North Atlantic Ocean and accumulates into a bottom current that flows towards and around the Antarctic continent. Some of the water is brought to the surface by wind-driven up-welling in the Antarctic divergence zone. The remaining deepwater is circulated into the Indian and Pacific Oceans where it is brought to the surface by wind-driven up-welling and tidal mixing. Its gravity decreases due to heating by solar radiation and addition of freshwater.

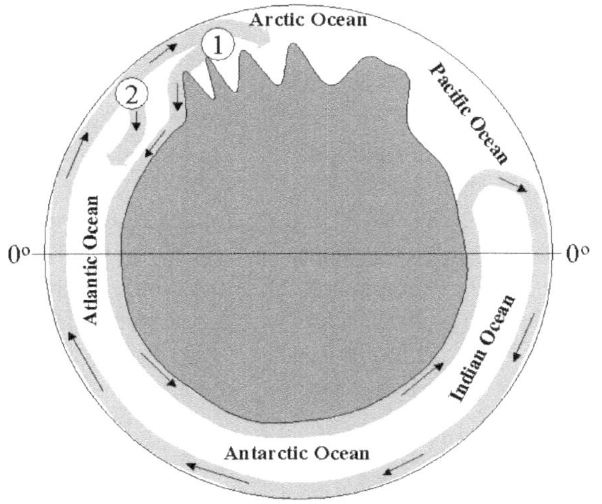

Figure 2. Idealised presentation of the World Ocean's Thermo-Haline Circulation with the two centres of deep-water formation: 1) Greenland Sea. 2) Labrador Sea.

The light surface water flows from the Pacific and Indian Oceans into the southern Atlantic Ocean. It turns northeast-wards while it traverses the Equator and then enters the Gulf of Mexico. The Gulf Stream conveys the surface water northwards along the east coast of USA and feeds it into the wind-driven North Atlantic Drift. Parts of this Atlantic surface water flows into the Norwegian Sea where it is known as the Norwegian Atlantic Current. Its role is to compensate for the seawater that is lost by sinking of Greenland Sea water. However, some is also fed into the Arctic Ocean where it mix with river water and generate the polar surface water that is eventually discharged into the Labrador Sea by the East Greenland Current.

Theoretical scenarios take into consideration effects of Global climate change on the surface salinity of the Arctic Ocean and the Greenland Sea, which poses a major concern. During the last Ice Age there was no inflow of Atlantic water into the Norwegian Sea, possibly because there was no sinking of surface water in the Greenland Sea. Thus, the stability of the water column in the waters to the north of Jan Mayen Island is a central scientific focal point.

Oceanographic fieldwork in the Greenland Sea has mainly been organised as brief research vessel operations in the summer season. Very little work has been made in the winter season when the deepwater formation is probably at its maximum. That is due to navigational problems, safety precautions and general working conditions related to the hazards of research vessel operations in drifting sea ice and turbulent open waters. Jan Mayen is safely moored on the Mid-Atlantic Ridge and its shore and shelf is in direct contact to the Greenland Sea in W-NE sector. To the SW the island is in direct contact with the Norwegian Sea and its Atlantic influence. It is therefore very suited for studies of Global processes by short excursions from its shores or by moorings on any flank of its shelf.

2.2 The Regional Scope

Jan Mayen Island is affected by the circumpolar low-pressure belt that undulates at about N60°, which is characterised by a series of low-pressure centres around the northern hemisphere. There, cold and dry air that sinks from the upper polar troposphere meets warm and humid air from the south (Fig. 3), causing fog and precipitation. Jan Mayen is situated within the marginal zone of the polar airflow system.

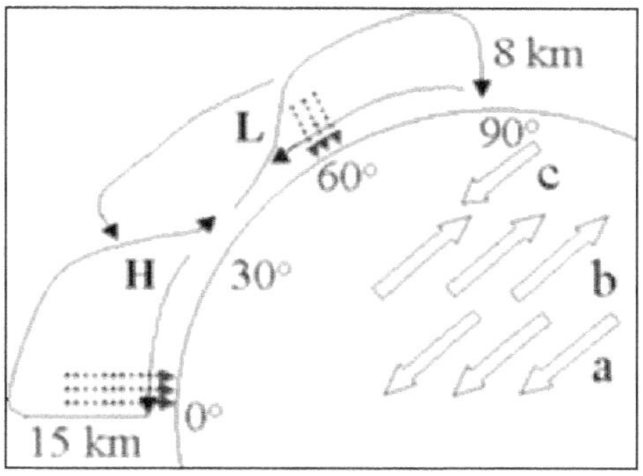

Figure 3. Idealised air circulation in the Earth's troposphere (curved arrows). Open arrows indicate the general near-surface latitudinal flow of air over the North Atlantic Ocean, being deflected by the rotation of the Earth (Coriolis' force): a) North-East trade wind. b) North Atlantic westerlies. c) polar airflow. L: Low pressure trough. H: High pressure ridge. Dashed arrows: Precipitation due to the lifting of humid air.

The geographical centres of atmospheric high and low pressure in the North Atlantic shifts their position with time and the pressure gradient between them is subject to inter-annual fluctuation. This is evident in the pressure difference between meteorological stations in Iceland and Portugal. When the pressure differences are large (NAO positive), the frequency and the strength of the westerly storms increases and humid Atlantic air causes large precipitation on the Norwegian west coast (Fig. 4). When the pressure differences are small, the weather is less violent and the transport of humid air is directed towards southern Europe.

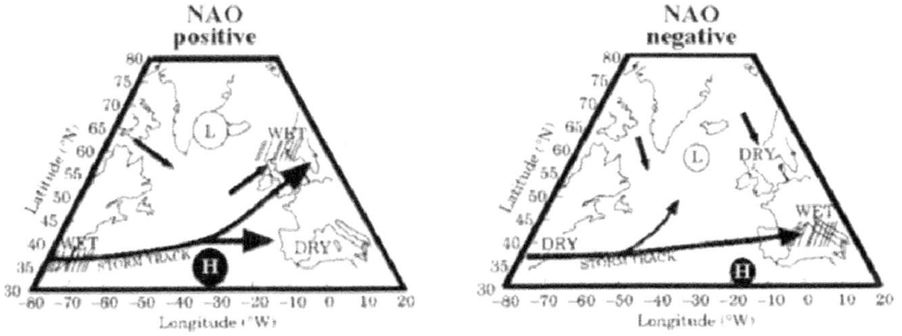

Figure 4. Latitudinal shifts in westerly winds and precipitation caused by changes in the pressure gradient between the Azores high and the Icelandic low pressure regions (North Atlantic Oscillation). NAO positive: Above average. NAO negative: Below average (Greene and Pershing 2000).

The interaction between the sea and the atmosphere is increasingly important in World climatic studies, being fundamental in prognostic modelling of geophysical processes that interfere with safety precautions, economical enterprises, and everyday conditions for human populations.

The inter-annual signal observed in NAO records is also detected in different time-series of scientific data, ranging from precipitation and sea surface temperature to zooplankton and fish populations. Thus, NAO appears to be a simple indicator of geophysical processes that generate complex biological responses of importance to the exploitation of living resources in the North-East Atlantic.

Different benthic invertebrates, fish and marine mammals that live in the Nordic Seas are subject to exploitation, and there is a large potential for harvesting of zooplankton resources. The spatial extension of each population system spans over wide geographical areas, resulting from the large-scale geophysical dimensions of their habitats. Within each population, different life-stage groups occupy particular habitats that are often far apart. Thus, many living resources perform large-scale migrations between different habitats where exploitation occurs, with particular reference to the reproduction and feeding habitats of adults.

Jan Mayen Island, its shelf and the waters around it has a central position within a basin-scale arena that involves geophysical forcing factors and production of resources. Information from the island and its surroundings is of direct importance to the work of people at sea and to human populations in northern Europe. Its resources support national economies and international trade with staple products. The Jan Mayen region therefore calls for scientific activities that may reveal processes governing human well-being.

2.3 The Local Scope

Jan Mayen Island features some particular and even unique conditions that are due to its geoposition and its remoteness. Beerenberg, a 2277m high active volcano with slopes that rises directly from the sea level, is the most stunning feature of the landscape. The slopes continue to fall into the abyss of the Nordic Seas. The volcanic structure is part of the Mid-Atlantic Ridge but features local tectonic processes that are unique to the area. That is also the case with the Beerenberg glaciers that accumulate precipitation drained from tropospheric air masses directed towards it. The glaciers provide archives of climatic information that remains to be studied.

The distances from Jan Mayen to other land-masses challenge the ability of plants and animals to populate the island on land and in lakes. The species diversity of the island's communities is therefore rather low, despite new species may have been established by anthropogenic introduction during ancient whaling and fox trapping enterprises, early scientific expeditions, previous war-fare activities, and the more recent service of meteorological and navigational infrastructure (Figs 5 and 6).

Jan Mayen's nature is very much influenced by the marine foodweb. Cliff-breeding sea-birds visit the island during the summer months to hatch and feed their juveniles. In this period they bring on shore marine biomass that fertilises not only their colonies, but also the landscape below their migration routes. The waters around Jan Mayen is also the reproduction and mating habitat for two seal populations, and is for this reason visited by Norwegian sealers in late winter. Fishing expeditions are also performed to exploit resources of pelagic fish and benthic invertebrates. Tourists show increasing interest for Jan Mayen. Private yachts and cruise ships visit the island and some of the landings are unauthorised. Amateur landings are risky and hazardous activities like snow-boarding from the top of Beerenberg have occurred.

All human activities on Jan Mayen Island may harm the vulnerable biota. The Norwegian authorities have therefore established an environmental action plan to monitor the development (Anon. 2000). It has not been followed up as scheduled because of logistical problems, and lack of national priority. The establishment of a scientific station on the island may be expected to facilitate environmental monitoring and protection.

Figure 5. The meteorological station. Photo: T Høyvik (Original on http://www.jan-mayen.no).

Figure 6. The Jan Mayen Island Loran-C navigational station. Photo: PE Dalen (Original on http://www.jan-mayen.no).

3. LEGAL STATUS

Jan Mayen Island is not a part of the Svalbard archipelago where many nations have signed a treaty that grants them certain rights to explore and exploit natural resources. The island is therefore not under the jurisdiction of "Sysselmannen" who is the Norwegian Governor on Svalbard.

Jan Mayen Island is governed by "Fylkesmannen i Nordland", the County Governor of Nordland who holds office in Bodø, northern Norway. The Governor enforces Norwegian civil authority that is delegated to the local Commander of the island's Loran-C station. As part of the administrative organization the Governor has established an advisory board that discusses effects of human presence, and instalment of adequate measures for protection of the nature and the cultural heritage. It provides the management with background information and gives advice on priorities regarding monitoring and maintenance.

4. LOGISTICAL ACCESS

Each year, several scientific institutions and individual scientists apply to Norwegian authorities for logistical support to study processes or objects on Jan Mayen Island, but with negative results. The area is rather inaccessible to research vessels in winter, due to the formation and drift of sea ice. The large distance from European harbours even makes research efforts in summer rare and brief. Both marine and terrestrial studies require high-tech logistics for field operations, as well as supply systems and support from a home-base in Norway. There is an air-field on the island (Fig. 7) and most of the transport of personnel and cargo by plane and ships occur from Bodø. However, foul weather and low visibility due to fog at Jan Mayen are frequent obstacles that make landings from air and sea random and some times risky.

Figure 7. C-130 Hercules during take-off from the airstrip on Jan Mayen Island. Beerenberg in the background. Photo: G Utkilen (Original on http://www.jan-mayen.no).

Figure 8. Sea ice front (curve) and thickness (increasing with darker tone) 28 February 2004 (Redrawn from: http://www.natice.noaa.gov).

Despite the described difficulties, the general logistics for support of research operations on Jan Mayen Island are already there. Housing, an airfield and landing systems for loading and unloading of ships are established and have been operated successfully for several decades. The problem is that the transport systems are costly to operate and national priorities do not allocate economical means that may support Jan Mayen research, be it national or international.

Several logistical difficulties have to be overcome, if Jan Mayen Island and its waters are to be studied in some detail. The scientific output per invested unit of capital can probably be optimised if dedicated research logistics were installed on the island itself. It would enable land-based monitoring and short sea excursions by hovercraft and helicopter operations when the sea is ice-covered (Fig. 8) and by moderately sized ships when the ice retreats. Automated measuring and sampling may be achieved in remote and dangerous localities on land, and by anchored buoys and remotely operated or self-navigated underwater robots in the sea.

5. SCOPE OF A NATO ARW

In scientific terms Jan Mayen Island is certainly not in the middle of nowhere. It has a unique position in relation to the Global climate regime, and to human ecology that addresses basin-scale physical and biological processes of economical importance in northern Europe. These processes attract the interest of scientists, not least because of the remoteness from continental systems and processes. Jan Mayen represents unique challenges for science that may contribute to unravel many of the large unknowns on Earth and benefit the Global human society.

A NATO ARW organised in Oslo 11-15 November 2003 was initiated to convince both Norwegian authorities, international scientific organisations, and the scientific funding agencies that research on and around Jan Mayen Island should gain higher priority than at present.

The legitimacy of the incentive is in particular, but not only, related to large international programmes like for instance GLOBEC (Global Ocean Ecosystem Dynamics) and CLIVAR (Climate Variability and Predictability). Norway has made only small contributions to such programs, and the Norwegian research community seeks ways to come into grips with the ongoing and proposed international activities. One possibility is to offer participation in the running of a research facility that will be geographically located in an area that is in focus of the international scientific community.

According to responses from the NATO Panel and the invited scientists, a meeting that discussed the feasibility of a joint international scientific observation facility on Jan Mayen Island seemed very appropriate. The NATO ARW was expected to ignite a process that would eventually meet the needs of the international scientific community to do research on and around the remote, but scientifically very interesting part of the Globe. The assessment of scientific options and the requirements for doing actual research on Jan Mayen Island will hopefully be wished welcome by Norwegian authorities who govern the area.

REFERENCES

Anon. 2000. Miljøhandlingsplan for Jan Mayen år 2000-2005. Fylkesmannen i Nordland, Rapport 2/2000. 45 pp (in Norwegian).

CLIVAR Scientific Steering Group 1998. CLIVAR Initial Implementation Plan. World Climate Research Programme, Report No 103. 314 pp.

Greene CH and AJ Pershing 2000. The response of *Calanus finmarchicus* populations to climatic variability in the North Atlantic: basin scale forcing associated with the North Atlantic Oscillation. *ICES Journal of Marine Science* 57, 1536-1544.

Chapter 2 NATO Advanced Research Workshop

Geology of Jan Mayen Island and Surroundings

An Overview

Krzysztof Birkenmajer
Institute of Geological Sciences, Kraków, Poland
E-mail: ndbirken@cyf-kr.edu.pl

Abstract: Jan Mayen is an Arctic island situated in the middle of the North Atlantic Ocean, at the northernmost tip of the submarine aseismic Jan Mayen Ridge just southward of the seismically active Jan Mayen Fracture Zone. Volcanic rocks of Quaternary age predominate on the island. Its northern part (Nord-Jan) is formed by the large Beerenberg stratovolcano (2,277m). The oldest historical records of its activity are from 1732 and 1818 AD. In September 1970, strong parasitic eruptions of lava and ash from northeastern part of the volcano gradually declined towards the beginning of 1971. The most recent, slight volcanic activity was observed on Jan Mayen Island in 1985 and 1986.

Key words: Arctic, Jan Mayen Island, Beerenberg volcano, geological structure, volcanic activity, volcanic hazards

1. LOCATION

Jan Mayen Island lies in Arctic part of the Northern Atlantic Ocean, between 70-72°N and 8-9°W, 550km northeast from Iceland and 500km east of Greenland. The island is 53.6km long and 2.5-15.8km wide, its total land area amounts to 373 km^2 (Berg *et al.* 2000, Barr 2003). It is elongated NE-SW, parallel with the active submarine Kolbeinsey Ridge - part of the spreading Mid-Atlantic Ridge (Fig. 1). The island sits at the northern tip of a narrow aseismic submarine Jan Mayen Ridge, once part of East Greenland, which runs N-S along eastern margin of the extensive submarine Icelandic Plateau. The Jan Mayen Ridge terminates in the north at the Jan Mayen

S. Skreslet (ed.), Jan Mayen Island in Scientific Focus, 13-26

Fracture Zone, a system of active transform faults that bound the Icelandic Plateau from the north (Johnson and Heezen 1967, Johnson 1968, 1975, Johnson *et al.* 1970, Campsie *et al.* 1990).

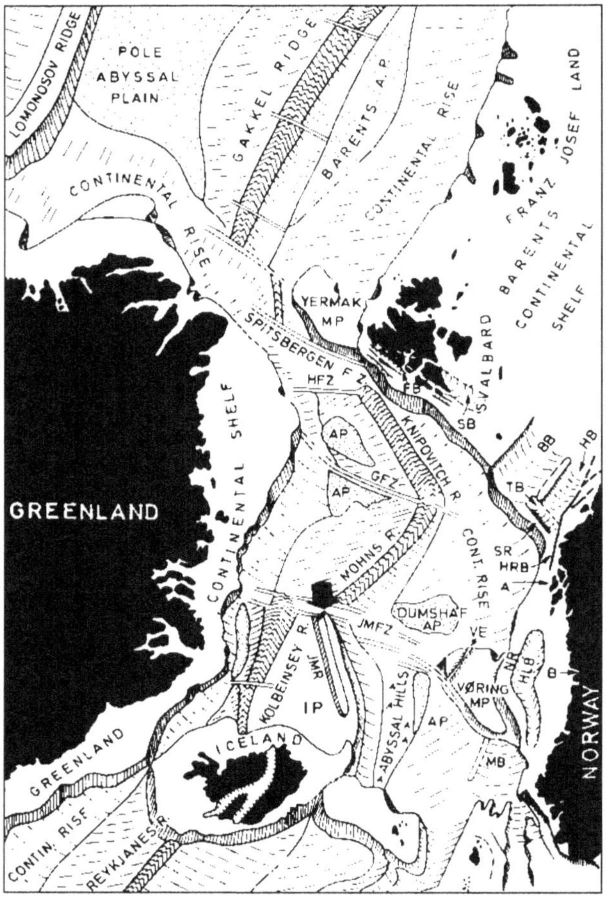

Figure 1. Position of the Jan Mayen Island (arrowed) in plate-tectonic structure of the North Atlantic Ocean basin. Compiled by Birkenmajer (1981) from Johnson and Heezen (1967), Johnson (1975) and Ronnevik *et al.* (1975). Abbreviations: A - Andøya; AP – Abyssal Plain; B - Beistadfjorden; BB - Bjørnøyrenna Basin; FB - Forlandsundet Basin; GFZ - Greenland Fracture Zone; HB - Hammerfest Basin; HFZ - Hovgaard Fracture Zone; HLB - Helgeland Basin; HRB - Harstad Basin; IP - Icelandic Plateau; JMFZ - Jan Mayen Fracture Zone; JMR - Jan Mayen Ridge; MB - Møre Basin; MP - Marginal Plateau; NR - Nordland Ridge; SB - Main Spitsbergen Basin; SR - Senja Ridge; TB - Tromsø Basin; VE - Vøring Escarpment.

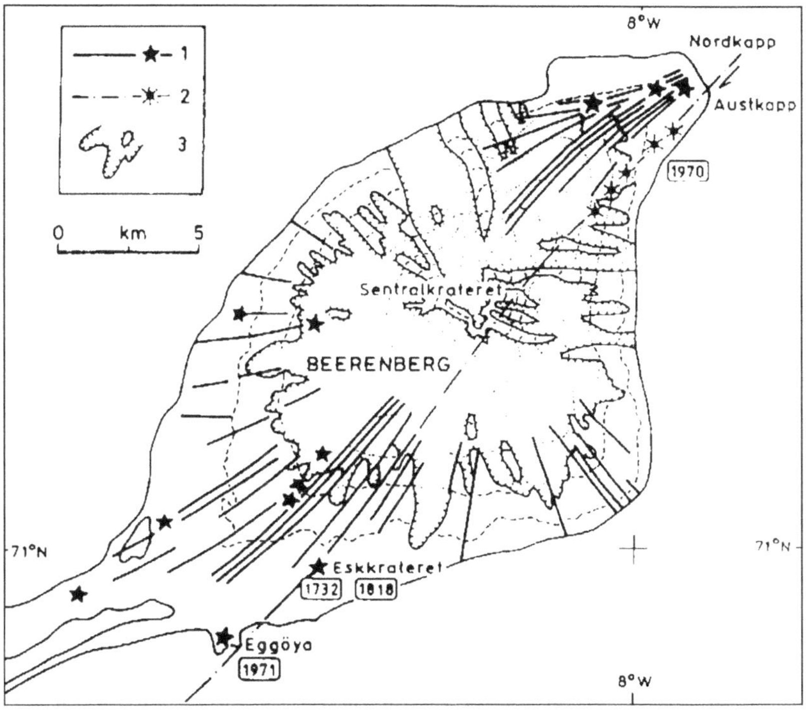

Figure 2. Main linear volcanic features on Nord Jan. Sites of historical volcanic activity from 1732 to 1971 marked (from Birkenmajer 1972). 1) Older fissures and volcanic veins with main extinct parasitic craters (asterisked). 2) Parasitic craters active in 1970 (asterisked) and the line of supposed deep-seated dextral strike-slip fault. 3) Ice cap and glacier margin.

2. TOPOGRAPHY

Jan Mayen Island shows contrasting topographies of its southern (Sør-Jan) and northern (Nord-Jan) parts. Sør-Jan, rising to about 800m above the sea (Rudolphtoppen, 769m), is much lower than Nord-Jan which is formed of the large Beerenberg volcano (2,277m). The volcano is about 15 km in diameter at sea level, and rises to about 5000m above the ocean floor. Firn fields, ice caps (Kronprins Olavs Bre and Kronprinsesse Märthas Bre), and about 20 steep, Alpine-type glaciers which radiate from the ice caps and from the central crater (Sentralkrateret) cover most of the volcanic cone (Fig. 2).

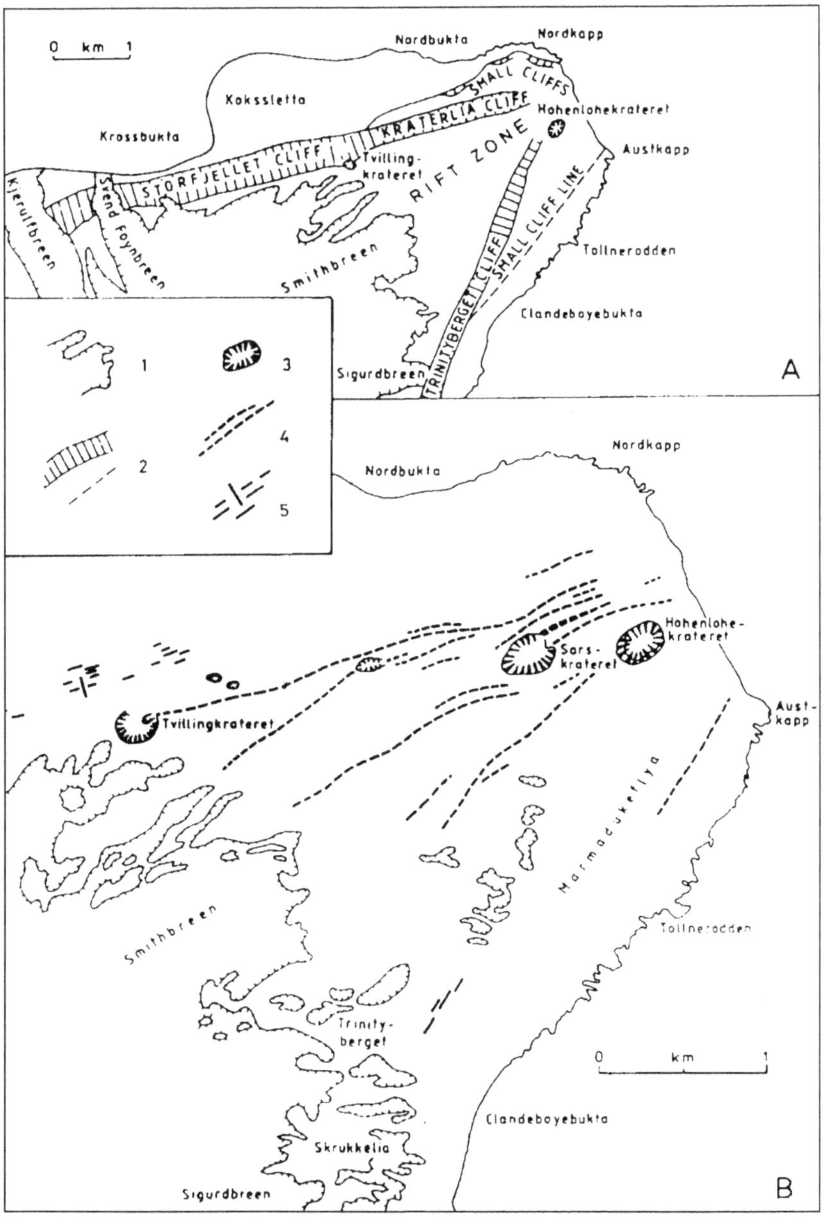

Figure 3. Nord-Jan, northeast part. (A) Main morphologic features and (B) main structural elements (adapted by Birkenmajer 1972, from Fitch 1964, and Roberts and Hawkins 1965). 1) Glacier margin. 2) Fault-controlled cliffs. 3) Extinct, pre-1970 volcanic craters. 4) Volcanic dykes.

Table 1. Geological history of the Beerenberg Volcano (after Fitch *et al.* 1965a, simplified). Age in years B.P.

<table>
<tr><td rowspan="3">Nordkapp Group</td><td>Smithbreen Formation</td><td colspan="2">Small parasitic eruptions up to historical times</td></tr>
<tr><td>Koksslettta Formation</td><td colspan="2">2,500 – 3,500</td></tr>
<tr><td>Tromsøryggen Formation</td><td colspan="2">4,000 – 5,000</td></tr>
<tr><td>Beerenberg high cone</td><td>Sentralkrateret Formation</td><td rowspan="3">Atlantic</td><td>6,000 – 7,000</td></tr>
<tr><td rowspan="6">Kapp Muyen Group</td><td>Nordvestkapp Formation</td><td rowspan="2">7,000 – 8,000</td></tr>
<tr><td>Havhestberget Formation</td></tr>
<tr><td>Storfjellet Formation</td><td colspan="2">Boreal c. 9,000</td></tr>
<tr><td>Kapp Fishburn Tillite</td><td rowspan="4" colspan="2">Greater than 10,000</td></tr>
<tr><td>Krossbukta Formation</td></tr>
<tr><td>Hidden volcanic formations</td></tr>
<tr><td>Submarine foundation of Jan Mayen</td><td>Submerged volcanic formation</td></tr>
</table>

Typically volcanic landscape features, such as numerous, well preserved extinct craters, lava flow fields, volcanic ash cones, fault-controlled ridges and scarps, basaltic dykes etc., are recognizable all over the island (Fig. 3). Most of the island's coastline is made up of steep rugged lava cliffs which are being continuously remodelled by strong marine abrasion.

3. ROCKS AND AGE OF THE ISLAND

Jan Mayen Island is built mainly of volcanic rocks and, to a lesser extent, of glacigenic and marine coastal deposits. Weathering cover, subject to recurrent freezing and thawing, gives way to landslides active during summer. Due to volcanic nature of the island, its small size and mid-ocean location, permafrost zone is probably thin and discontinuous.

Volcanic rocks exposed above sea level are fairly young, Quaternary (Pleistocene and Holocene: some 2 million years before present to the present) in age. Still older rocks (Tertiary in age), may form submarine basement of the island (Fitch 1964, Fitch *et al.* 1965ab, Roberts and Hawkins 1965, Hawkins and Roberts 1972).

Above the sea, the volcanic complex is divided in two parts by glacigenic deposits distinguished as the Kapp Fishburn tillite (Table 1). Below the tillite, basalt lavas occur interbedded with fossil-scree deposits - the Krossbukta Formation. The lavas have been radio metrically (K-Ar) dated at 500,000 years (mid-Pleistocene) through 30,000 years (late Pleistocene).

Above the Kapp Fishburn tillite, there occur in the succession (Table 1): the Storfjellet Formation (basalt lavas and cinder accumulations); the Havhestberget Formation (basaltic pumice tuff, subordinately lava flows and fissure ridges); the Nordvestkapp Formation (centrally erupted basalt lava flows - the main dome of Beerenberg); the Sentralkrateret Formation (plagioclase basalts); the Tromsøryggen Formation (basalt lavas and pyroclastics forming great fissure ridges, cones and lava fields); the Koksslетта Formation (plagioclase basalts forming fresh fissure/cone systems, fissure-fed lava fields and coastal lava platforms); finally, the Smithbreen Formation (explosive volcanic accumulations, moraines, outwash, torrent-river gravels, beach and lagoon deposits, scree and avalanche fans (see Fitch 1964, Fitch *et al.* 1965ab, Roberts and Hawkins 1965, Hawkins and Roberts, 1972).

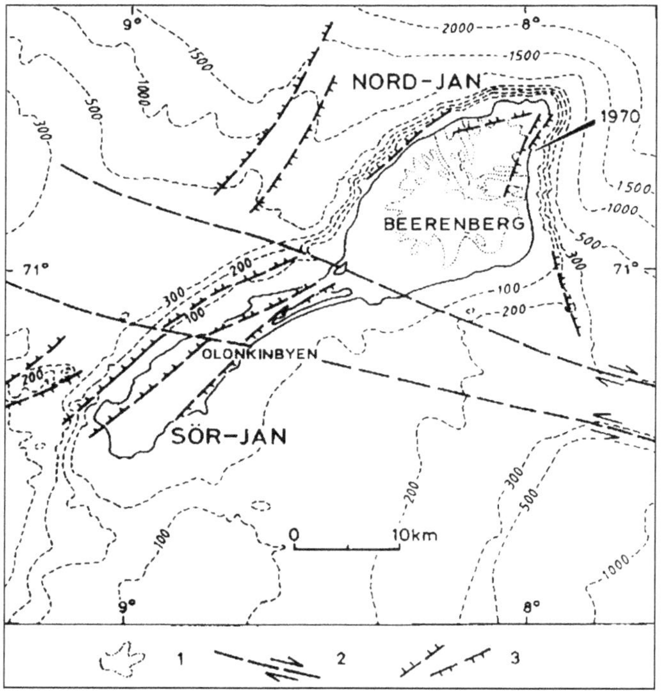

Figure 4. Supposed major fault systems on Jan Mayen (Birkenmajer 1972). 1) Limit of the present glaciation. 2) Strike-slip faults parallel to the Jan Mayen Fracture Zone. 3) Step-faults parallel to the Mid-Atlantic Ridge (Kolbeinsey Ridge). Submarine contours based on *Topografisk kart over Jan Mayen, 1:50,000* (Norsk Polarinstitutt, Oslo, 1959).

4. STRUCTURAL FEATURES

The majority of linear features such as volcanic fissures, basaltic dykes and fault-controlled cliffs on Nord-Jan strike NE-SW (Figs 2, 3) sub-parallel with the Mid-Atlantic Ridge (Kolbeinsey Ridge - Fig. 1). Another system of large-scale fractures, which parallel the submarine Jan Mayen Fracture Zone, is recognizable at a junction between Nord- and Sør-Jan (Fig. 4).

5. VOLCANIC ACTIVITY PRIOR TO 1970

Since its discovery in 1614, the Beerenberg volcano was considered to be extinct or nearly extinct, as only slight volcanic activity had been observed on the island.

A small eruption of flames and black ashes from the south-eastern slope of the volcano that lasted for 28 hours was observed on 17 May 1732, by a German sailor Jan Jacobsen Laab from his ship lying at anchor close to Beerenberg. The site of the eruption could be that of Eskkrateret (Fig. 2). An eruption of smoke rising every 3-4 min. to a height of about 4000feet was reported on 29 April 1818, from approximately the same site (Fig. 2) by William Scoresby Jr. These reports were, however, subject to some doubt, since neither of these eruptions had produced lava (Orvin 1960, Gjelsvik 1970).

There were no reports of volcanic activity on Jan Mayen between 1818 and 1970, save for exhalations of steam which escaped from small cracks near the Eggøya crater (Fig. 2).

6. PARASITIC ERUPTIONS OF 1970

The strong eruption of the Beerenberg volcano in September 1970 was quite unexpected. According to the British Meteorological Office, based on a study of ESSA 8 satellite images showing the extent and height of the volcanic dust and vapour, the eruption had begun by 9a.m. GMT on 18 September 1970. The main event might have been at the time of the strongest earthquake, reported about five hours earlier that day by the U.S. Coast and Geodetic Survey from the vicinity of Jan Mayen Island. The quake had a magnitude of 5.1 on the Richter scale. "It seems possible, owing to the spread of the dust cloud by the 21st that there were further eruptions which put out material into the stratosphere between that time (the 18th) and the morning of 20th" (Hahn 1971, p. 17). Volcanic dust, sunset colours and prolonged twilight were noticed in southern England on some of these days.

The eruption got a good coverage in Norwegian press, and several volcanologic excursions were organized by tourist offices in Oslo, on 2, 4, 9 ad 11 October 1970 (Naterstad 1970, Oftedahl 1970ab, Birkenmajer 1970, 1971). See Birkenmajer (1972) for a summary of reports in the Norwegian press. More factual information on the event was soon provided in official reports by the Norwegian Polar Institute (Gjelsvik 1970, Smithsonian 1970). Analysis of volcanological, geotectonic and petrological aspects of the eruption followed suit (Birkenmajer 1972, Siggerud 1972, Weigand 1972).

The eruption started along a line of fissures that opened in north-east part of Beerenberg (Figs 5, 6), producing basaltic lavas and ejecta (ash and explosion bombs), causing rapid melting of nearby glaciers (the so-called Jökullhlaup phenomenon). The lava poured out, or erupted in fountains up to about 200m high, while bombs of lava were ejected into the air, presumably up to 1000m or so. Temperature of the lava reached 1030°C near the eruption site, falling to 975°C as the lava streams reached the sea. Brown pumice belts were soon floating in the sea. The lava velocity on mountain slope was about 4 m/s. A cloud of vapour and volcanic dust smelling of hydrosulfur had risen to 7,000-10,000m on 20 September, decreasing to about 5,000m the following day, and to 2,500-2,800m on 9-11 October.

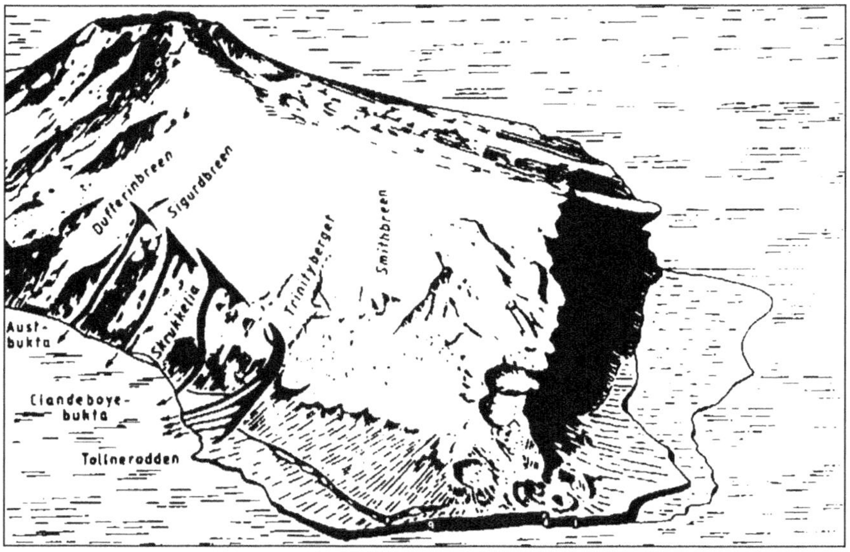

Figure 5. Beerenberg Volcano on Jan Mayen. Sketch from an aerial photo showing sites of the 1970 eruption (Birkenmajer 1998).

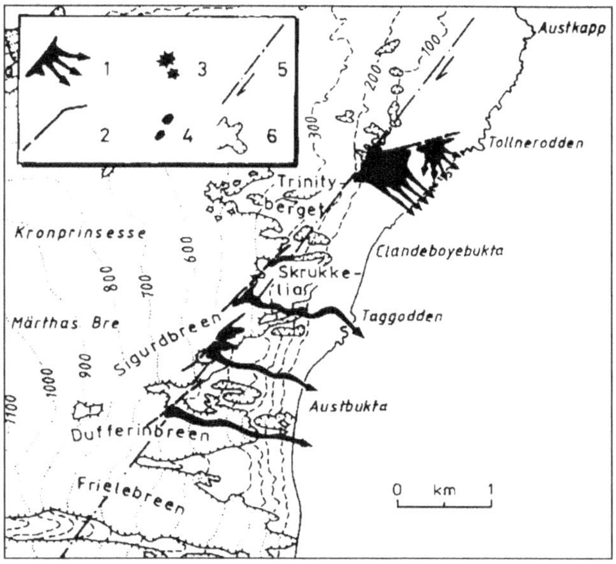

Figure 6. Sites of the Beerenberg eruption, northeast Nord-Jan 20-21 September 1970 (after Oftedahl 1970ab, and Birkenmajer 1972). Explanations to figures 6-8: 1 - lava streams; 2 - main eruption fissures; 3 - active craters; 4 - extinct craters; 5 - probable course of deep-seated dextral strike-slip fault; 6 - glacier margin.

The lava heated the sea-surface water which is normally about 1°C. On September 20, temperature of 13°C was recorded (by the RNAF telemetric devices) about seven miles offshore, rising to 19°C the following day. Some days later, temperatures as high as 30 to 39°C were directly measured at Clandeboyebukta and Austbukta, about 500m outside the lava front. Large numbers of seabirds were seen swimming in warm water, but no dead fish were observed.

From 23 September on, five major craters began to form along the pre-existing fissures (tension-gashes). Originally, some of them had 2-3 vents. Towards the end of September, the craters apparently enlarged to form single vents: the Tollnerodden-, the Trinityberget-, the Skrukkelia-, the Sigurdbreen-, and the Dufferinbreen craters. Simultaneously, the four initial fissures (tension-gashes) had disappeared probably due to filling up with products of eruption, while secondary fissures had opened along a line about 6 km long interconnecting the craters. This line extended southwards as far as Frielebreen, climbing up to about 1000m above the sea (Fig. 6).

Parasitic volcanic activity on north-eastern slope of Beerenberg was strongest during the first four days of eruption (18-21 September 1970). It gradually diminished towards the end of September (Fig. 7). On 9-11

October, only two craters were still active (Fig. 8). At the beginning of 1971, it seemed that even these craters had ceased to produce lava.

On 16 March 1971, the crew of the Meteorological Station on Jan Mayen, again observed smoke and vapour over northeast part of the island. The renewed volcanic activity had also been preceded by an earthquake. On 23 March 1971, another strong earthquake was felt at the Station: it was registered by the Seismological Institute in Uppsala at 10.26,19 h GMT, having a magnitude of 5.7 on Richter scale. The same day, black and yellow patches appeared on snow at Eggøya, southeast corner of Nord-Jan (see Fig. 2), apparently caused by volcanic dust and sulfuric exhalations from opening fissures.

7. MORPHOLOGICAL EFFECTS OF THE 1970 ERUPTION

As a result of aggradation by lava, the coastline of Nord-Jan at Clandeboyebukta and Austbukta has shifted eastwards by about 1km, thus increasing the land area of the island by about 4km^2 (Fig. 8). Echo-sounding performed from M/S "Polarbjørn" revealed that in Austbukta the sea bottom had risen by about 100m at the previous 650m isobath, about 2.5km east of the previous coastline.

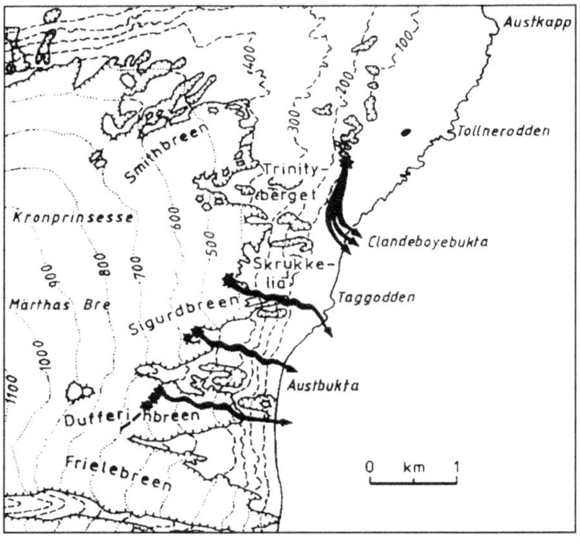

Figure 7. Sites of the Beerenberg eruption, northeast Nord-Jan 23 September 1970 (after Oftedahl 1970ab, and Birkenmajer 1972). For explanations see Fig. 6.

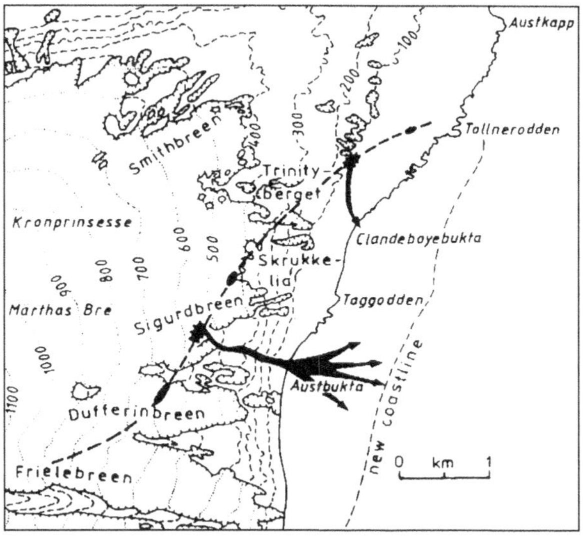

Figure 8. Sites of the Beerenberg eruption, northeast Nord-Jan, beginning of October 1970 (after Oftedahl, 1970ab, and Birkenmajer 1972). For explanations see Fig. 6.

8. GEOTECTONIC ASPECTS OF THE 1970 ERUPTIONS

The renewed volcanic activity (in 1970-71) of the Beerenberg volcano, associated with strong earthquakes registered on Jan Mayen Island, gives clues as to the state of geotectonic regime in this part of the Atlantic Ocean. The Beerenberg stratovolcano itself sits at an aseismic submarine ridge consisting of thinned continental-crust (Jan Mayen Ridge). However, it is very close to the seismically active Jan Mayen Fracture Zone, JMFZ (transform fault), along which the Mid-Atlantic Ridge had been relocated from its previous position near Greenland (at Kolbeinsey Ridge) by about 200km east to its present position - the Mohn's Ridge (see Johnson *et al.*, 1970; Campsie *et al.*, 1990; and Fig. 1). Moreover, the volcano lies within intersection zone of two geofracture systems: the first one trending WNW-ESE (Fig. 4), roughly parallel with the JMFZ (which stretches NW-SE - see Fig. 1), and the second one trending NNE-SSW (Fig. 2), parallel with the Kolbeinsey Ridge (western part of the Mid-Atlantic Ridge).

The latter geofracture system, well expressed on Nord-Jan, seems to have been tectonically active during the 1970 eruptions (Fig. 2), probably being the epicenter of strong earthquakes which preceded/accompanied the volcanic eruptions. From a geological perspective, its importance is also augmented by similarly oriented older fissures and volcanic vents and fault-controlled cliffs on Nord-Jan (Figs 2, 3). Prolongation of this geofracture towards the SSW matches Eskkrateret, the supposed site of eruptions in 1732 and 1818, and Eggøya, where slight volcanic activity was observed in March, 1971 (see Birkenmajer 1972, 1998). Towards the NNE, the discussed geofracture coincides with some extinct, pre-1970, craters near Austkapp (Fig. 2).

9. VOLCANIC AND EARTHQUAKE HAZARD ON JAN MAYEN ISLAND

Since the 1970-71 eruptions, slight renewed volcanic activity at the Beerenberg volcano was noted several times. In 1972, a large, apparently volcanic vapour-cloud was observed to rise from Sentralkrateret. This phenomenon seems to have been more or less permanent ever since (e.g., reports from January 1985, and August 1986), along with vapour exhalations at the Eggøya crater.

Permanent seismic observations carried out on Jan Mayen Island since autumn of 1972, have provided evidence for numerous, mainly small, but also some quite strong, quakes with epicentres located both on land and at sea-bottom nearby. In January 1985, a series of earthquakes up to magnitude 5 Richter scale were registered. On 7 January 1985, a volcanic eruption apparently smaller than that of 1970, was observed from a SAS commercial flight at the northernmost part of Nord-Jan (Barr 2003, p. 252). This again indicates a close interrelation between earthquakes and renewed volcanic activity on the island.

Location of the renewed volcanic activity (in 1985 and 1986), at Sentralkrateret and in northern part of Nord-Jan, might indicate that parts of the Beerenberg volcano other than that of the 1970 eruption sites, are currently being tectonically reactivated.

The volcanic hazard on Jan Mayen Island thus calls for a continuous registration of earthquake frequency, magnitude and trends, as well as for determination of quake epicentre location with respect to geologically recognized active fracture zones and faults. Geodaetic surveys (nivellation and GPS positioning), repeatedly performed on fixed points and/or along field transects, might also give insight into dynamic state of the Beerenberg volcano, its possible volume and shape changes. Recording of temperature and determination of chemistry of volcanic exhalations at selected sites, on a regular basis, are recommended as further elements which may help predict timing, location and scale of future volcanic eruptions on the island.

Hazards to people working on Jan Mayen Island, are posed by possible renewal of stronger volcanic activity on the island lies, besides obvious dangers such as hot lava, ash fall, toxic gases etc., in rapid melting of ice caps and glaciers, resulting in formation of torrential streams and appearance of fast-moving lahar-type (mixed water-ice-volcanic material) landslides. These hazardous events may be restricted to lava eruption sites only but, in case of considerable rise of ground temperatures and the resultant thawing of permafrost, they might spread out over large areas of the island causing a regional disaster.

REFERENCES

Barr S 2003. Jan Mayen. *Norges utpost i vest. Øyas historie gjennom 1500 år.* Schibsted & Norsk Polarinstitutt, Oslo, 266 pp (in Norwegian).

Barr I, L Marstrander, JPH Hansen, JR Hansen, and S Skreslet 2000. Miljøhandlingsplan for Jan Mayen. År 2000 til 2005. Fylkesmannen i Nordland Miljøavdelingen, Bodø (in Norwegian).

Birkenmajer K 1970. New Polar record. *Problemy* (Warszawa) 12, 745-748 (in Polish).

Birkenmajer K 1971. Beerenberg eruption on Jan Mayen Island. *Wszechswiat* (Kraków) 2 (2088), 32-37 (in Polish).

Birkenmajer K 1972. Geotectonic aspects of the Beerenberg Volcano eruption in 1970, Jan Mayen Island. *Acta Geologica Polonica* 22 (1), 1-15.

Birkenmajer K 1981. The geology of Svalbard, the western part of the Barents Sea, and the continental margin of Scandinavia. Pp 265-329 in AEM Nairn, M Churkin Jr, and FG Stehli (eds) *The Ocean Basins and Margins* 5. Plenum Publishing Co. New York.

Birkenmajer K 1998. The geology of Jan Mayen: an outline. *Circumpolar Journal* (Groningen) 13, 10-21.

Campsie J, MH Rasmussen, LC Kovacs, F Dittmer, JC Bailey, NO Hansen, J Laursen, and L Johnson 1990. Chronology and evolution of the northern Iceland Plateau. *Polar Research* 8, 237-243.

Fitch FJ 1964. The development of the Beerenberg Volcano, Jan Mayen. *Proceedings of the Geological Association* (Colchester) 75, 133-165.

Fitch FJ, RI Grasty, and JA Miller 1965a. Potassium-argon ages of the rocks from Jan Mayen and an outline of its volcanic history. *Nature* 207, 1349-1351.

Fitch FJ, AEM Nairn, and C Talbot 1965b. Palaeomagnetic studies on rocks from North Jan Mayen. *Norsk Polarinstitutt Årbok* 1963, 49-60.

Gjelsvik T 1970. Volcano on Jan Mayen alive again. *Nature* 228, 352.

Hahn J 1970. Of volcanoes - satellites - whales - jealousy and cold feet (The eruption of a long dormant volcano on the Arctic island of Jan Mayen provides some historical reflections). *Oceanus* (Woods Hole) 16 (1), 15-21.

Hawkins TRW and B Roberts 1972. The petrology of the volcanic and intrusive rocks of Nord-Jan, Jan Mayen. *Norsk Polarinstitutt Årbok* 1970, 19-41.

Johnson GL 1968. Marine geology in the environs of Jan Mayen. *Norsk Polarinstitutt Årbok* 1966, 105-111.

Johnson GL 1975. The morphology and structure of the Norwegian-Greeland Sea. Ph.D. Thesis, University of Copenhagen. 157 pp.

Johnson GL and BC Heezen 1967. The morphology and evolution of the Norwegian-Greenland Sea. *Deep-Sea Research* 14, 755-771.

Johnson GL, JS Freytag, and JA Pew 1970. Structure of the Norwegian Basin. *Norsk Polarinstitutt, Årbok* 1969, 7-16.

Naterstad J 1970. Tilleggs-guide til Aftenposten/Bennetts tur til Jan Mayen 9.10. og 11.10.1970. Oslo (Norwegian).

Oftedahl C 1970a. Guide for Bennetts tur til Jan Mayen 2.10. og 4.10.1970. Oslo (in Norwegian).

Oftedahl C 1970b. Landet som glødet. *Aftenposten A-Magasinet* 1970 (44), 12-14 (in Norwegian).

Orvin AK 1960. The place-names of Jan Mayen. *Norsk Polarinstitutt Skrifter* 120.

Roberts B and TRW Hawkins 1965. The geology of the area around Nordkapp, Jan Mayen. *Norsk Polarinstitutt Årbok* 1963, 25-48.

Ronnevik H, EI Bergsager, A Moe, O Øvrebø, T Narvestad,and J Stangenes 1975. The geology of the Norwegian continental shelf. Pp 117-129 in AW Woodland (ed.) *Petroleum and the Continental Shelf of Northwest Europe, I. Geology.* Applied Scientific Publications, London.

Siggerud T 1972. The volcanic eruption on Jan Mayen 1970. *Norsk Polarinstitutt Årbok* 1970, 7-18.

Smithsonian 1970. Smithsonian Institution, Center for Short-lived Phenomena: event 89-70 (23 Oct. 1970, Cards Nos. 1035 & 1036), Beerenberg volcanic eruption.

Weigand PW 1972. Bulk-rock and mineral geochemistry of recent Jan Mayen basalts. *Norsk Polarinstitutt Årbok* 1970, 42-52.

Chapter 3 NATO Advanced Research Workshop

The Potential of the Beerenberg Glaciers for Climate Studies

Jon OM Hagen
Department of Geosciences, University of Oslo. Norway
E-mail: j.o.m.hagen@geo.uio.no

Abstract: The glaciers in Jan Mayen are found on the slopes of Beerenberg. The total glacier area is about 117 km^2, or about 30 % of the whole island. Three different possible programs can be outlined. 1) Annual glacier front variations can give smoothed long-time information about the net mass balance of the glaciers. 2) Mass balance monitoring can be done as measurements of altitude variations of winter accumulation and summer melting and give information about the glaciers immediate response on current climate changes. 3) Ice core studies can possibly be conducted in the central crater of Beerenberg and provide information about the climate history and development of contaminants deposited over the last few hundred years.

Key words: glacier mass balance, ice cores, paleoclimate

1. INTRODUCTION

The glaciers in Jan Mayen are found on the slopes of Beerenberg (Fig.1). The total glacier area is about 117 km^2, or about 30 % of the whole island. The glacier can be separated in 20 separate glaciers or ice streams (Hagen *et al.* 1993, Orheim 1993). The glaciers are steep as can be seen in Figs 1 and 2. Some of them go from sea level to 2100m a.s.l. over less than 8 km length. The mean glacier elevation is 850m a.s.l. The highest precipitation is in north-west and three of the glaciers facing this direction extend to sea level and calve in the sea. The equilibrium line altitude is also lower in the

S. Skreslet (ed.), Jan Mayen Island in Scientific Focus, 27-36

north-west, about 600m a.s.l. and up to 900m a.s.l. in the southward glaciers (Orheim 1976, Anda 1980). There are very few measurements of snow accumulation, but a study on Sørbreen indicated that the winter snow accumulation there varies usually between 1-2 m water equivalents (3-6m of snow). However, there are large local variations in the snow accumulation due to wind redistribution and local topography.

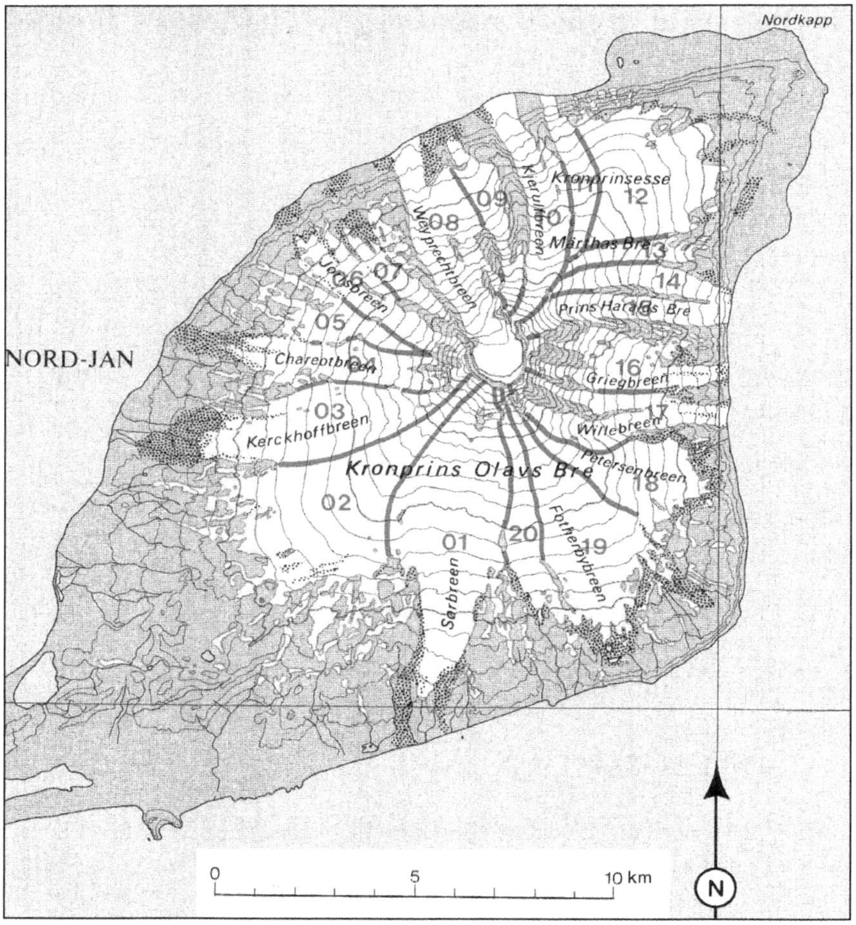

Figure 1. The glaciers of Beerenberg, with boundaries and numbers as given in "Glacier Atlas of Svalbard and Jan Mayen". .Glacier fronts are from 1975. Lines represent the latitude of N71° and the longitude of W8°. Base map compiled from Jan Mayen 1:20000 (modified from Hagen *et al.* 1993).

Figure 2. Weyprechtbreen facing north-north-west and calving in the sea (Photo: Norsk Polarinstitutt).

There are very few studies of the glaciers in Jan Mayen, and since early 1980ies no studies have been conducted. The available information concerns the Holocene variations of the glaciers and some information about the mass balance of the glaciers. In this paper a review of the available information will be given and future possibilities for using the glaciers for climate information will be discussed.

2. HOLOCENE GLACIER VARIATIONS

University of London made expeditions to Jan Mayen in 1959 and 1961 and have published the most extensive descriptions of the glaciers and the Holocene glacier variations (Kinseman and Sheard 1963). There has probably been three Holocene periods of glacier expansion; the first may have taken place around 4000 B.P., the second about 2500 B.P. and the last during the Little Ice Age maximum extension around A.D. 1850 (Anda *et al.* 1985). The glacier advances have been reconstructed based on remnants of old moraine ridges, the oldest advance is uncertain and can also be from last main glaciation. The glaciers have been oscillating since the end of the LIA, retreating after 1850, but advanced in early 1900 with a maximum in 1910. A glacier retreat started then after 1910, corresponding well to the start of retreat of glaciers in Svalbard. The minimum extension was around 1950, and an advance occurred again in the early 1960ies, culminating in 1965.

Since then a general retreat has occurred, but no systematic observations of glacier fronts have been done over the last twenty years. During maximum advances several of the glaciers were calving in the sea. The glacier fronts were never more that 1-2km further out than to-days glacier fronts. This can be illustrated by the frontal position at different times on Sørbreen, the most well studied glacier in Jan Mayen (Fig. 3).

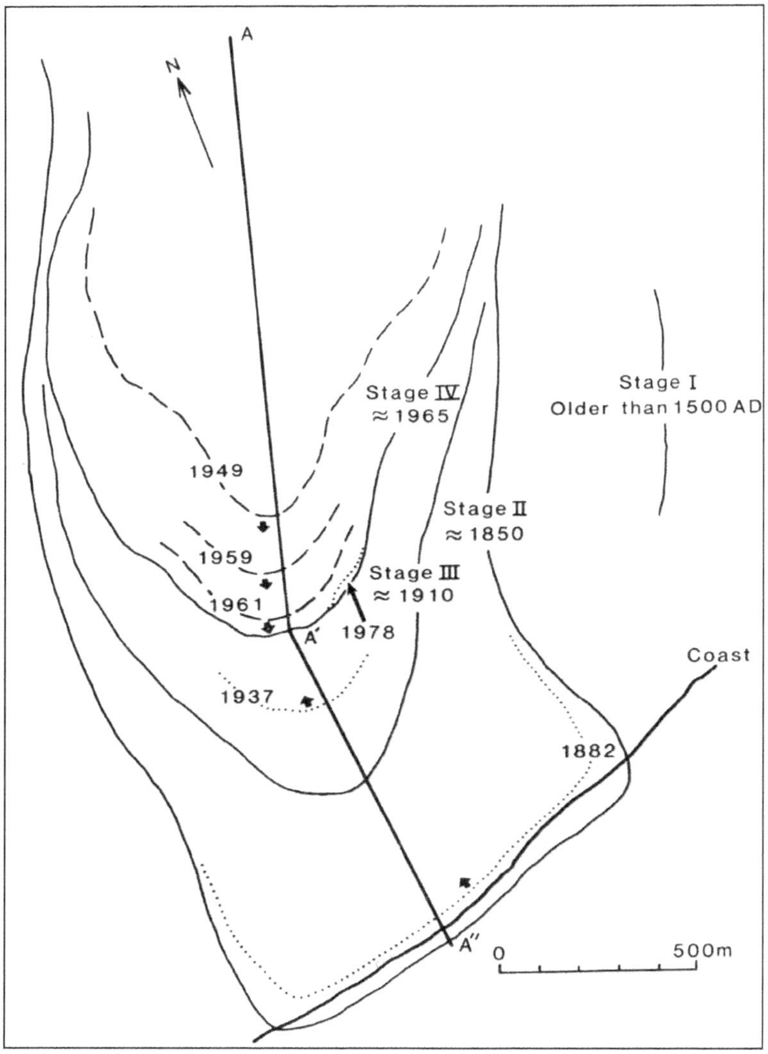

Figure 3. Front position variations at Sørbreen since Little Ice Age (from Anda *et al.* 1985).

3. PRESENT MASS BALANCE

The Norwegian Polar Institute carried out a limited mass balance program on Sørbreen in the years 1972-1974 (Orheim 1976) and later a thesis work from University of Bergen (Anda, 1980) gave some additional data from 1976. The observations indicated that the equilibrium line altitude is about 600 m a.s.l. in north-west and up to 900m a.s.l. in the southward facing glaciers. The study on Sørbreen indicated that the winter snow accumulation varies between 1-2m in water equivalents (3-6m of snow). There are large local variations in the snow accumulation due to wind redistribution and local topography. As the average yearly precipitation at the meteorological station close to sea level is about 660 mm it is likely that the snowfall on Beerenberg glaciers many places are higher than what was observed on Sørbreen. However, large amounts of the snow fall on the steep slopes of Beerenberg is redistributed and probably partly transported out to the sea by wind. The measurements of ablation showed an abnormal pattern with increase of ablation with altitude in the altitude range 200-400m a.s.l. (Fig. 4). This is probably caused by long-lasting fog in these lower parts which reduce incoming radiation and lower the temperatures. The fog creates an inversion layer and gives very special energy balance conditions on the glacier (Fig. 5).

A future program on mass balance of Beerenberg glaciers should be conducted on Sørbreen where some data already is available (Fig. 4). The surface mass balance of the glaciers is sensitive to both changes in the winter snow accumulation and the summer ablation (summer above zero temperature). The special meteorological conditions requires that mass balance measurements should be monitored in altitude intervals, not necessarily covering the whole altitude span of the glacier, but rather conducted at the same elevation year by year. A minimum program should cover the area above and below the mean fog level of about 400 m a.s.l. Such program would give information about the immediate response of the glaciers on a change in the climate, but also interesting information about the energy balance processes on ablation on the glacier surface.

4. ICE CORING

Ice coring in glaciers, where the purpose is to obtain information about the past climate from annual and seasonal variations of meteorological parameters derived from layers stored in the glaciers, requires that two main conditions are fulfilled:

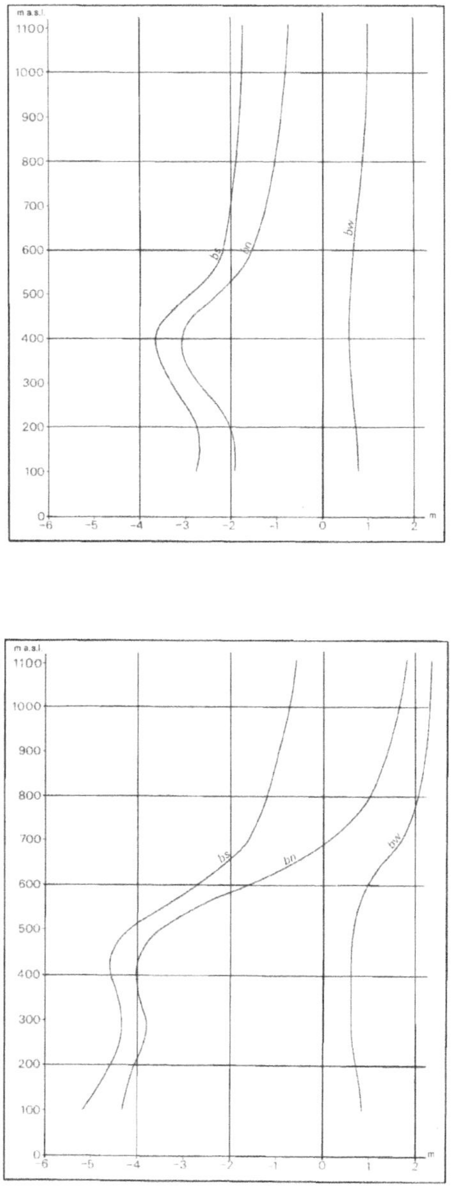

Figure 4. Mass balance curves on Sørbreen from the seasons 1972/73 (above) and 1973/74 (below). The curves show altitude variations of specific mass balance given in meter water equivalents, bs: summer balance (melting), bw: winter balance (snow accumulation) bn: net balance (bn = bw - bs). Modified from Orheim (1976).

Figure 5. Fog in lower part of Beerenberg creating inversion layer (from Anda 1980).

1) The annual layers must not be completely disturbed by summer melting. On temperate glaciers the summer melting destroys the layering as meltwater penetrates down and quickly warm up the snow pack to zero by release of latent heat as the water refreeze. On these glaciers the meltwater partly refreeze in ice layers and partly the winter snow accumulation melt away. The possibility to use oxygen isotopes to detect the seasonal layering is then destroyed and impurities in the snow layers are washed out. However, in subpolar or polythermal glaciers, where the summer season is short and surface melting is not able to warm up the snow pack completely to the melting point, some of the layered structure can be preserved and climatic information can be found in the cores. This has been shown in several Svalbard glaciers (Isaksson *et al.* 2003) and Canadian ice caps (Koerner and Fisher 1990, Koerner 1997).
2) The core site must be at a place where the dynamics of the glacier does not destroy the annual layers. On ice caps and ice sheets all cores are taken close to the summit where the surface velocity and internal deformation is as low as possible.

The mean annual temperature at sea level on Jan Mayen is –0.8°C. If a standard lapse rate of –0.6°C/100m is used the mean annual temperature at the top of Beerenberg should be about –14°C which clearly indicate permafrost conditions and cold ice at temperatures below zero. Even if the lapse rate in periods are lower and even if high summer sun radiation may give surface melting and thus refreezing of meltwater that release latent heat and warms up the snow/ice even in the highest parts of Beerenberg it is still likely that the ice temperatures on top of Beerenberg is well below zero. The average temperature in the warmest summer month, August, is only 5.2°C at sea level, which indicate that the precipitation on top of Beerenberg falls as snow also during the summer.

On Jan Mayen glaciers the only place that could be stable enough to give any chance to preserve the layering is at the summit, in the crater of Beerenberg. However, even there it might be difficult to find a place where many years of accumulation could be found undisturbed. The accumulation rate is large, probably about 5 m of snow every year, and the crater is fairly small, about 2x2km only. This accumulation will then gives ice flow out of the crater and down in the glacier Weyprechtbreen. The crater is a part of the accumulation area of this glacier.

A volcanic eruption would melt away all ice, but the central crater of Beerenberg has not erupted in historical time (Siggerud 1972). The most recent eruption on Beerenberg took place in September-October 1970 when five crater groups were formed along a 6 km long fissure zone on the north-east slope from about 1000 m a.s.l. down to sea level. Lava flow built up approximately 4 km^2 land into the sea and the volume was probably more than 0.5km^3 (Siggerud, 1972). The eruptions were quite explosive and lava fountains were observed hundreds of meter into the air. Thus, also tephra and ash was thrown in the air and deposited in the sea and on land. The heat from the eruptions could not melt any of the glacier ice in higher parts of Beerenberg. The draught of cold air flowing down the snow/ice covered slopes of Beerenberg probably also prevented thick sediment layers to be accumulated on the glaciers, but some deposits must have settled all over the mountain. It is therefore likely that a layer of ash is preserved in the ice on top of Beerenberg. If the ice is stable enough this layer can be used as a dating reference horizon in ice cores drilled on the summit crater. In addition other reference horizons may be found in an ice core in the crater, most likely layers from volcanic eruptions in Iceland and the more recent radioactive fallout from the atomic bomb-test explosions in 1962/63 and the Tsjernobyl accident in 1986 can be detected.

These reference horizons can be used to obtain information bout the net accumulation history on Beerenberg. Oxygen isotope studies can in addition be used to get information about the annual accumulation rate and also about the temperature variations. However, nothing is known on either the thickness of the ice in the central crater or the dynamics/flow velocity of the ice there.

The ice cores may also provide data about the history and development of contaminants deposited over the last few hundred years. Analysis of the layers, heavy metals and chemical compounds, may give valuable information about the history of long-transported fallout back to before the industrial revolution in Europe.

Before it is possible to evaluate the potential for an ice coring, information on the summit crater conditions must be investigated. This includes current precipitation rate, temperature conditions, ice thickness and flow pattern. This can in principle easily be done by shallow cores, Ground Penetrating Radar and flow measurements on poles on the surface of the ice. It is, however, probably a greater challenge to organize the logistics to get to the place.

If the conditions are suitable, an ice core from the summit of Beerenberg could give very valuable information about the climatic conditions in this part of the Atlantic Ocean over the last few centuries, covering the Little Ice Age. Core data could also give information about the pre-industrial long-transported fallout history.

5. CONCLUSIONS

Jan Mayen glaciers are logistically difficult to access. Three different programs are most likely possible to be conducted, given here in increasing level of complexity.

5.1 Frontal Dynamics

The measurement of glacier front variations is a fairly simple monitoring program that can easily be established and even operated by the permanent staff on the island. The front variations give smoothed long-time information about the net mass balance of the glaciers. However, it does not give any information about whether the changes are caused by summer melting or snow accumulation. This can only be detected by measurements of the individual components of the mass budget on the glaciers.

5.2 Mass Balance Monitoring

Even if a mass balance program that covers the whole area of a glacier on Beerenberg is very difficult it is still possible to run a good monitoring program on mass balance per altitude variations in the lower part of a glacier, preferably Sørbreen where some data already exists. A monitoring of mass balance in some altitude intervals can be as good as a program covering the overall area of the glacier and give information about the glaciers response on current climate changes.

5.3 Paleoclimatic Studies from Ice Cores

This program requires more logistics and is much more difficult to conduct as it requires work on the ice in the central crater of Beerenberg. A pre-site survey is necessary in order to investigate the ice thickness, temperature conditions, accumulation rate and flow pattern. A full evaluation of the potential for ice core drillings can not be done until this information is available. However, there is a potential then that cores can provide information about the climate over the last 100-300 years in this part of the Atlantic Ocean. It is so important to understand and obtain more data about the development of the north-Atlantic climate that that a pre-site survey should be recommended.

REFERENCES

Anda E 1980. En glasiologisk og glasialgeologisk undersøkelse på Jan Mayen. Unpublished Cand. Real. Thesis, University of Bergen. 65 pp + 65 pp maps and figures.

Anda E, Orheim, O and J Mangerud 1985. Late Holocene glacier variations and climate at Jan Mayen. *Polar Research* 3 n.s., 129-140.

Hagen JO, O Liestøl, E Roland and T Jørgensen 1993. Glacier Atlas of Svalbard and Jan Mayen. *Norsk Polarinstitutt Meddelelser* 129, 1-141.

Isaksson E, M Hermanson, S Hicks, M Igarashi, K Kamiyama, J Moore, H Motoyama, D Muir, V Pohjola, R Vaikmäe, RSW van de Wal, O Watanabe 2003. Ice cores from Svalbard - useful archives of past climate and pollution history. *Physics and Chemistry of the Earth* 28, 1217-1228.

Kinsman DJJ and JW Sheard 1963. The glaciers of Jan Mayen. *Journal of Glaciology* 4, 439-448.

Koerner RM, and DA Fisher 1990. A record of Holocene summer climate from a Canadian high-Arctic ice core. *Nature* 343 (6259), 630-631.

Koerner R 1997. Some comments on climatic reconstructions from ice cores drilled in areas of high melt. *Journal of Glaciology* 43 (143), 90-97.

Orheim O 1976. Bremålinger på Jan Mayen. *Norsk Polarinstitutt Årbok* 1974, 249-252.

Orheim O 1993. Glaciers of Jan Mayen, Norway. Pp 153-164 in RS Willimas and JG Ferrigno (eds.) Satellite Image Atlas of Glaciers of the World. *U.S. Geological Survey Professional Paper 1386-E.*

Siggerud T 1972. The volcanic eruption on Jan Mayen, 1970. *Norsk Polarinstitutt Årbok* 1970, 5-18.

Chapter 4 NATO Advanced Research Workshop

Meteorological Observations on Jan Mayen

Øystein Hov
Norwegian Meteorological Institute, Oslo
E-mail: oystein.hov@met.no

Abstract: The climate on Jan Mayen is Polar Marine most of the time. The average annual temperature was 0.8°C for the period 1921-1996, with no clear trend. The average annual precipitation for 1921-1996 was 664.4mm, highest in December (64.8mm), lowest in June (31.5mm) The annual average number of overcast days was 242.0 and the average cloud cover 84.6% for 1922-1996. The annual frequency of fog was 13.7 for the 1956-1996 period. The lowest cloud height was below 100m 6.7% of the time 1956-1996. In the period 1971-1981 the mean sea ice freezing up date was 8 January and melting date 21 April for more than 0 octas.

Key words: temperature, precipitation, clouds, wind

1. INTRODUCTION

The Norwegian Meteorological Institute (met.no) and Jan Mayen modern history are related in that the meteorological station was established in September 1921 at the same time as the institute occupied the island. Met.no owns and operates the meteorological station. Housing infrastructure is joint with the operator of Loran-C. The station is administratively and technically under the branch office of met.no for North Norway in Tromsø. At present there are four met.no employees on Jan Mayen, being one radio sonde leader, one technician and two meteorological officers. The met service had up to 14 employees there in the past, including the responsibility for Jan Mayen Radio. At present the met.no personnel carries out radio soundings

S. Skreslet (ed.), Jan Mayen Island in Scientific Focus, 37-40

twice/day and weather observations every three hours. The measurements are distributed globally via the WMO network.

2. CAPABILITIES

Issues that can be addressed by appropriate research observational capabilities on Jan Mayen are tele-connections in regional or global weather (how can a perturbation in the physical state in one place act to change weather patterns in other, seemingly non-connected, regions?), climate change, coupling atmosphere-ocean-cryosphere, high impact weather, ocean circulation, polar stratospheric ozone, biogeochemical cycles, sudden chemical events, and the global distribution ("distillation") of toxic species. The Jan Mayen radio soundings are considered to be crucial for the quality of numerical weather prediction in the North Atlantic region. It is important to take into account that the resources for Arctic science need to focus on the use of existing infrastructure (e.g. Ny Ålesund on Svalbard).

3. LOCATION

The meteorological station has always been located in the middle of the island, and except for the period of World War II either at the NW or SE shores. It was established in September 1921 on the SE shore. In September and October 1940 the station was hit and destroyed by war actions but already in March 1941 re-established. During the period from 1941-1949 the station was located in five different places. From 5 September 1949 to 1 October 1962 it remained in the same position on the NW shore. 1 October 1962 the station was moved again to the SE shore, 11.3km to a more southern position than where it started in 1921. On 25 August 1969 there was a minor relocation of the station to its present position (Førland *et al.*, 1997).

4. OBSERVATIONS

The following summary is taken from Førland *et al.*, 1997. The temperature series from Jan Mayen has been tested against reference series in Svalbard (Bjørnøya and Svalbard Airport) and Greenland (Ammassalik and Scoresbysund) and a significant break in summer season was found in 1939. Its magnitude was -0.6°C; i.e. the first part of the series was too warm compared with the latter one. In the precipitation series there were inhomogeneities in 1932, 1949 and 1946, the latter two linked to several relocations of the precipitation gauge.

The average annual temperature was 0.8°C for the period 1921-1996, maximum in August (5.2°C), minimum in February (-5.2°C). The lowest annual mean was -4.1°C (1968), the highest 0.7°C (1984). The lowest single temperature ever measured was -28.4°C in February 1969. Average annual temperature updated to 2003 is presented in Fig. 1.

The annual average relative humidity was 84.6 for the 1921-1996 record; highest in July (89.6) and lowest in April (82.7).

Average annual precipitation for 1921-1996 was 664.4 mm, highest in December (64.8mm), lowest in June (31.5mm). Number of days in a year with more than 10mm precipitation was 14.8 as an average for the 1931-1996 period.

Number of clear days (defined as sum cloud cover <20%) per year was on the average 4.2 for the 1922-1996 period, with 0.0 as the average in October and 0.6 in May as the "clearest" month. The annual average number of overcast days was 242.0 and the average cloud cover 84.6% for 1922-1996.

The annual frequency of fog was 8.2% for the 1956-1996 period; lowest in December (3.0%), and highest in July (20.6%). The horizontal visibility was less than 1 km 12.1% of the time as an average 1958-1996; 7.9% of the time in October and 20.4% of the time in July. The last numbers include the contribution from weather situations with drifting snow.

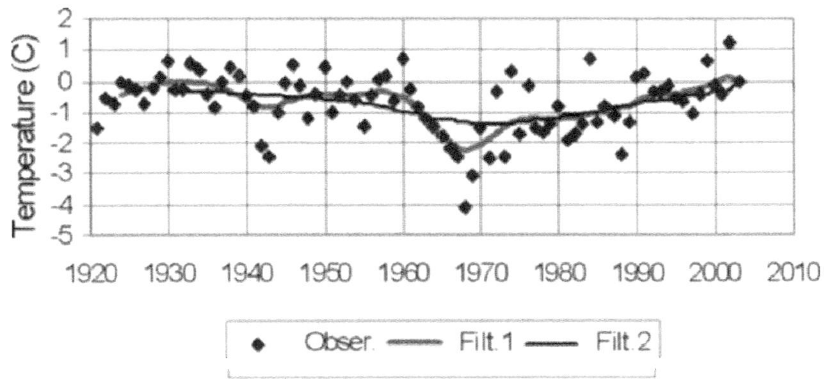

Figure 1. Observed annual average temperature Jan Mayen 1921-2003. With filters 1 and 2 the annual average temperatures were smoothed by a low pass Gaussian filter where the two filters are suitable to identify decadal and thirty year variability, respectively. The low pass Gaussian filter and the weighting function used is defined in Førland *et al.* (1997). Figure provided by Øyvind Nordli, Norwegian Meteorological Institute.

The lowest cloud height was below 100m 6.7% of the time 1956-1996, that is, 6.7% of the observations had lowest cloud height lower than 100m; 20.4% of the time in July as the highest frequency and 1.6% in December as the lowest.

In 1937-1996 the max wind force was above 6B (12m/s) on 200.8 days as an annual average, with 22.8 days in January and 9.7 days in July as the highest and lowest monthly averages, respectively.

In the period 1971-1981 the mean sea ice freezing up date was 8 January for more than 0 octas of ice and 22 January for 7 octas or more of ice; the mean melting date was 21 April (more than 0 octas) and 26 April (7 or more octas). The mean number of days with sea ice in a season was 51, the highest 117, and the lowest 0.

Steffensen (1982) classified the climate on Jan Mayen according to Køppen's system of classification and concluded that it fluctuates from Polar Tundra climate in relatively cold decades and to Polar Marine climate in relatively mild 10-year periods.

ACKNOWLEDGEMENT

Eirik Førland, Inger Hanssen-Bauer, Knut Harstveit and Øyvind Nordli, all at the Norwegian Meteorological Institute, contributed to this summary.

REFERENCES

Førland, E.J., I. Hanssen-Bauer and P. Ø. Nordli (1997) Climate statistics & long term series of temperature and precipitation at Svalbard and Jan Mayen. Norwegian Meteorological Institute report Klima No 21/97. 72 pp.

Steffensen, E.L. (1982) The climate at Norwegian Arcitc stations. Norwegian Meteorological Institute report Klima No 5/82. 44 pp.

Chapter 5 NATO Advanced Research Workshop

North Atlantic Influence on Lithuanian Climate

Effects of Large-Scale Atmospheric Circulation

Gintautas Stankunavicius
Department of Hydrology and Climatology, Vilnius University, Lithuania
E-mail: gintas.stankunavicius@gf.vu.lt

Abstract: The North Atlantic Oscillation (NAO) and the Arctic Oscillation (AO) have a strong influence on the European climate and reflect opposing variations of barometric pressure near Iceland and Azores and the strength of circumpolar vortex respectively. The dynamics of winter NAO indices causes long-term fluctuations of air temperature and hydrothermal regimes in Lithuania. The warm (cold) winters in Lithuania are related to the positive (negative) NAO phase. The correlation between the different NAO/AO indices and precipitation in Lithuania is weak. In Lithuania Hurrell's NAO index appears to be the best temperature indicator during the cold season, except for November and a fairly good indicator for precipitation in January and March. The AO index becomes significant for local precipitation in it's negative phase for January-February.

Key words: North Atlantic oscillation, Arctic oscillation, temperature and precipitation anomalies

1. INTRODUCTION

1.1 The Importance of the North Atlantic Region

The Northern European water basin connecting Atlantic and Arctic regions is the most sensitive area to climatic fluctuations all over the European-Atlantic sector. Deep sea convection, sea ice boundary and enhanced heat fluxes effects are of primary importance for climate change

S. Skreslet (ed.), Jan Mayen Island in Scientific Focus, 41-54

monitoring. High frequency atmospheric eddies lose their energy and become quassi stationary vortices after passing the North Atlantic. They maintain high values of the absolute vorticity. The end tail of Atlantic storm tracks is in this region, causing high variability of geopotential height and sea level pressure. The wintertime first empirical orthogonal function of 1000 hPa heights shows that pressure changes around Iceland-Greenland region are in phase with Northern Eurasia (Fig. 1). Therefore, an expansion of the Siberian High to the west and northwest is closely connected to the phase of the North Atlantic Oscillation (NAO) and to the strength of the polar vortex – Arctic Oscillation (AO).

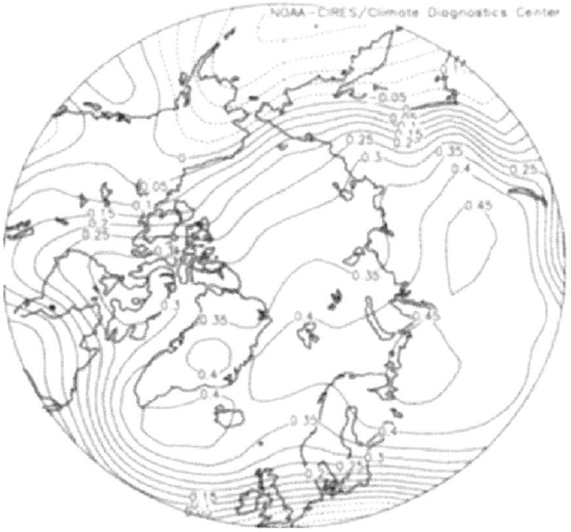

Figure 1. The 1st interannual EOF (Empirical Orthogonal Function) of December-March 1000 hPa geopotential height in the Northern Hemisphere. Image provided by NCEP/NCAR Reanalysis.

1.2 Data and Methods

Air temperature and precipitation data from 11 Lithuanian meteorological stations were used in this study. For comparison additionally near surface air temperature and precipitation rate gridded data were taken from the NCEP/NCAR Reanalysis project (Kalnay *et al.* 1996). Additionally sea level pressure, geopotential height, zonal wind and relative vorticity data from this source were used. The station based monthly NAO index was taken from the NCAR climate analysis center and the principal component based NAO index – from the Climate Prediction Center (NOAA CPC). The monthly AO index defined by Thompson and Wallace (2000) was taken from the NOAA CPC web site (http://www.cpc.ncep.noaa.gov/). Daily NAO and AO values were taken from NOAA CPC ftp site.

Cyclone trajectories were extracted from the atlas of extratropical storm tracks (http://www.giss.nasa.gov/). And, finally, a joint time-frequency analysis was applied to daily NAO and AO indices in order to segregate the active NAO/AO periods of less than one month timescale. Here the software package "Signalogram" was used, which was developed in the Dept. of Radiophysics, Vilnius University (Aleksiejunas and Ivaska 2001).

1.3 North Atlantic Oscillation and Arctic Oscillation: Connection to Regional Temperature and Precipitation Patterns

The North Atlantic Oscillation (NAO) is a measure of the flow over the Atlantic basin and over Europe. Typically NAO index values are computed using mean sea level pressure (SLP) departures over Iceland and the Azores. Periods of low NAO have been shown to be connected to cold temperatures in Northern Europe and eastern North America, whereas periods of high NAO are connected to dry conditions in the Mediterranean and the Middle East and mild winters in Northern and Central Europe (Jones *et al.* 1997). The Arctic Oscillation is also known as the Northern Hemisphere annular mode. The AO is defined as the first empirical orthogonal function (EOF) of the Northern Hemisphere (20°-90°N) winter SLP data. It explains 23% of the extended winter (December-March) mean variance, and it is clearly dominated by the NAO structure in the Atlantic sector. The AO is an almost annular (or zonally-symmetric) structure that reflects a more hemispheric-scale meridional seesaw in SLP between polar and middle latitudes. Detailed description is presented in Kutzbach (1970), Wallace and Gutzler (1981), Trenberth and Paolino (1981) and Thompson and Wallace (1998, 2000).

Long lasting wintertime weather anomalies in Lithuania depend on large scale circulation features over the North Atlantic and Europe. Mid-tropospheric zonal wind fields over central and eastern North Atlantic, which strongly depend on NAO phase, are responsible for maritime air mass transportation deep into the European continent. Long term blocking episodes over North Atlantic and Western Europe are associated with negative NAO phase. Anomalous northerly and north easterly flow over the Baltic region during such periods brings cold Arctic or continental polar air masses and causes large negative surface temperature anomalies. Negative temperature anomalies are also possible under positive NAO phase but with blocking over Eastern Europe. Cloud cover and local factors (snow cover, soil moisture, Baltic SST and other internal factors) play a primary role in temperature anomaly formation in such periods (Stankunavicius 2000).

Zonal wind in the middle troposphere over the southern Baltic is also a good indicator of winter temperature anomalies. Zonal winds over the North Atlantic and over Lithuania are often in phase during zonal circulation periods and almost uncorrelated during meridional and divergent circulation periods (Fig. 2b). The influence of zonal winds on temperature anomalies is stronger in Eastern Lithuania than close to the Baltic Sea coast of Lithuania. Precipitation anomalies are well separated in Europe concerning extremely negative and positive NAO and/or AO phases. The zero correlation line lies a little southward from Southern Baltic and seems to not depend on the selected index. Only the largest positive correlation coefficients are shifted northward in the AO case: Scandinavian mountains precipitation is closer related to an AO index change than to a NAO index change (not shown). The positive (negative) phase is associated with above (below) normal precipitation in Northern Europe (particularly close to the western coast of Norway) and below (above) normal precipitation over the Mediterranean and over Central Europe (Fig. 2a). During negative NAO/AO phase the storm tracks position are shifted southward and the path of extra-tropical cyclones lies within the Mediterranean. Storm tracks in the positive phase are shifted northward and high frequency synoptic disturbances travel from the North Atlantic to north-eastern Europe and Arctic (Bardossy and Caspary 1990). This is clearly visible in the map of standard deviation of precipitation over Europe (Fig. 2a). Cyclone activity is maintained by large temperature gradients (both air and sea surface temperatures) between the northern and southern parts of the North Atlantic and dry cold air outbreak in the rear of the cyclones. Frontal activity in winter is higher over Northern Europe during the positive NAO phase (Rogers 1997). Abundant frontal precipitation patterns in Lithuania and surroundings are connected to orographically enhanced fronts and damped fronts by downstream blocks (Stankunavicius and Bartkeviciene 2003).

Although correlation between winter air temperatures in Lithuania and NAO is high, the precipitation amount does not directly depend on sign and intensity of this large scale phenomenon. The main factors accompanying increased precipitation under positive NAO phase are large positive relative vorticity (RV) and negative SLP anomalies. The correlation between NAO index and SLP over Lithuania is weak (0. 09 during the cold season). Between NAO index and RV at 500hPa level over Lithuania the correlation is also small (-0.25). Only the correlation between RV and SLP is rather large (-0.50) on a monthly timescale. Even during extreme positive NAO index periods ($I \geq 1.5$, 26 months from a total of 208) there are only few months with significant positive RV anomalies over Lithuania.

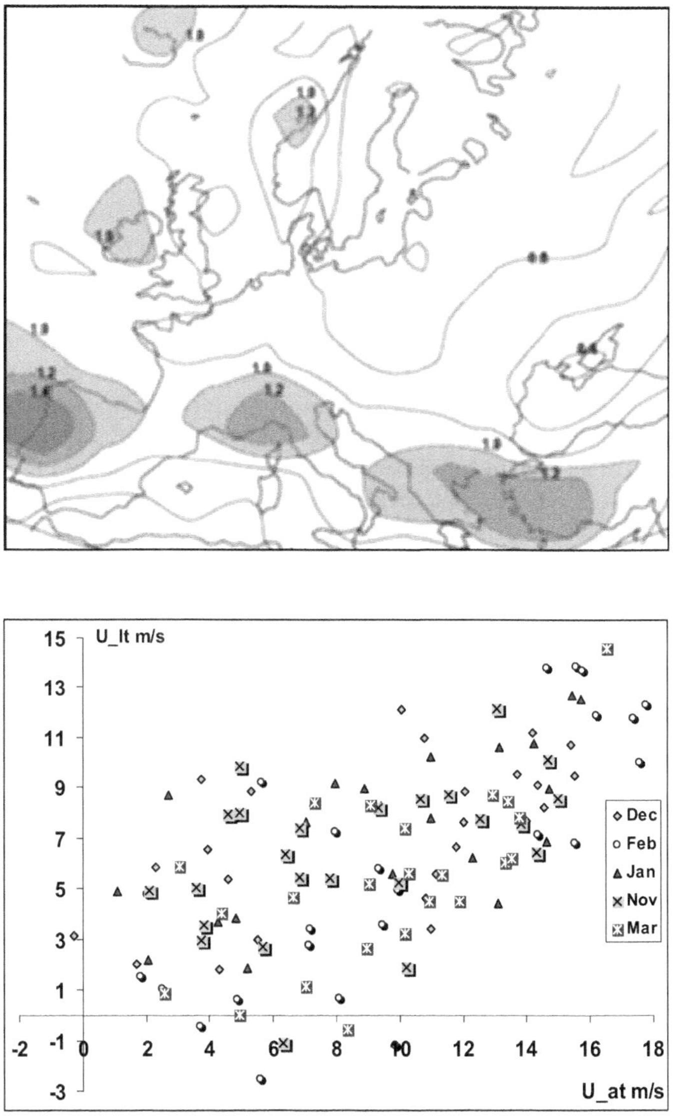

Figure 2. Interannual standard deviation of precipitation rate (mm day^{-1}) for December-March season. Values higher than 1 are shaded (upper). Relation between zonal wind speed at 700hPa level over Northeastern Atlantic (U_at 30°W–10°E, 45°N–60°N) and over Lithuania (U_lt 21°E–26°E, 54°N–56°N). Negative values show easterly winds (lower).

2. CLIMATE CHANGE AND NAO

According to Ulbrich and Christoph (1999) the NAO index, based on SLP fluctuations over the North Atlantic region, increases in comparison to a control run only moderately in a 240 year scenario run with increased greenhouse gases. The long-term trend exceeds the variability of the control climate only at the end of the scenario simulation. In contrast, the steadily growing storm track activity over north-west Europe already surpasses the standard deviation derived from the control run after about 160 years. This effect is associated with a change of the NAO pattern.

During the scenario run, empirical orthogonal functions for sequential 10-year periods show a systematic north-eastward shift of the NAO's northern variability centre from a position close to the east coast of Greenland, where it is also located in the control run, to the Norwegian Sea (Fig. 3). So the simulated sub-polar low is shifted towards Jan Mayen.

Other authors report a statistically significant increase in the mean NAO index (at the 95% confidence level) in their scenario runs with increased greenhouse gases, especially during late summer/autumn and in winter, suggesting more westerly and typically milder weather over Europe during the cold season (May 1999, Hu and Wu 2004).

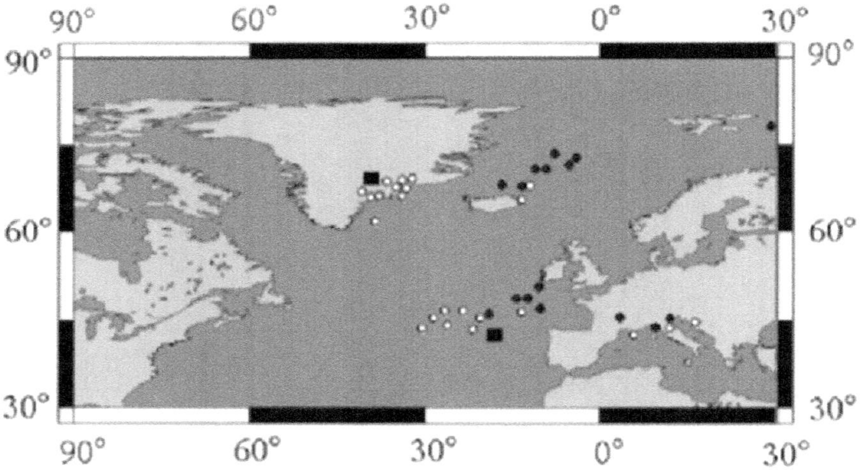

Figure 3. Locations of NAO centres from the coupled Global model ECHAM4/OPYC developed by MPI/DKRZ. The average winter positions presented as black boxes (from control run), consecutive decades in the scenario run (individual months) before 2020 (white dots), and after 2020 (black dots). Transformed from IPCC (2001).

Winter months with largest north-eastward shift of NAO centres (particularly northern centre – sub-polar low) show that the spatial distribution of air temperature anomalies is very similar to those of the positive NAO/AO phase. This was also proven even using two different Global surface (near surface) temperature data-sets: i) the NCEP/NCAR reanalysis (N/N R) with spatial resolution of 2.5° latitude x 2.5° longitude, and ii) University of Delaware (U_D) - land surface stations 2 m temperature analysis on a 0.5° latitude x 0.5° longitude grid (Willmott and Robeson, 1995). Largest positive N/N R temperature anomalies are concentrated over north-western Russia and U_D anomalies over Poland. Note a distinguishable spot of positive anomaly around the Jan Mayen Island (Fig. 4). It is strongly influenced by observations of the Jan Mayen meteorological station and surrounding areas experienced a significant interpolation impact.

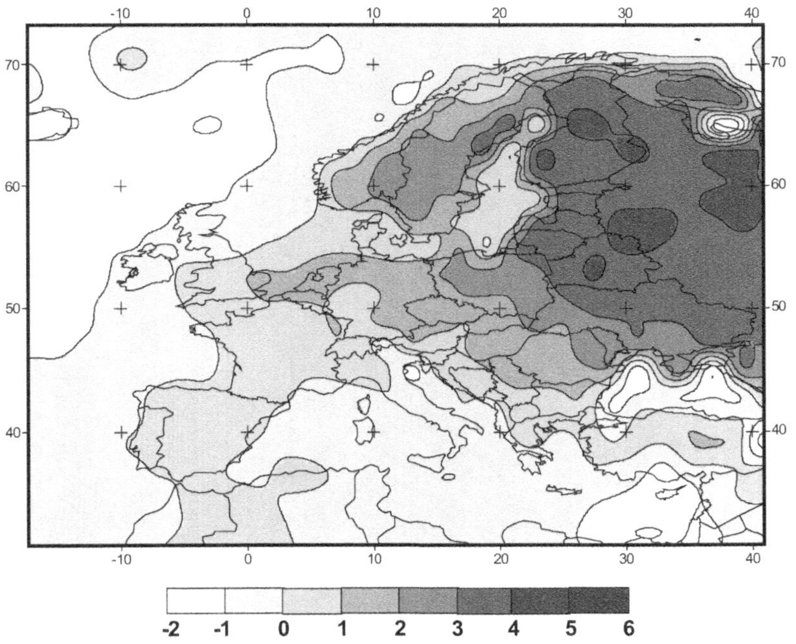

Figure 4. Air temperature composite anomaly (° C) for January-March seasons when NAO centres have the extreme north-eastern position (NCEP/NCAR reanalysis data).

Precipitation anomalies were analysed using four different datasets: N/N R, U_D, CMAP (CPC merged analysis of precipitation) and MSU (microwave sounding unit precipitation). These last two data-sets are available from the Climate Diagnostics Center, Boulder, Colorado web-site. Shifted NAO centres to north east are related to significant precipitation increase over northeastern Atlantic, Norwegian Sea, Scotland and the Alpine and Scandinavian mountains. Largest differences between data-sets are found over Northern Scandinavia, Finland, the northeastern Baltic and over the Mediterranean. Over the Southern Baltic differences are small, although the zero anomaly line having most northern position is only in U_D data (Fig. 5).

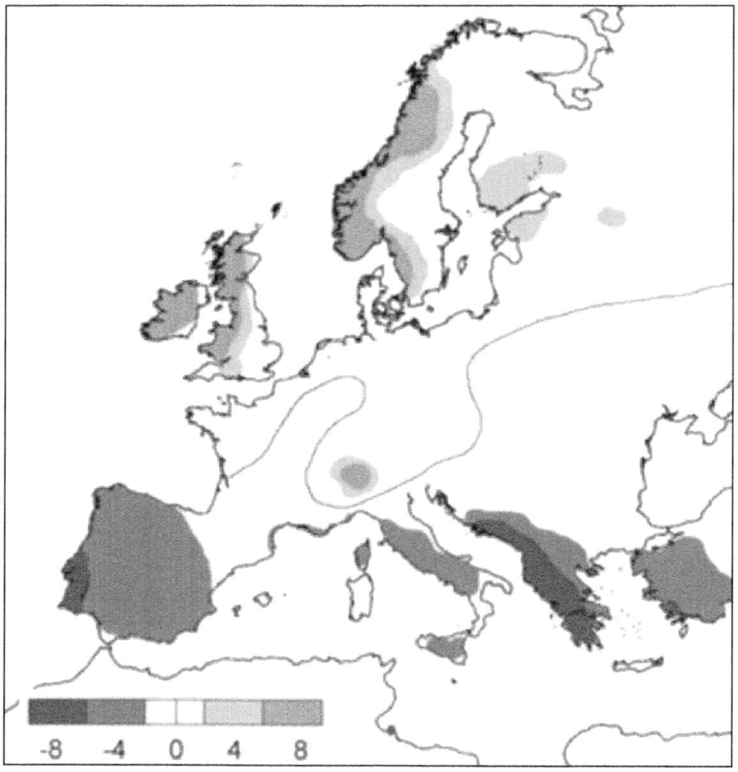

Figure 5. Precipitation composite anomaly (cm month^{-1}) in January-March seasons when NAO centres have the extreme north-eastern position (University of Delaware land surface observations dataset). Solid curve indicate zero anomaly.

3. DOMINANT CYCLONE TYPES IN WINTER

Prevailing cyclone trajectories over Europe extend from west-south-west to east-north-east. Six types of cyclones having influence on Lithuanian weather can be distinguished. The most frequent of them are sub-polar and British-Faeroe cyclones (about 30 % in cold season). Also they are most intensive but their centres are too far from the southern Baltic. Only atmospheric fronts associated with such cyclones have an impact on Lithuanian weather. Such cyclones are related to predominant zonal flow over Europe and intensification of centres of action in the North Atlantic. The frequency of sub-polar cyclones is highly correlated with AO/NAO, and the frequency of British-Faeroe cyclones is correlated with NAO.

Norwegian Sea and Barents Sea cyclones are less frequent types, but they bring significant weather changes in Scandinavia and in the Baltic states. The first type is related to a stationary wave's ridge over the Northeast Atlantic and western Europe, and a trough over eastern Europe, and almost no Atlantic cyclone is allowed further east or southeast. The second type occurs when the upper ridge is shifted towards the east compared to the first type position.

Norwegian Sea type cyclones make up less than 5% of all cold season cyclones influencing Lithuanian weather. Developing in the area of a meridionally extended upper frontal zone and fast moving they directly transport air masses from the Norwegian and Greenland Seas to the Baltic region without significant transformation (Fig 6). Also, there are typical accompanying weather phenomena: strong wind, blizzards, intensive short term snowfall, and rapid temperature change behind the cold front (2–10°C). Other types the southern (Mediterranean and Black Sea) cyclones rarely reach Lithuania with their central parts. They are characterised by large horizontal temperature and humidity gradients in their frontal zones, causing sudden change of weather conditions. Inverse temperature distribution (warmer in the eastern and colder in the western part) is frequent under such circulation conditions in winter.

4. SHORT TERM NAO/AO VARIATIONS AND A CASE STUDY

The connection of monthly NAO/AO variations with low frequency atmospheric variability of temperature, wind stress, precipitation and other is enough covered by an increasing number of scientific papers dealing with these large scale phenomena in recent years (e. g. Hurrell 2003). In contrast, the daily NAO index, which is constructed by properly projecting the daily

500 mb height anomalies over the Northern Hemisphere onto the loading pattern of the NAO has less substantial relations with surface temperature and precipitation patterns across the Europe. A correlation between this daily index and temperature in Lithuania is weak and usually statistically insignificant for periods longer than 20 but shorter than 40 days. The correlation decreases for shorter periods because of the daily NAO nature: it depends on high frequency eddies particularly maintaining the NAO northern centre (sub-polar low). Similar relations were found for deep slow moving cyclones.

The daily AO index is constructed by projecting the daily near surface pressure anomalies onto the loading pattern of the AO (NCEP 2003). The sign of the daily AO is predetermined mostly by the presence of cyclones in the Arctic area or by strong continental anticyclones. Also AO values could be strongly influenced by processes only in the Pacific-North American sector. Nonetheless, daily AO and NAO in some periods changes simultaneously and are highly correlated.

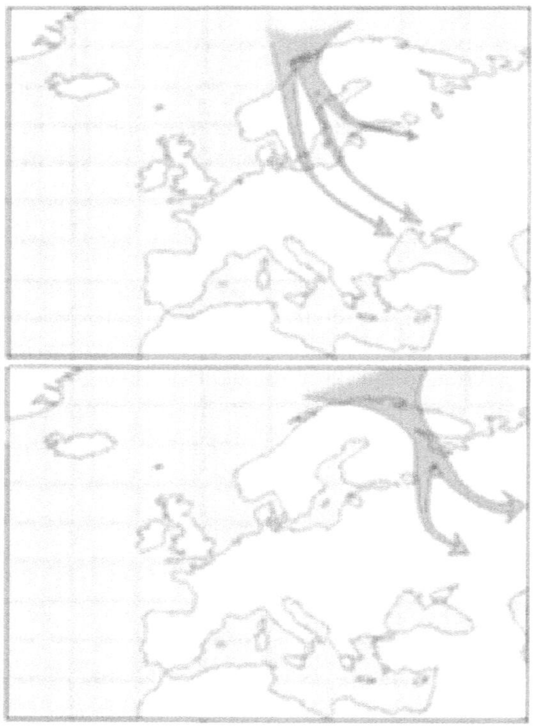

Figure 6. Prevalent trajectories of Norwegian Sea (upper) and Barents Sea (lower) type cyclones in cold season. Recomposed from Stankunavicius (2001).

Using a joint time frequency analysis active NAO/AO periods shorter than 1 month were extracted. This analysis shows that NAO/AO short term cycles range from 6 days to 1.5 month. The highest power spectra values, which are statistically significant, have periods of around 18-27 days. AO and NAO cycles are almost synchronous but power spectra may differ. Best matching between these indices is found when the polar vortex is shifted to the North Atlantic side. Daily temperature variations (in various Lithuanian sites) usually are weakly connected to NAO/AO changes in such active periods. A 10 years analysis shows that the correlation increases when NAO/AO index leads for 3-6 days, and these coefficients become statistically significant.

In the first part of December 2002 very cold and dry weather was advected to the Southern Baltic region. Mean daily air temperatures in Vilnius gradually decreased while the NAO index became extremely raised. This short 8 days positive NAO phase was only a short deviation from the longer negative NAO/AO phase lasting since October. Note that the AO index all the time remained negative because of the anti-cyclonic circulation over the Arctic. The deep Atlantic cyclone was blocked in the NE Atlantic, determining the sign of the NAO index. Temperature in Vilnius rather closely follows the AO index with a lag of 2-4 days. The blocking ridge over Scandinavia was the primary cause of low AO index values (Fig. 7). During this period Vilnius daily temperature correlation with NAO was negative. The correlation with AO was large and positive, but with a 3-day lag of temperature behind AO.

5. DISCUSSION

The area north of Iceland is not directly related to high frequency weather variability in the Baltic region although it is of great importance for low frequency variations of circulation intensity in the North Atlantic European sector. In winters with a substantial shift of the Iceland Low to north-east, temperature and precipitation anomalies in Europe as well as in Lithuania have the same sign and almost the same magnitude as in extremely positive NAO phases. Primary data taken from either land stations, or reanalysis or satellite measurements show almost the same result. The existing positive NAO/AO and Lithuanian temperatures trend does not mean that winter conditions become milder. The rising NAO trend reflects only an increasing frequency of zonal winds over Central and Northern Europe. But in other periods when the zonal flow is not dominant (better visible in daily NAO/AO variations), the sign of temperature and precipitation anomalies could be reversed or close to normal.

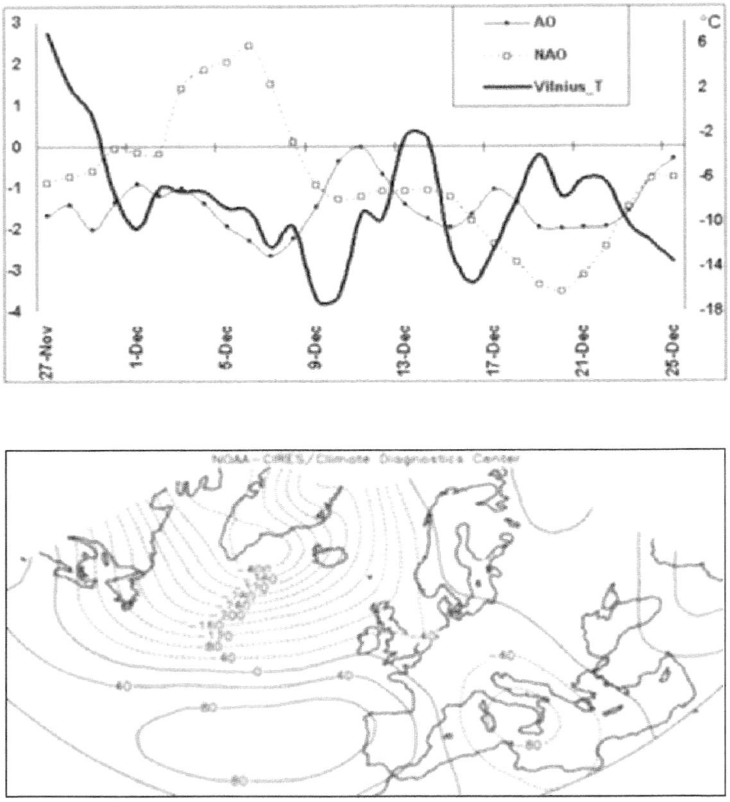

Figure 7. Upper figure: The change of NAO and AO indices, and Vilnius air temperature during December 2002. Lower figure 1000mb eddy height for 3-8 December 2002.

The dominant positive NAO/AO phase during last two decades indicates an increasing extreme behaviour of climate. Additionally there are substantial differences between these indices. The areas which experience the maximal AO influence on temperatures are located 10-15° northward of those with maximum NAO impact. Also AO reflects circulation anomalies over a larger region than NAO. The best link between NAO/AO and Lithuanian temperatures exists in the JFM season, in December it is much weaker and in November the correlation become negative due to frequent anticyclonic circulation over the Baltic region during positive NAO phases.

Precipitation tendencies for increasing NAO/AO are better visible in Northern and Southern Europe. There is no significant trend in Lithuania, only winter precipitation will slightly increase in north-western part (frontal activity) and decrease in the central part of Lithuania (lee-side).

REFERENCES

Aleksiejunas R and V Ivaska 2001. Geometric phase estimation using time-frequency analysis. *Journal of Physics A: Mathematical and General* 34, 8835-8850.

Bardossy A and H J Caspary 1990. Detection of climate change in Europe by analyzing European atmospheric circulation patterns from 1881 to 1989. *Theoretical and Applied Climatology* 42, 155-167.

Hu Z Z and Z Wu 2004. The intensification and shift of the annual North Atlantic Oscillation in a global warming scenario simulation. *Tellus A*, 56 (2), 112-124.

Hurrell JW 2003. Climate variability: North Atlantic and Arctic Oscillation. Pp. 439-445 in J Holton, J Pyle, and J Curry (eds.) *Encyclopedia of Atmospheric Sciences*. Academic Press, London.

IPCC 2001. *Climate Change 2001: The scientific basis*. Cambridge University Press

Jones PD, T Jónsson, and D Wheeler 1997. Extension to the North Atlantic Oscillation using early instrumental pressure observations from Gibraltar and south-west Iceland, *International Journal of Climatology* 17, 1433–1450.

Kalnay E, M Kanamitsu, R Kistler, W Collins, D Deaven, L Gandin, M Iredell, S Saha, G White, J Woollen, Y Zhu, M Chelliah, W Ebisuzaki, W Higgins, J Janowiak, KC Mo, C Ropelewski, J Wang, A Leetmaa, R Reynolds, R Jenne, and D Joseph 1996. The NCEP/NCAR Reanalysis 40-year Project. *Bulletin of the American Meteorological Society* 77, 437-471.

Kutzbach, J. E., (1970). Large-scale features of monthly mean Northern Hemisphere anomaly maps of sea-level pressure, *Monthly Weather Review* 98, 708–716.

May W 1999. A time-slice experiment with the ECHAM4 A-GCM at high resolution: The experimental design and the assessment of climate change as compared to a greenhouse gas experiment with ECHAM4/OPYC at low resolution. Scientific report 99-2, Danish Meteorological Institute, Copenhagen.

NCEP 2003. http://www.cpc.ncep.noaa.gov/products/analysis_monitoring/bulletin/.

Rogers JC 1997. North Atlantic storm track variability and its association to the North Atlantic Oscillation and climate variability over Northern Europe. *Journal of Climate* 10, 1635-1647.

Stankunavicius G 2000. Synoptic conditions of the long term air temperature anomalies in Lithuania. (Ph.D. thesis) Vilnius University Press.

Stankunavicius G 2001. Synoptic climatology: wintertime cyclone types; practical applications. CD-ROM NOMEK/EUMETSAT courses (CALMET 01 Nordic Training Courses for Meteorologists) Vilnius 18-24 April 2001.

Stankunavicius G and G Bartkeviciene 2003. The extreme atmospheric circulation conditions over North Atlantic: the air temperature and precipitation anomalies, *Geographical yearbook*, 36 (1), 18-33. (in Lithuanian).

Trenberth KE, and DA Paolino 1981. Characteristic patterns of variability of sea level pressure in the Northern Hemisphere, *Monthly Weather Review* 109, 1169–1189.

Thompson DWJ and JM Wallace 1998. The Arctic Oscillation signature in the wintertime geopotential height and temperature fields. *Geophysical Research Letters* 25, 1297–1300.

Thompson, DWJ and JM Wallace 2000. Annular modes in the extratropical circulation, Part I: Month-to-month variability, *Journal of Climate* 13, 1000–1016.

Ulbrich U and M Christoph 1999. A shift of the NAO and increasing storm track activity over Europe due to anthropogenic greenhouse gas forcing. *Climate Dynamics* 15, 551-559.

Wallace JM, and DS Gutzler 1981. Teleconnections in the geopotential height field during the Northern Hemisphere winter. *Monthly Weather Review* 109, 784–812.

Willmott CJ and SM Robeson 1995. Climatologically aided interpolation (CAI) of terrestrial air temperature. *International Journal of Climatology* 15, 221-229.

Chapter 6 NATO Advanced Research Workshop

Northern Hemispheric Air Circulation

Vladimir F Romanov
Arctic and Antarctic Research Institute, St.-Petersburg, Russia
E-mail: romanov@aari.nw.ru

Abstract: New theoretical basis is developed and is applied to diagnosis of the climatically prevailing atmospheric transports using the time mean climate data. Balance equations are decomposed into the continual mixture of interacting circulation patterns in the ranges of the synoptic eddy (SE) variability and climate background flow (BG) including the upscale fluxes and sources of the SE formation and decay. Second correlations are modeled including different transport mechanisms. Constant coefficients are evaluated through the data assimilation. Adaptation to the Arctic is through the parameterization of the planetary boundary layer and interaction with Arctic surfaces describing the main Arctic features. Comparisons with data show agreement. Transport paths are focused in the chess-type system of the eddy-advective jets over the hemisphere.

Key words: diagnosis, climate, atmospheric circulation statistics, transports, upscale fluxes, planetary boundary layer, parameterisation, surface exchange

1. INTRODUCTION

The ability to diagnose hemispherical CPAT (climatically prevailing atmospheric transports) is important to many human activities, political and economical decision-making concerning climate and ecology, and to prevent harmful anthropogenic effects on the Global environment.

An important problem concerns the evaluation of paths and the intensity of air transport to the Arctic from the industrial mid-latitudes. This is because the Arctic climate system of atmosphere, land, ice, ocean,

S. Skreslet (ed.), Jan Mayen Island in Scientific Focus, 55-66

vegetation, etc. is very sensitive to any anthropogenic forcing. For example, even short-range transport of soot is able to decrease the surface albedo of the vast Arctic and sub-Arctic regions. It intensifies the radiation heating and melting of ice, activates heat fluxes from the ocean and possibly creates multi-annual and self-reinforcing positive feedbacks that may change the Arctic climate. The result may be gradual changes in the atmospheric meridional contrasts, which could influence the Global circulation system and the World climate.

The Global circulation and climate models (GCM) are widely used in the study of atmospheric circulation statistics and CPAT. They are based on the non-averaged primitive equations and describe explicitly the weather-scale fluctuations. In order to detect the distribution of CPAT it is necessary to accumulate the representative family (climatic ensemble) of the daily resolved circulation fields (GCM output). Averaging and filtering these fields (using different subtle statistical procedures) could provide estimation of the prevailing and dispersed components of circulation statistics (e.g. different transport parameters). These procedures are very computer demanding because distributions that are highly resolved in space and in time must be converted to time averages, which eliminates most of the detailed GCM information. Usually the computed prevailing features represent the small prevailing differences of the relatively large synoptic-scale fluctuations, and thus the evaluation is not accurate. For example, intense northward/southward prevailing eddy transport jets (which cover the latitudinal belt in different longitudes) are mutually compensated. The resultant meridional transport is small and is balanced through the mean meridional advection where the mean velocity is less than 1 m/sec while the typical wind is around 10-15 m/sec. Therefore it is tempting to apply the averaged equations and time averaged data to reduce very computer-demanding reproduction of long-term momentary resolved circulation fields to diagnose the circulation statistics and CPAT. However, this causes a complicated system of equations and difficult parameterizing of the unknown second correlations (e.g. eddy fluxes).

The CPAT represent essential features of the self-organizing atmospheric planetary climate circulation structures (APCS) (Glansdorff and Prigogine 1971, Nicolis and Prigogine 1977, Haken 1993, Mokhov and Petukhov 1989) where the synoptic eddy variability is concentrated during climatic periods. These APCS show very contrasting regions, the climate pressure lows and highs, frontal zones and eddy transport jets. Synoptic eddy (SE) variability (especially the cyclones) generates most of the kinetic energy on account of the baroclinic instability. As a result, this prevailing SE variability forms the upscale fluxes, which feed the mean circulation and support the APCS. The kinetic energy spectra follows here a "-3" power law (Boer and Shepherd 1983). Therefore the eddy flux modeling can not be based on the diffusive hypothesis where the energy spectra follows a "-5/3" power law. Evidently the SEs over the Northern Atlantic transport pollutants, which are absorbed near the eastern US regions. Pollutants accumulate when these cyclones are focused in the Icelandic low, causing concentrations to be higher than in the surroundings. Subsequent cyclones that come into the low-pressure region will also transport pollution. Therefore this transport follows the prevailing velocity instead of the diffusive transport against the mean gradient of concentration. Additionally, the prevailing synoptic vorticity reflects the antisymmetry of the stress tensor and can not be kinematically determined through the curl of the mean wind. This vorticity satisfies the additional balance equation of the redundant angular momentum that determines the non-symmetrical tensor components.

This paper attempts to derive and test a new theoretical basis of averaged mass balance equations, which model the climatically prevailing transport as the statistical mixture of two components, i.e. the back-ground (BG) flows and the climatic family of the SEs. These equations and variables have been smoothed and averaged over climate scales. The input information for this problem represents very available time averaged climate data and data characterizing the cyclone frequencies. Therefore, the application of this technique is very economical.

2. DIAGNOSTIC METHOD BASED ON THE AVERAGED BALANCE EQUATIONS

The main idea is that the synoptic eddy dynamics (especially cyclone) is very different from the surrounding small-gradient flows and essentially non-linear. The SEs are responsible for most of the kinetic energy generation in the range of the SE variability, and the SE dynamics produces the upscale fluxes from the range of the SE variability to the APCS (Romanov *et al.* 1987). This supports the prevailing features of the atmospheric circulation. Therefore, the momentary relevant variables and the primitive basic balance equations are decomposed into the SE and BG regions. In connection with the traditional concept accepted in synoptic meteorology the SE regions are defined as bordered by the outermost closed isobar within the lower troposphere. The SE vertical extension covers the layer within the family of the outermost closed isobars at different height levels, taking into account the typical SE size (the Rossby radius, the radius of curvature of the synoptic isobars, etc.).

Table 1. Acronyms used in the text.

Acronym	Full text
APCS	Atmospheric Planetary climate Circulation Structures
BG	Back-Ground
CPAT	Climatically Prevailing Atmospheric Transports
GCM	Global Circulation and climate Models
NCEP	National Centers for Environmental Prediction (USA)
PBL	Planetary Boundary Layer
SE	Synoptic Eddy
UDF	Universal Dimensionless Functions

As a result of this decomposition, all the momentary variables (density, momentum, vorticity, energy, etc.) are described (as the continual mixture) both in the SE and in the BG regions through different balance equations. Summation of those equations restores the primitive total balance equations.

In order to get the continuous variables and differential operators these equations are spatially smoothed and time averaged. The smoothing scale and the time averaging period are chosen so that the mean variables characterize the prevailing features in climatic families of the SEs and BG. Variability that is the subgrid to these new resolution scales from smoothing and averaging represents the subsynoptic variability. It satisfies a "-5/3" power law (Romanov *et al.* 1987) and is modeled through the turbulent diffusion concept which connects the second correlations with the gradient of averages.

The smoothing and averaging procedures are mathematically correct and lead to the averaged balance equations describing the total climatic ensemble of the atmospheric circulations as the continual mixture of the SE and BG. These components naturally reflect the prevailing and dispersed features of the SE and BG dynamics. They are proportional to the entirely new variable, the climate SE concentration describing the SE climatic probability. The concentration describes the number of eddies having their typical size and life-period within the smoothing region during the averaging period. The concentration is expressed through the traditional SE frequency, known from synoptic climatology. Balance equations for the SE and BG represent the exact integrals of the basic primitive three dimensional balance equations and contain all the governing processes as the corresponding terms, which are expressed through the mean variables and second correlations.

Previous diagnostic SE studies (Romanov *et al.* 1987) have suggested an approach that expresses the second correlations through the time averages calculated from the usual climate data. As a result of this modeling the total climatic ensemble of the current circulations is described through the coupled balance equations for the SE and BG components. This system reflects the next main governing processes: 1) The mean advection (through the prevailing mean climatic circulation), 2) The prevailing eddy advection (due to the prevailing SE drift and prevailing eddy circular, rotary winds), 3) The dispersed, turbulent (synoptic and sub-synoptic scale) eddy transient and rotary motions, 4) The downscale eddy fluxes (due to the barotropic instability), 5) The upscale eddy fluxes (account of the SE energy generation in connection with the baroclinic instability) and effect of the prevailing vertical eddy circulation as well as the SE generation (cyclogenesis, anticyclogenesis) and decay (cyclolysis and anticyclolysis).

This kind of modelling reflects that the drifting SE transports is only a part of the total mass that moves with the phase velocity of the drifting pressure pattern (bordered by the outermost isobar). Another part of mass constantly flows through the SE volumes. Modelling has led to new and unknown coefficients, which scale the describing processes numerically. In order to evaluate these coefficients and to calibrate the model, a special procedure has been developed. It represents the climate data assimilation in the coefficient values to satisfy the balance and to minimize deviations between the modeled imbalances and data.

The resultant balance equations represent a diagnostic technique that allows analyzing different governing processes and their responsibility for the total climatic circulation. The input includes only the usual climate data (pressure, temperature, geopotential and the SE frequencies). In order to provide the diagnosis, these data fields should be decomposed into the SE and BG components. The developed method of decomposition strictly follows all the theoretical derivations. The procedure is based on the hypothesis of the ergodicity using the possible rearrangement of the linear operators of spatial smoothing and time averaging. The spatial smoothing scale is evaluated following the classical approach (Lorenz 1967, Peixoto and Oort 1992) where means represent the zonal and time means, but deviations from these means are the eddy perturbations of all the smaller scales. The total kinetic energy here represents a sum of the energy components including zonal circulation, stationary and transient eddies. The procedure that evaluates the scale of smoothing is organized as a filtering of the transient eddies and guarantees accuracy of this estimation from above and from below.

The 8-year long series of the mean monthly NCEP data has been used to verify the diagnostic technique and reproduce the climate circulation statistics. For simplicity only the lower troposphere has been modelled. Although the 850mb height level varies over the hemisphere and has been approximated as a constant value, H=1457m following Oort (1983), the model reconstructs the basic features of the northern hemispheric circulation. The largest deviations between the modeled imbalances and data cover a small region and do not exceed 1 mb per 8-year mean month, which demonstrates accurate modelling of the total mass redistribution.

3. ADAPTATION OF THE LOWER TROPOSPHERE MODEL TO ARCTIC CONDITIONS

The Arctic and sub-Arctic regions are exposed to different solar radiation and surface cooling by infrared radiation. The planetary boundary layer (PBL) where the small scale turbulence dominates is characterized by the typical temperature inversion and highly stable stratification. The near surface baroclinity due to surface horizontal inhomogeneity is typical. The regular vertical circulation and quasi-stationary changes in the boundary parameters (snow and ice melting, ice growth, time change of the PBL thickness, etc.) as well as the salinity stratification and baroclinity in the ocean PBL are also significant. All these horizontal irregularities could impact very sharp horizontal changes, which are stable in time (ice edge and polynias, land surface irregularities, etc.). These irregularities could support very different surface exchanges, forming the atmospheric baroclinic structures and influencing circulation and transports.

To adapt the diagnostic model to Arctic conditions, in contrast to the Monin-Obukhov similarity which is commonly used in the GCMs PBL parameterization schemes (e.g. Boer *et al.* 1984, Schubert *et al.* 1993, Kiehl *et al.* 1998), the model includes a special similarity (Romanov 1977) that reflects all the important peculiarities. All local turbulent variables (turbulent fluxes and profiles) are expressed through boundary parameters undisturbed by the local turbulence. In the atmospheric PBL the temperature and humidity stratification is modeled. The ocean PBL similarity includes the temperature and salinity stratification (Romanov 1976b). The boundary parameters and turbulent variables are connected through the universal dimensionless functions (UDF) that depend on two dimensionless numbers. Baroclinity and vertical motions are described through additional UDFs (the dimensionless perturbations) and depend on two dimensionless numbers (Romanov 1976b, 1977). All the UDFs have been computed as the numerical universal solution of the atmospheric and oceanic PBL nonlinear multilevel models where formulation corresponds to the present similarity theory (Romanov 1976ab]. As a result, the computationally intense solutions of the numerical models have been converted to much more economical approximation of the UDFs.

This scheme of parameterizations is closed through the surface heat/mass balance (Romanov 1976c). It guarantees the local conservations and a realistic reproduction of the surface inversions because the heat balance describes the near surface energy sinks of infrared radiation and evaporative cooling. The approach also supports correct evaluation of compensating turbulent fluxes, which cause inversions and stable stratification. The surface infra-red radiation is parameterized using the empirical method (König-Langlo and Augstein 1994) that is adapted to the Arctic features. The scheme describes correctly the quasi-stationary changes in boundary parameters, reproduces the local parameters, describes very special Arctic features, and is quite economical. Parameterization has been widely tested against different data and shows good agreements (Dethloff *et al.* 2000, Makshtas *et al.* 2002).

4. EXAMPLE OF THE ATMOSPHERIC TRANSPORTS DIAGNOSIS

In order to calibrate the diagnostic model (evaluating the model coefficients) and to analyze the circulation (transport) statistics the NCEP 8-year mean monthly data have been used over the Northern hemisphere (surface pressure, 850 mb, geopotential and temperature averaged between the surface and 850 mb). The SE frequencies and drift velocities have been calculated by Lagun and Yazev (1994) using the twice daily surface NCEP pressures where the outermost isobar (i.e. the border of the every cyclone) was determined for every time step. The 8-year mean SE frequencies and SE concentrations were calculated for every location and month. The Lagrangian coordinates of the SE mass centers for successive moments have been used calculating the SE drift speeds (vector and scalar mean) for every grid point and month. Data characterizing the period 1981-88 include the sufficiently representative climatic ensemble of the SE patterns. Additional input has been used to adapt results to the Arctic features. The water temperature and salinity below the ocean PBL at a depth of 50m is given as the climatic and zonal averages (Stepanov 1974]. The surface albedo is used according to Kukla and Robinson (1980). Thickness of the ice and snow cover is taken from the Treshnikov (1985) atlas. The surface macro-scale roughness comes from the classification by Zilitinkevich (1970). The surface OPBL flow velocity is used as a constant for the hemisphere. Because these data are from various sources, they may not reflect the same spatial and temporal averaging and, thus, need not be internally consistent.

The modelled month-to-month mass imbalances have been converted to local surface pressure time tendencies and compared with corresponding NCEP tendencies. Minimisation of those deviations through corresponding variation of the model coefficients has evaluated the model coefficients as hemispherical and annual stable constants for every 8-year mean month.

The 8-year mean monthly surface pressure local time changes (from month to month) are in close correspondence to NCEP. The largest deviations occurred in the small region and do not exceed 1 mb per 8-year mean month. Producing the 8-year mean statistics including the heat and mass exchanges between the atmosphere and surface, and using only the climate mean data and physical constants, this technique has used only around 20 minutes of the processor time with Pentium 4 (instead of usually applying super-processor CRAY). Large volumes of output information have been produced. We show here only some illustrations as examples (Fig. 1).

Dimensionless SE concentration and prevailing eddy pressure anomaly show the two main APCS, the Aleutian and the North Atlantic lows where most of the SEV is focused during 8-years. These climatic structures represent the planetary regions of the SE concentration including the prevailing SE tracks and regions where the most of SEs lose their mobility and stop near the sub-polar frontal zone. Comparison of the prevailing climate eddy pressure anomaly with the mean square SE anomaly shows similar horizontal structures, but more modest values of the prevailing anomaly. For example, the prevailing anomaly over the Barents and Kara Seas is around 2 mb, whereas the mean square anomaly is deeper. This brings a qualitative conclusion concerning the dispersed properties of the SEV distribution and reflects the chaotic, turbulent component of the SE situations. The prevailing SE vorticity is focused in the SE concentration centers. Counter clockwise (cyclonic) wind rotation is focused over the Barents and Kara Seas, in the Aleutian low, and over the Atlantic low. Maximum values of vorticity reach $1.0 \cdot 10^{-7}$-$2.5 \cdot 10^{-7}$ sec^{-1}, except over Iceland, where they exceed $4.0 \cdot 10^{-7} sec^{-1}$. This prevailing vorticity is 10 times smaller than the momentary vorticity.

Mean BG winds have the maximal zonal values over eastern Asia (6-7m sec^{-1}), over the northern and southern parts of the Aleutian low (2-4 and 5-6 6-7m sec^{-1}) and over the southern area of the Atlantic low (6-7m sec^{-1}). Meridional BG winds have lower intensity (around 1.5-2.5m sec^{-1}) but reach 4.0-5.5m sec^{-1} over the western and eastern regions of the Aleutian and North Atlantic lows. The two most pronounced tracks of the mobile SE are located in the Aleutian and North Atlantic lows. Zonal (5-7m sec^{-1}) and meridional (2-4m sec^{-1}) drift speeds are maximal in these areas (south of 55-65°N). The prevailing circular winds are mainly focused in the peripheral areas of the SE concentration seats. The most intense zonal rotary component (0.8-1.0m sec^{-1}) occurs in the southern and northern areas of the Atlantic low. The meridional rotary component is highest over the eastern and western parts of the Aleutian and Atlantic lows in the fashion of the concentrated long jets. Well pronounced streams over the eastern and western peripheries of the Atlantic low start at latitudes 50-55°N and extend far into the Arctic up to 70-75°N, having maximal speeds of 1.0-1.2m·sec^{-1}. The prevailing vertical motions are most intense over the central parts of lows (upward) and highs (downward).

In order to analyze the main transport mechanisms, these transports have been converted to the mass imbalances and to corresponding local pressure tendencies. Comparison of those tendencies with the total mass imbalance evidently shows the corresponding input of different transport processes. An important role belongs to BG advection in forcing the pressure changes, as it is similar to changes following from data. However, the meridional BG winds are not intense enough to support the sufficiently active mass exchanges of the Arctic and mid-latitudes. The SE climatic family produces the main meridional transports. Narrow northward/southward eddy jets are mutually compensated over the latitudinal belts showing a small total prevalence. The mean BG meridional advection maintains the total mass balance. The spectral transformation from SE to BG is responsible for this compensation. The regulated spectral flux from the SE supports the compensating BG motions. It is easy to pick out the SE eddy transport, and these could be crucial for forming the total picture among all the simulated effects. The prevailing SE transports are more effective than stochastic turbulent transports through the synoptic and sub-synoptic scale eddies. These transports cause the mass redistribution and only smooth the spatial inhomogeneity that is produced by the prevailing eddy effects.

Ensemble of the drifting SEs supports the systematic transport that is focused in the longitudinally concentrated patterns extending from 30-50°N up to the sub-polar zone 65-75°N. Although the pressure changes are not high (around 0.01mb month^{-1}), most of the transported mass only passes the points without sensible inflow. Most of the mass carried (97-99%) comes from SE to BG due to the SE-BG exchange. Inasmuch as most cyclones coming to lows are stopped and do not drift further to the north, this transport is limited, and the prevailing SE drift could not be considered as a reason for the exchange between polar and middle latitudes. Over the areas where the most of drifting eddies are stopped the well pronounced circular winds occur and support the intense further transport across the frontal zone. This stimulates very active spectral upscale transformation of the air mass anomalies from SE to BG (around 90% of the incoming mass). SE rotary motions are focused at the eastern and western peripheries of the lows and support the transfrontal fluxes. The combination of SE drift and rotary wind transports supports the intense meridional redistribution from the middle latitudes to the sub-polar frontal zone and further exchange between the middle and polar latitudes (Figure 1).

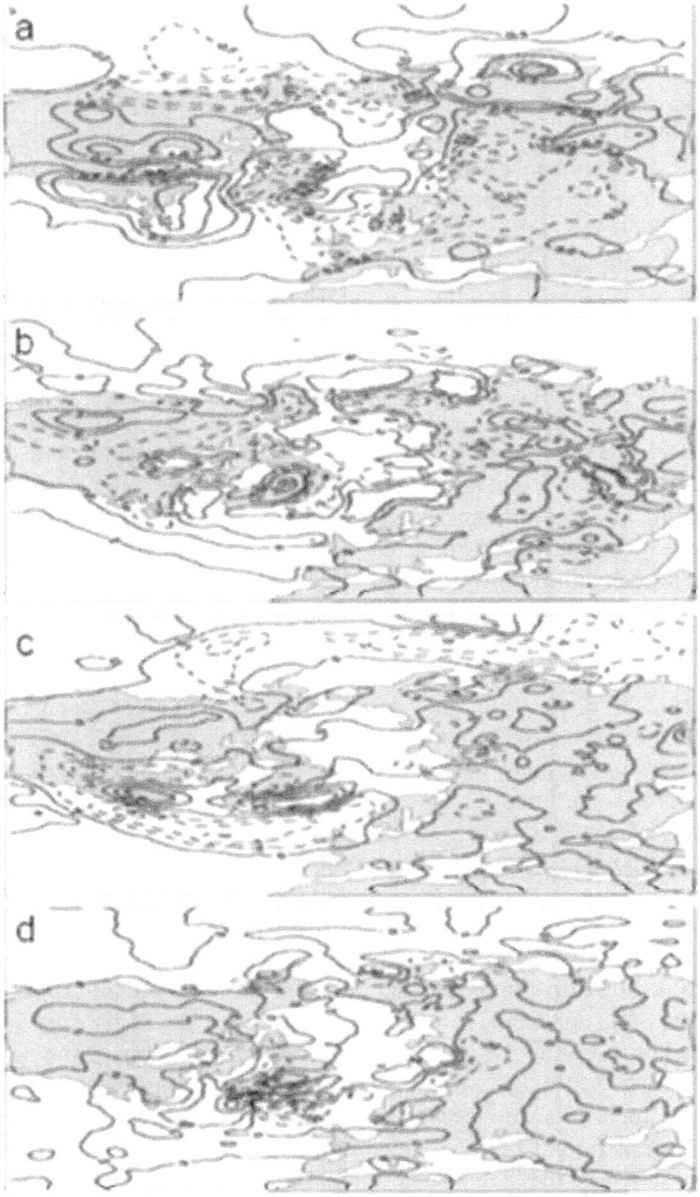

Figure 1. Local mass change presented as local surface pressure change (mb·month^{-1}) in the surface-1457m layer accounting for a) mass BG advection, b) the SE drift, c) circular eddy winds, and d) synoptic scale turbulence. Dashed lines show negatives. Dark fields show continents and islands.

Results show that the low pressures APCS are surrounded by high-pressure areas. Total field represents an intermittent eddy-pattern system where the northward transports (over the eastern parts of lows) extend over western peripheries of the highs. So, the prevailing circular eddy jets form the so-called "chess-type" organized cells covering the whole hemisphere. Additional output from this technique characterizes the lower troposphere (wind, temperature, humidity profiles, turbulent fluxes, potential, internal, kinetic and turbulent kinetic energy balance components, the every surface heat and mass balance component, etc). This information depicts a full picture of the interactive processes between the atmosphere and the underlying surfaces and represents the sufficiently complete basis to diagnose the processes that are responsible for the precipitation of the transported contaminants. Through this set of parameters it is also possible to detect the response of the Arctic climate system to contaminants transported from the mid-latitudes.

5. RELEVANCE OF JAN MAYEN ISLAND TO MODEL DEVELOPMENT

Through the study of various parameters related to the air transport we detected that the Jan Mayen region is characterized by prevailing transports typical for the North Atlantic APCS where the intense mean and eddy transports prevails. We conclude that observations at Jan Mayen could be important for long term monitoring of the total sub-Arctic transport system.

ACKNOWLEDGEMENTS

The author is thankful to Prof. Z. Kowalik from the Institute of Marine Science, University of Alaska for the great help with this manuscript.

REFERENCES

Boer GJ and TG Shepherd 1983. Large-Scale Two-Dimensional Turbulence in the Atmosphere, Journal of Atmospheric Science 40 (1), 164-184.

Boer GJ, NA McFarline, and R Laprise 1984. The climatology of the Canadian Climate Centre general circulation model. Atmosphere and Ocean 22, 430–473.

Dethloff K, C Abegg, A Rinke, I Heberstadt, and VF Romanov 2000. Sensitivity of Arctic climate simulation to different boundary layer parameterizations in a regional climate model. Tellus 53A, 634-659.

Glansdorff, P and I Prigogin 1971, Thermodynamic Theory of Structure, Stability and Fluctuations. Wiley Interscience, New York.
Haken, H 1993. Advanced Synergetics: Instability Hierarchies of Self-Organizing Systems and Devices. Springer-Verlag, New York.
Kiehl JT, JJ Hack, GB Bonan, BA Boville, DL Williamson, and PJ Rash 1998. The National Center for Atmospheric Research Community Climate Model: GCM3. Journal of Climate 11, 1131–1149.
König-Langlo G and E Augstein 1994. Parameterization of the downward long-wave radiation at the Earth's surface in polar regions. Meteorologische Zeitung 3, 343-347.
Kukla G and D Robinson 1980. Annual cycle of surface albedo. Monthly Weather Review 108, 56–68.
Lagun VE and AI Yazev 1994. The global distribution and time-variability of the synoptic-scale disturbances in atmosphere. USSR Academy of Science Reports 334 (5), 642-645 (in Russian).
Lorenz EN 1967. The nature and theory of the general circulation of the atmosphere. WMO, Geneva, 259 pp.
Makshtas AP, VF Romanov, and SV Shoutilin 2002. Sensitivity of modeled sea ice to external forcing and parameterizations of heat exchange processes, Ice in the Environment: Proceeding of the 16th IAHR International Symposium on ice, Dunedin, New Zealand, 2-6 December 2002, 76-85.
Mokhov II and VK Petukhov 1989. Space-time climatic structures. Preprint, USSR Academy of Sciences, Institute of Atmospheric Physics, Moscow, p.1 and p 2, 99 pp (in Russian).
Nicolis G and I Prigogine 1977. Self-Organization in Nonequilibrium Systems: From Dissipative Structures to Order through Fluctuations. John Wiley & Sons New York.
Oort AH 1983. Global atmospheric statistics 1958-1973, US Dep. Of commerce, NOAA Professional Report 14, 1-180 pp.
Peixoto JP and AH Oort 1992. Physics of Climate. American Institute of Physics, 520 pp.
Romanov VF 1976a. Vertical motions and turbulence in the planetary boundary layer, Izvestia. Academy of Scences of USSR Atmosheric and Oceanic Physics 12, 478–484 (in Russian).
Romanov VF 1976b. Turbulence in the ocean boundary layer under the ice cover, Oceanology, 16, 32–40 (in Russian).
Romanov VF 1976c. Parameterization of the nonlinear small-scale air-sea interaction with allowance for ice cover, Meteorology and Hydrology 12, 39-56.
Romanov VF 1977. Similarity theory for baroclinic planetary boundary layer, Izvestia. Academy of Scences of USSR Atmosheric and Oceanic Physics 13, 739–743 (in Russian).
Romanov VF, NV Ariskina, VF Vasiliev, and VE Lagun 1987. Energetics of the atmosphere in polar regions. Gidrometeoizdat, Leningrad, 296 pp (in Russian).
Romanov VF 1988. Parameterization of the synoptic eddy processes for the climate models of the general atmospheric circulation (in Russian), USSR Academy of Science Izvestiya, Atmospheric and Oceanic Physics, 24 (4), 367-380 (in Russian).
Schubert SD, RB Rood, and J Pfaendtner 1993. An assimilated data set for Earth science applications, Buletin of the American Meteorological Society 74, 2331-2342.
Stepanov V 1974. The World Ocean. Knowledge, Moscow 255 pp. (in Russian).
Treshnikov AF (ed.) 1985. Atlas of the Arctic. Main Office for Geodesy and Cartography of the Council of Mineralogy of the U.S.S.R., Moscow. 264 pp. (in Russian).
Zilitinkevich SS 1970. The Atmospheric Boundary Layer Dynamics. Gidrometeoizdat, Leningrad. 291 pp. (in Russian).

Chapter 7 NATO Advanced Research Workshop

Natural and Anthropogenic Aerosols in Arctic Areas

Marko Vana
Institute of Environmental Physics, University of Tartu, Estonia
E-mail: mvana@ut.ee

Abstract: Some properties, sources and effects of natural and anthropogenic atmospheric aerosols in Arctic areas are discussed in this paper. Long-range transport of anthropogenic aerosols, and local biochemical sources that are thought to lead to aerosol formation and growth are also discussed. Current knowledge and questions remaining to be answered clearly indicate that the future research and measurements of atmospheric aerosols are needed in the Arctic. An important issue is to reduce the uncertainties in assessments of the impact of climate change.

Key words: polar aerosols, Arctic haze, particle formation and growth

1. ROLE OF ATMOSPHERIC AEROSOLS

Aerosols, consisting of solid and/or liquid particles suspended in a gas, play a very large part in nature and the life of man (Fuchs 1964). Atmospheric aerosol particles come from a wide variety of natural and anthropogenic sources and their size extends from one nanometer up to hundreds of micrometers. The geographically localized sources and sinks and relatively short lifetime (from few minutes to several weeks) give aerosols an extreme spatial and temporal in-homogeneity in the atmosphere, despite of the turbulent mixing (Haywood and Boucher 2000). An understanding of the properties of aerosols is of great practical importance, since aerosol particles are ubiquitous in the Earth's atmosphere and influence our quality of life in many different ways. (Hinds 1999).

S. Skreslet (ed.), Jan Mayen Island in Scientific Focus, 67-77

On a mass basis, two populations of particles, with characteristically different sources, sinks, sizes and chemical compositions, generally constitute the tropospheric aerosol (WMO/GAW 2003). The sub-micrometer fraction originates from condensation sources and from atmospheric gas-to-particle conversion processes. It is primarily removed by precipitation and dry deposition processes. The super-micrometer fraction is produced by mechanical processes (wind-driven soil erosion and seawater bubble-bursting) and is mainly removed by sedimentation, impaction and interception at the Earth's surface. It has considerably shorter atmospheric residence time than the sub-micrometer fraction.

Atmospheric aerosols are thought to be an important factor influencing the evolution of the Global climate through two primary mechanisms (Seinfeld and Pandis 1998). The first is the direct effect, whereby aerosols scatter and absorb solar radiation, therefore increasing the albedo of the earth and cooling the climate system. The second is the indirect effect, aerosols are also the nuclei around which cloud droplets form. Enhanced concentrations of aerosol particles modify cloud properties and increase the albedo of the earth's atmosphere. IPCC (2001) estimates the average direct radiative forcing, through scattering of incoming radiation, at –0.4, –0.2, and +0.2$W{\cdot}m^{-2}$, for sulphate, biomass and fossil fuel black carbon aerosols, respectively, and the indirect forcing, through cloud condensation nuclei effects, at 0 to –2.0$W{\cdot}m^{-2}$ for all aerosols. The effect of radiative forcing by anthropogenic aerosols is one of the largest sources of uncertainty in climate predictions. It is difficult to distinguish anthropogenic from natural aerosols as even individual particles can have both natural and anthropogenic components.

At elevated levels aerosol particles affect not only visibility and climate, but also our health and quality of life. Our lungs are exposed to a variety of respirable particles from the outdoor environment, work places and our homes (Stone and Donaldson, 1998). The most recent World Health Organization (WHO) review based on a number of recent studies has concluded that fine particles (PM2.5) are most hazardous in terms of mortality, cardiovascular and respiratory endpoints. It was therefore recommended to use PM2.5 as the indicator for health effects induced by particulate pollution. However, there are evidences that not only particulate mass, but also particle number, size, surface and chemical composition may be relevant characteristics for health effect studies (AMAP 1998).

Pollutants emitted to the atmosphere such as SO_2 and NO_x are oxidized to sulphate and nitrate through gaseous and aqueous phase reactions (AMAP 1998). These particle species are removed by both dry and wet deposition to the earth's surface leading to effects such as acidification and eutrophication (mostly due to fertilizers and particle deposition). The deposition of sulphate and nitrate particles is dependent on their size that is controlled by the aerosol dynamic processes in the atmosphere. Eutrophication is becoming a serious threat to coastal environments and seems to be a Global problem in the next decades.

2. AEROSOLS IN THE ARCTIC

2.1 Arctic Haze

Due to the climatic peculiarities polar aerosols have a special character. Arctic atmosphere is not a pristine environment (ACIA 2000). The Arctic Ocean is surrounded by industrialized continents that contribute gaseous and particulate pollution. The Arctic provides a unique laboratory to study the atmospheric processes of a great number of elements and compounds from natural and anthropogenic sources in a new light and under quite different conditions than in the majority of studies conducted elsewhere on the globe (Sturges and Barrie 1988).

In the Arctic, the concentration of pollution has a strong seasonal variation characterized by a summer minimum and a winter maximum. In the clean conditions, the monitoring of Arctic aerosols reveals the aged character of particles and the concentrations are very low (Seinfeld and Pandis 1998). During the winter and early spring the Arctic aerosol has been found to be influenced significantly by anthropogenic sources. The phenomenon is commonly called Arctic haze. During this period, the aerosol number concentration can increase over $200 cm^{-3}$, which is 10-20 times higher than in summer.

During the 1970s, it was discovered that even remote parts of the Arctic were influenced by air pollution (AMAP 1998). Part of this is due to human activities at lower latitudes. Arctic haze is anthropogenic aerosol that reaches the Arctic mostly as submicron aerosol (includes particles with diameter of 0.1-1μm) and trace gaseous constituents (Shaw 1995). The pollution is transported to the Arctic over spatial scales of several thousands of kilometers. In extent, this generally polluted air mass system can be about as large as the African continent. Horizontal visual range in Arctic haze may be restricted to 5–8km by the haze (Schnell 1984). The Arctic haze is primarily a winter phenomenon, and the transport from Eurasia is controlled by the low-pressure systems which intensify over the North Atlantic in late winter and early spring (Ottar 1989). During the transport across the Arctic, the concentration level is maintained, as there is no significant dry or wet deposition, and very little transformation of SO_2 to sulfates (SO_4^{2-}).

The haze was first noted by Murray Mitchell in 1956. However, it was not associated with anthropogenic air pollution and considered unimportant. The importance of this phenomenon was recognized and the active research was arranged during 1970s and 1980s.

Through chemical fingerprinting it has been found that mainly European and Asian emissions effect to the pollutant concentrations in the Arctic, while the North American sources have less important role (Shaw 1995). North American sources evidently contribute only in a minor way to the Arctic pollution burden. Polluted air masses originated from the east coastal region of North America usually travel east across the Atlantic Ocean and experience rapid removal by storms within the Icelandic cyclone system. The major Eurasian sulfur dioxide emission sources are also 5-10° higher in latitude in comparison to those in North America.

Arctic haze in mid-winter mainly consists of sulfur dioxide produced in the burning of fossil fuels and smelting of sulfide ores (Shaw 1995). The gaseous form converts to particulate sulfates when the sun rises in spring. Heavy metals and organic compounds (about $1 \mu g \cdot m^{-3}$) have also been measured in Arctic air pollution. But sulfate has been found to be the predominant chemical compound in the haze particles (up to $2 \mu g \cdot m^{-3}$) representing roughly 40% of the mass concentration. Concentrations of black carbon are also elevated (about $0.3–0.5 \mu g \cdot m^{-3}$), consistent with the combustion that takes place in inefficient coal power plants of the type used in the former Soviet Union.

There are transport and chemical processes in the sulfur cycle that are strongly latitude-dependent (AMAP 1998). The lack of sunlight in the Arctic for large parts of the year limits the production of the OH radical and H_2O_2, which slow the sulfate production cycle and increase the SO_2/SO_4^{2-} ratio observed in the Arctic. This photochemical mechanism is critical to the timing of the Arctic haze maximum. The seasonality of SO_2 oxidation to sulfate is important in prolonging the presence of sulfate aerosols in the Arctic into April and May.

One favourable factor for the occurrence of the Arctic haze is that during the winter season, the Arctic air mass occupies, in addition to the polar region, a large fraction of the snow-covered continental land mass of North America and Eurasia. This allows polluted air to move into the Arctic at ground level, as the temperature in the source areas is approximately the same as in the Arctic (Ottar 1989). The Arctic air mass is a unique meteorological feature of the troposphere characterized by: (i) subzero temperatures much of the year; (ii) little precipitation (most of which falls during the warm half of the year, some parts of the Arctic meet the definition of a desert, with an annual precipitation of less than 100 mm at many Arctic sites); (iii) stable stratification that prevents strong vertical mixing; and (iv) low level of solar radiation especially during winter (Barrie 1986).

The reasons why Arctic haze happens are thought to be as follows. Firstly, strong temperature inversions cause great atmospheric stability. Lack of sunlight and snow-covered surface reflects that little sun radiation back to space results in a cooling of the surface layers and causes strong surface-based temperature inversion in the polar night. Secondly, low water vapour pressure in the cold air results in thin and sparse cloudiness, and small precipitation. This causes slowdown in the removal rate of pollutants in the cold polar air. As a consequence, under conditions of great atmospheric stability and low water content, the residence time of aerosols can increase up to several weeks. So, it is mostly because of the increase in residence time, rather than simply large air masses of injected pollutants, that the Arctic atmosphere becomes generally contaminated.

In the past decade or so, the occurrence of Arctic haze has become weaker and less frequent, perhaps due to break-up of the Soviet Union and associated decrease in heavy industrial emissions, though systematic changes in meteorological patterns may also be involved (Shaw 2003). As a consequence, the further studies of the properties of anthropogenic particles and contaminant gases present in the Arctic atmosphere are needed.

2.2 Formation and Growth of Natural Arctic Aerosols

The central Arctic Ocean in summer provides a limited influence of continental and anthropogenic sources resulting in low particle concentrations (>80nm diameter between 20 and 65cm^{-3}) (Covert *et al.* 1996). In summer, the central Arctic is a specially favorable region for studying the natural sulfur cycle, as the open water surrounding the pack ice are the only significant sources of dimethyl sulfide (DMS) excreted by most species of phytoplancton. The natural sulfur cycle in the troposphere involves release of the marine biogenically produced DMS from the ocean surface, which is transferred to the atmosphere and is oxidized photochemically, and eventually converted in part to sulfate compounds in the particulate, and therefore can exert a significant Global climatic control.

Nucleation mode aerosol particle (3-25nm in diameter) concentration bursts in the atmosphere have been observed at several places around the world (Kulmala *et al.* 2000). During the bursts, the intensive gas-to-particle conversion process occurs due to the presence of precursor gases in the atmosphere. The relevance of natural particle formation bases on the fact that the recently formed particles can grow to the accumulation mode sizes (0.1-1μm in diameter), where they are considered to be an important factor influencing the evolution of the Global climate.

In summer Arctic, a measurable number of particles smaller than about 25nm in diameter is present for a surprisingly large proportion of the time (70%) (Leck *et al.* 2001). Due to their relative short lifetime (typically few hours), the presence of ultrafine particles, especially at the lower limit of measurement range, is an indication of recent production (Covert *et al.* 1996).

The questions are: what are the formation mechanisms of these nucleation mode aerosol particles and how do particles grow to sizes large enough to act as nuclei for the formation of cloud drops through water condensation? It has been found that in summer Arctic there are a more or less continuous low level production of particles 3-25nm together with sporadic periods of enhanced concentrations (Leck *et al.* 2001). The indications suggest that these particles were not produced at the point of the observation but were transported there from layers above, possibly from the free troposphere. The low surface area of pre-existing aerosol in the free troposphere is considered to make it a far more likely place for nucleation to occur than in the boundary layer. Further growth of the particles to the much more stable Aitken mode (25-80nm in diameter) by condensation of H_2SO_4 is assumed. However, calculations of growth by condensation and coagulation suggest that this process requires more than a week and few of particles should survive that length of time. Organic species may also be involved. The measurement of organic species in atmospheric particles requires substantial development. So, the studies of particle formation and growth in Arctic areas have retained many problems in need of further research.

Particles in the diameter range of 25nm–1μm constitute about 85% of the total number concentration (Covert *et al.* 1996). The sources of these particles in the Arctic marine boundary layer in the summer and autumn appear to be a combination of two processes. On some, few occasions this aerosol may be present due to long-range transport, largely aloft, from distant anthropogenic sources on the continents. More often it seems to be due to production and transformation from naturally occurring gas phase precursors within the Arctic region and the marine boundary layer. The chemistry of accumulation mode particles has showed to have major components of sulphate, ammonium (NH_3) and methane sulphonate (MSA) (Leck *et al.* 2001). It is therefore certain that the oxidation products of DMS and their reaction with NH_3 contribute strongly at some stage to the growth of particles. Much of the MSA is thought to be acquired during the pre-CCN stage. It seems therefore that particle growth processes in the Arctic may differ from those elsewhere and further investigation of the evolution of particles and associated chemistry is needed.

3. MONITORING OF ARCTIC AEROSOLS

The Arctic aerosol has been a subject of active research since 1970 (Shaw 1995). There are several monitoring programmes initiated to monitor the level of pollutants and their effects in all compartments of the Arctic environment. The European Monitoring and Evaluation Programme (EMEP) includes several observational sites in the Arctic, where acidifying and eutrophying compounds, ozone, heavy metals, persistent organic pollutants, particulate matter (PM) and volatile organic compounds (VOC) are measured (EMEP 2003). The EMEP database contains monitoring data on a rather large number of chemical and physical parameters of aerosols, but the sampling sites are in general distributed inhomogeneously.

The World Meteorological Organization (WMO) established the Global Atmosphere Watch (GAW) programme in 1989 to address the lack of coordination among the research and monitoring networks. In recent years satellite programmes have produced important measurements of atmospheric compounds and related parameters that complement the GAW network measurements (WMO/GAW 2003). The present network of GAW Global stations consists of 22 stations. Four of them are located in the Arctic (Point Barrow, Alert, Ny Alesund, Pallas-Sodankylä). Aerosol optical depth, mass concentration and major chemical components in two size fractions, aerosol light scattering and absorption coefficients are the main aerosol parameters, which are measured in the GAW stations.

The Arctic Monitoring and Assessment Programme (AMAP) is a program integrating both monitoring and assessment activities in order to produce integrated assessment reports on the pollution status and trends of the conditions of Arctic ecosystems and to identify possible causes for changing conditions. Within the AMAP area, 10 key areas have been identified that are a special focus for coordinated and harmonized monitoring and research activities. Jan Mayen Island is situated near the Svalbard and East Greenland area, and therefore, is potentially suitable measurement site to complement AMAP.

As concerns the measurements of aerosol properties, despite of these monitoring programs the observational network in the Arctic is still sparse and inhomogeneous.

Figures 1 and 2 illustrate the detection of the polluted air inflow to the Finnish Arctic by the measurements of atmospheric aerosol size distribution at Värriö (Finland). Värriö measurement station (67°46'N, 29°35'E) is located to Eastern Lapland close to the Russian border and not far from the Kola Peninsula (north-west Russia), which is the most industrialized area in the Arctic (see Figure 1). The nickel smelter in Nikel, the Cu-Ni-ore roasting plant in Zapoljarnij and the nickel smelter and refinery in Monchegorsk are among the world's largest SO_2 emitters (Kashulina *et al.* 2003). During the easterly wind pollution from Russian industry areas may reach Värriö station. Particle number concentrations were measured with Electrical Aerosol Spectrometer (EAS) designed and built at University of Tartu, Estonia (Tammet *et al.* 2002). EAS measures aerosol size distribution in the diameter range of 3 nm-10 µm.

Figure 1 shows 96-hours backward air mass trajectories arriving at 500 m level at Värriö station on 30 April 2003. Air pollution is transported to the Finnish Arctic after the air mass movement over industrialized areas of the Kola Peninsula. Number concentration of the accumulation mode or submicron aerosol particles (0.1-1µm in diameter) is used as an indicator of the cleanliness of air masses. Figure 2 illustrates the temporal variation of the number concentration of submicron particles. We can see how concentration increased after polluted air inflow from approximately 100 particles cm^{-3} to almost 600 particles cm^{-3}.

This example indicates the need for continuous measurements of size distribution, chemical composition and other properties of the long-range transported anthropogenic aerosols at any place in the Arctic.

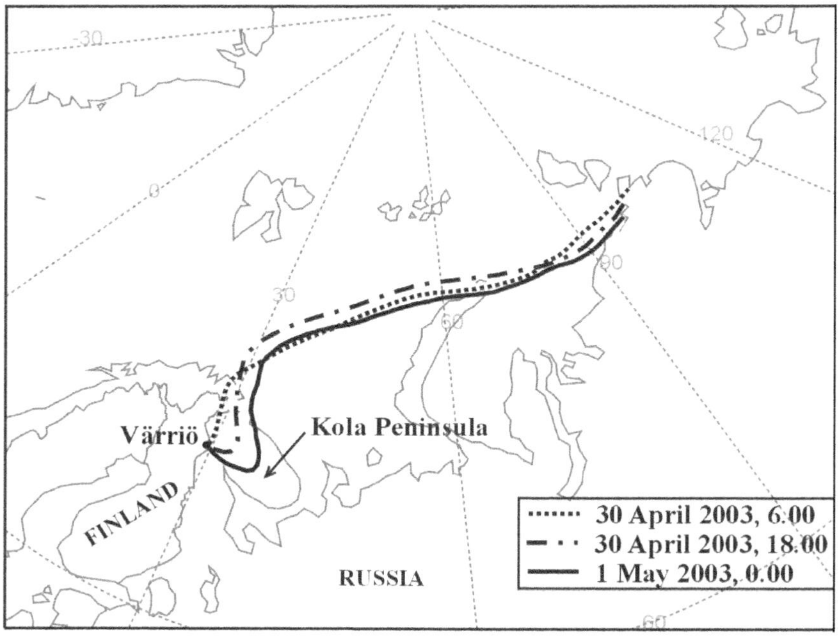

Figure 1. 96-hours backward air mass trajectories ending at Värriö.

4. CONCLUSIONS AND OBSERVATIONAL GAPS

Current knowledge about Arctic aerosols shows that we have come a long way towards understanding the occurrence, nature, and origin of Arctic pollution. However, much remains to be learned.

Since the atmospheric residence time of aerosol particles is relatively short (days to weeks in the troposphere and months in the stratosphere), a large number of measuring stations are needed.

Long-term monitoring of chemical properties of aerosols will be crucial in determining the role that aerosols play in climate, in documenting changes in the regional air quality and in providing a scientific basis for policy decisions regarding control strategies. Due to the complexity of aerosols, a successful network programme must address numerous issues in chemical measurement.

Aerosol impacts depend on particle size. Size-segregated measurements therefore allow the determination of the aerosol properties of fine and coarse particles so that their occurrence and hence effects can be evaluated separately. It would be highly desirable to supplement the aerosol monitoring programme, focusing on particle mass, with more detailed information about particle number distributions.

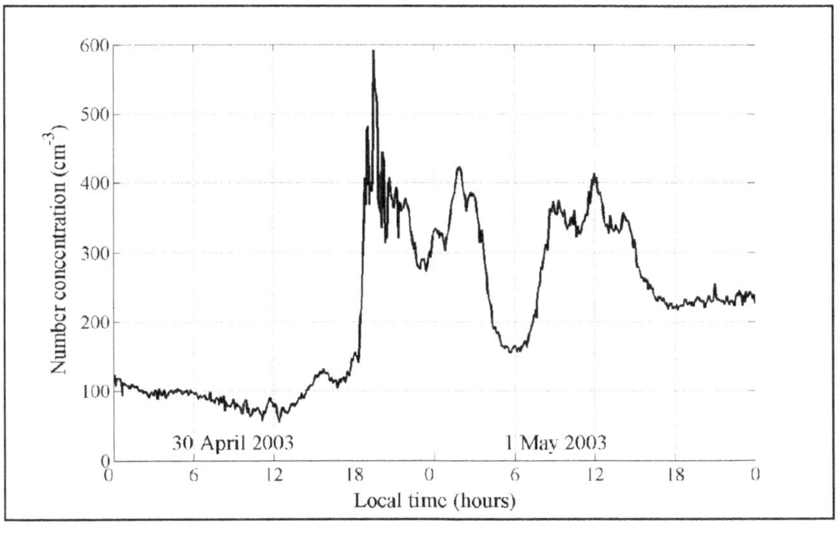

Figure 2. Temporal variation of the number concentration of the accumulation mode aerosol particles (0.1 - 1 μm) at Värriö.

There is a great challenge in adequately characterizing the nature and occurrence of atmospheric aerosols and in including their effects in models to reduce uncertainties in climate prediction. Satellite-based measurements of aerosols are a necessary but not sufficient component of an approach needed to acquire an adequate information base upon which progress in understanding the role of aerosols in climate can be built.

Nucleation is an unsolved problem in aerosol science. Bridging the gap between molecules and macroscopic (>3nm) particles remains a challenge to the scientific community. Formation and growth mechanisms of the ultrafine aerosol particles (3–25nm in diameter) and their contribution to the formation of cloud condensation nuclei in the Arctic atmosphere need to be elucidated.

The choice of study area to reduce the uncertainties of the calculated indirect aerosol-cloud forcing in assessments of the impact of climate change is important.

Measuring network in the Arctic is sparse. So, the new possibilities to found new measuring stations for the studies of the Arctic aerosols are necessary. For these reasons, the location of Jan Mayen Island is perfect for the long-term monitoring of the marine Arctic aerosols.

ACKNOWLEDGEMENTS

This research has in part been supported by the Estonian Science Foundation through grant 5387.

REFERENCES

ACIA 2000. An assessment of consequences of climate variability and change and the effects of increased UV in the Arctic region. Implementation plan. Prepared by the Assessment Steering Committee. 35 pp.

AMAP 1998. Assessment report: Arctic pollution issues. Arctic Monitoring and Assessment Program, Oslo.

Barrie LA 1986. Arctic air pollution: an overview of current knowledge. Atmospheric Environment 20, 643 – 663.

Covert DS, A Wiedensohler, P Aalto, J Heintzenberg, PH McMurry, and C Leck 1996. Aerosol number size distributions from 3 to 500 nm diameter in the Arctic marine boundary layer during summer and autumn. *Tellus* 48B, 197 – 212.

EMEP 2003. Transboundary particulate matter in Europe. Status report 4/2003. 85 pp.

Fuchs N 1964. *The Mechanics of Aerosols.* Pergamon, New York.

Haywood J and O Boucher 2000. Estimates of the direct and indirect radiative forcing due to tropospheric aerosols: a review. *Reviews of Geophysics* 38, 513 – 543.

Hinds WC 1999. *Aerosol Technology: Properties, Behavior, and Measurement of Airborne Particles.* J. Wiley & Sons, New York.

IPCC 2001. *Climate Change 2001: The Scientific Basis.* Cambridge University Press, UK. 944 pp.

Kashulina G, C Reimann, and D Banks 2003. Sulphuric in the Arctic environment (3): environmental impact. *Environmental Pollution* 124, 151 – 171.

Kulmala M, L Pirjola, and JM Mäkelä 2000. Stable sulphate clusters as a source of new atmospheric particles. *Nature* 404, 66 – 69.

Leck C, ED Nilsson, EK Bigg, and L Bäcklin 2001. Atmospheric program on the Arctic Ocean Expedition 1996 (AOE-96): An overview of scientific goals, experimental approach, and instruments. *Journal of Geophysical Research* 106, 32051 – 32067.

Ottar B 1989. Arctic air pollution: a Norwegian perspective. *Atmospheric Environment* 23, 2349 – 2356.

Schnell RC 1984. Arctic haze and the Arctic gas and aerosol sampling program (AGASP). *Geophysical Research Letters* 11, 361 – 364.

Seinfeld JH and SN Pandis 1998. *Atmospheric Chemistry and Physics: from Air Pollution to Climate Change.* J. Wiley & Sons, New York.

Shaw GE 1995. The Arctic haze phenomenon. *BAMS* 76, 2403 – 2413.

Shaw GE 2003. Arctic haze. Pp 155 – 159 in JR Holton (ed.) *Encyclopedia of Atmospheric Sciences*. Publisher, Place.

Sturges WT and LA Barrie 1988. Chlorine, bromine and iodine in Arctic aerosols. *Atmospheric Environment* 22, 1179 – 1194.

Tammet H, A Mirme, E Tamm 2002. Electrical aerosol spectrometer of Tartu University. *Atmospheric Research* 62, 315 – 324.

WMO/GAW 2003. Aerosol measurements procedures, guidelines and recommendations. *WMO TD* 1178, 1-67.

Chapter 8 NATO Advanced Research Workshop

On the Origin of Climatic Signals in the Nordic Seas

Vladimir I Byshev, Victor G Neiman, and Jury A Romanov
P.P.Shirshov Institute of Oceanology, Moscow, Russia
E-mail: romanov@ ocean.ru

Abstract: The Barents Sea was considered as a sort of the Nordic Basin`s key region to monitor short-term climate change manifestations. It is shown that the forcing impact like El Niño events can cause a heat flow pattern reconfiguration as in the Barents Sea so, perhaps, in whole Nordic Area. There is a good reason to assume that distant energetic connections in the sea-atmosphere climatic system are attributed to certain features of atmospheric circulation, which functions, in particular, as an immediate heat carrier.

Key words: Nordic Seas, climatic change, El Niño, atmospheric circulation

1. INTRODUCTION

The Nordic Seas that interconnect the Northern Atlantic and the Arctic Ocean, promptly reply to global climate disturbances (Byshev and Neiman 2000, Byshev *et al.* 2001). This is evident in observations of heat fluxes, convection intensity, and ice conditions, etc. Therefore, the Nordic Seas could be considered as an appropriate natural testing site for detection of climate change on a regional scale as well as with regard to the entire Northern Hemisphere.

One of the key areas for climate forcing is found in the Pacific region. It was Bjerknes (1966) who timely drew attention to a response of the atmosphere in the extra-tropical latitudes to the large-scale positive anomaly of the surface water temperature in the Pacific tropics. Since then it has become known worldwide as ENSO, i.e. El Niño – Southern Oscillation

S. Skreslet (ed.), Jan Mayen Island in Scientific Focus, 79-89

(IPCC 1995). Nowadays many students of climatology are enthralled with the quasi-cyclic El Niño (EN) phenomenon that appears to have a significant influence upon a global climate system. This interest has increased during the last two decades of the past century after EN expressed itself in the most severe manner in 1982-1983.

We have succeeded to observe effects the ENSO even in the Nordic Seas, which are surely very far away from the Pacific tropics. The available observation material for our case study was obtained from the regional data banks for the Barents Sea.

2. ENSO EFFECTS IN THE BARENTS SEA

The Barents Sea is affected by the relatively warm North Atlantic Current (NAC) and is known to be of extraordinary importance for the regional climatic subsystem of the Northern Europe. This is because of the intense heat flux from the sea surface to the atmosphere in this area (Adrov 1993) where the average diurnal heat flow is estimated to be 13.9MJ/m^2. The heat flow noticeably exceeds the corresponding specific values within, for instance, the Iceland Sea, and the Norwegian Sea (Birman and Pozdnyakova 1985).

Continuous advection of warm Atlantic waters is the source of the intense heat loss from the Barents Sea surface. The surface layer belongs to the Murmansk Current, while a warm intermediate layer is fed by recirculated Central Arctic water that is of the same Atlantic origin. It is natural to assume that the temporal variability in the sea component of this three-dimensional heat flux system should immediately affect the surface-atmosphere thermal balance of the Barents Sea, which consequently would change the weather and climate of the region. Therefore, an elucidation of the causes and regularities in the variability of the Barents Sea heat content could explain the physical nature of short-period fluctuations in the regional climatic system.

Environmental events of the past two decades have indicated that an effect of EN could be one of principal sources of such variability. To check this supposition an attempt was undertaken to separate the particular heat response of the Barents Sea to EN events from its general water temperature variability. Thus, we used the method of stochastic processes analysis, which obviously can be applied to ENSO phenomena that are characterized by a random phase, a variable intensity, and different frequency of occurrence, etc.

First we compiled two independent ensembles of phase states of the ENSO system (Sidorenko 1991), with low (EN) and high (La Niña – LN) indexes. Some representative periods of this oscillation characterized by maximal amplitudes are shown in Table 1. For each phase of secular observation series along the Kola longitudinal (standard hydrographic) profile (70°30' – 72°30' N) (Bochkov 1982, Yndestad 1999) the temperature within the upper 200m thick layer was compiled. Based on these data, annual series of monthly averaged seawater temperatures were calculated and ensembles of corresponding curves were constructed by detached manner for the EN and LN phases of ENSO.

In Fig. 1, areas of the inter-annual temperature variability within each of the two ensembles are outlined by thin lines. Average multi-annual temperature variations are marked by bold lines. According to an estimate of confidence intervals, the average phase state can be resolved with a probability of 0.95 for all months except January, for which this index is slightly lower because of the relatively smaller bulk of data presented in the statistical samples.

Noteworthy is the overlapping of opposite phase state areas owing to a superposition of short-period (2-8 year) climatic change, which is exited, for instance, by ENSO, and low-frequency (about 60 year) fluctuations, which reflect cold and warm epochs in the Earth climate (Byshev *et al.* 1997).

A comparison between the temperature characteristics of the upper layer of the Barents Sea calculated for the two principal phases of ENSO exhibits the following. The average monthly temperature variations reveal a sinusoidal annual wave related to a trivial variability of water heat content from season to season. In such a case, neither noticeable structural nor phase differences are observed between the two ensembles. At the same time, the 200m thick layer at the Kola meridian profile became colder by 0.5°C (confidence level 0.95) during a phase transition from LN to EN.

In spite of the relatively small absolute value of this difference, even a very rough estimate of the corresponding change in the total heat reserve in the sea's upper layer demonstrates that such a deviation from the "standard" value must substantially affect quantitative parameters of the sea-atmosphere heat exchange process. These variations, in their turn, inevitably lead to certain perturbations in the atmosphere's circulation, which bring about anomalies in the local and regional weather (and climate, ultimately).

This raises a problem: How much time is needed after the termination of the indirect impact of ENSO on the Barents Sea upper layer to recover its regular temperature, and for the regional sea-atmosphere heat fluxes to regain their "normal" state? This question is of fundamental importance in the climatic aspect. Its clarification undoubtedly requires a quantitative consideration of the heat influx to the Barents Sea owing to lateral advection versus the heat loss due to density convection.

Table 1. Low El Niño (EN) and high La Niña (LN) values of the Southern Oscillation Index (SOI).

Low SOI	Months	High SOI	Months
01.1899-12.1900	24	05.1909-08.1910	16
12.1901-02.1903	15	05.1916-01.1918	21
05.1904-09.1905	17	09.1937-11.1938	15
06.1911-05.1912	12	05.1942-04.1943	12
05.-10.1914	6	12.1949-12.1950	13
12.1914-05.1915	6	04.1955-07.1956	16
06.1918-05.1920	24	12.1959-12.1960	13
09.1939-03.1942	31	03.-10.1964	8
04.1972-04.1973	13	01.1967-07.1968	19
06.1976-04.1978	23	05.1970-03.1972	23
04.1982-09.1983	18	05.1973-05.1976	37
Total	189	Total	193

Some preceding estimates of the possible response of the sea-level atmospheric pressure in the Northern Hemisphere to a strong ENSO signal (Byshev and Lebedev 2000) have shown an average pressure growth by 2GPa at high latitudes (70° – 85°N), i.e. a significant reactivation of the polar anticyclone that appears in the Arctic during an EN phase. In this respect, the 1997-1998 EN, which was the most dramatic of all EN events that took place in the 20th century, has not been an exception. According to reported observations (Anon. 1997, 1998), the Barents Sea region in that period was dominated by an anticyclonic atmosphere circulation. It probably obstructed the ordinary annual cyclonic system of currents in the upper layer of the sea and hampered the westerly inflow of warmer Atlantic water into the Barents Sea area.

In order to verify this assumption we used available data on the discharge of Atlantic water (Fig. 2b) and the heat transfer through the Faeroe-Shetland Strait (Nikolaev and Alekseev 1989). Quantitative characteristics of these flows are indicators of the mass and heat exchange between the Atlantic Ocean and the Nordic Seas. Statistical processing of the data has demonstrated that the flow of warm Atlantic water into the Norwegian Sea during the EN phase decreased by approximately 30% (confidence level 0.99) as compared to the average regular state of the regional current system. In the same time, the heat advection through the Faeroe-Shetland Strait decreased by 29% with an almost analogous confidence level (0.95). In addition, the temporal variation of the Barents Sea ice conditions (Dmitriev and Gudoshnikov 1999), being considered as a climate severity indicator, shows (Fig. 2c) that during EN it increases at least by 10% compared to a LN phase.

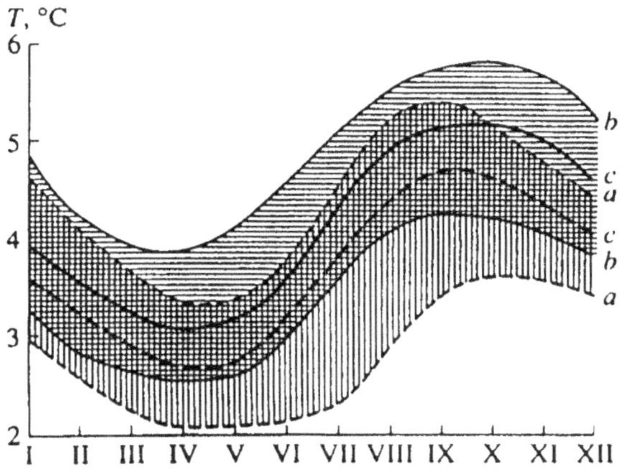

Figure 1. Seasonal temperature variation in the upper 200m thick water layer at the Kola meridian during the a) El Niño and b) La Niña phases. c) Averaged curves for each of the two assemblages.

Hence, a statistical analysis of the long-term series of independent data conclusively demonstrated an actual existence of the Nordic Seas' response to the ENSO signal. The most crucial component of this climatic response is, in our opinion, a significant decrease of the Barents Sea water heat content accompanied by a noticeable expansion of the sea ice conditions. All these actual distortions should eventually affect weather and climatic conditions over the sea surface areas and far beyond their boundaries. Furthermore, it is not unlikely that the shown perturbations inside the coupled system of the Barents Sea and the interactive atmosphere were responsible for the extremely severe winters in the European Arctic during 1997-1999 (Belinskii *et al.* 1999). Owing to the feedback, the observed telecommunication of climate forcing between remote areas of the Global Ocean could be realized immediately by means of fast heat fluxes in the atmosphere. Let us describe this type of feedback effect in the ocean-atmosphere climatic subsystem in some more detail.

Prior to the exertion of the above mentioned effects in the sea-atmosphere climatic sub-system it is necessary to emphasize our primary interest in the structure of modern climatic change in the Arctic Region. So, first of all, the attention was drawn to the diagnostic study of the Northern Hemisphere's atmospheric circulation, which drives the hydro-meteorological fields in global as well as regional scales.

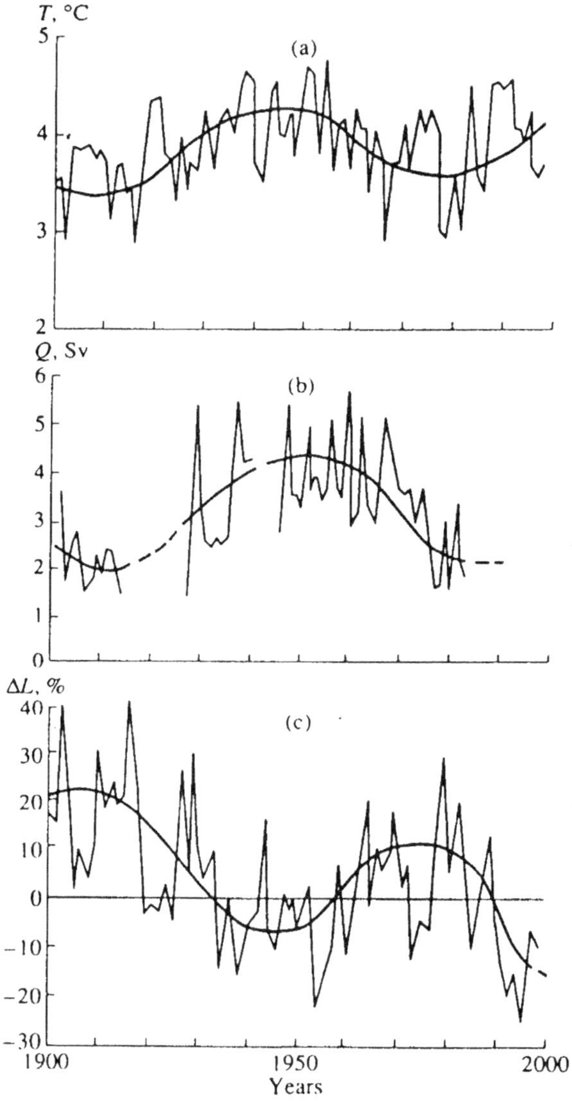

Figure 2. Temporal variations in a) average annual values of the temperature of the 200 m thick water layer at the Kola meridian (in °C). b) Discharge of Atlantic water in the Faeroe-Shetland Strait (in Sverdrups), and c) ice condition anomalies in the Barents Sea (in % of the climatic standard). Smoothed curves were obtained by means of low-frequency filtration of the initial data.

3. THE VARIABILITY OF ATMOSPHERIC CIRCULATION IN THE NORTHERN HEMISPHERE

The peculiarities of the atmospheric circulation and heat fluxes in the Northern Hemisphere in the 20th century were considered on the basis of the classification of atmospheric processes (Dzerdzeevskii 1968) and the calendar of the elementary circulation mechanisms (ECM) (Anon. 1987). According to this classification, the Northern Hemisphere includes 13 types and 41 subtypes of ECM, which correlate well with atmospheric circulation indices (Dzerdzeevskii and Monin. 1954).

The number of days for each of the 41 ECM was initially calculated from time series of acknowledged meteorological records from 1899 till 2000. Thereafter, the annual number of ECM day series were compiled and subjected to statistical probability and spectral analyses.

Figure 3 shows the time series and spectral densities calculated after prior filtration of the long-term trend approximated by a cubic polynomial for four basic groups of circulation. Both time-series and spectra exhibit well-expressed oscillations with periods of 2-8yr. In addition, the hints of cyclic variance with scale of some decades order are seen on all time series. A 20-yr periodicity is clearly distinct just in the fourth group spectrum (meridional southern circulation type). Unfortunately, the limited length of the available time series makes it impossible to resolve by standard methods with a fair degree of confidence the periods of climate change components much larger than the first decade. From this prudent point of view a long-term trend shown on Fig.3 (a) should be considered as well.

The spectra of atmospheric processes differ between the regional sectors of the Northern Hemisphere. Most of them are characterized by a significant decrease in the spectral density of energy after 5-10yr (Fig. 3b). Therefore, in order to separate the high- and low-frequency oscillations, 11yr running averaging of time series was performed. Longer-period cyclic variations were determined as the difference between the 11yr mean and a long-term trend value. The variations in each of the three scales can be studied individually for both the entire hemisphere and its separate sectors. In the latter case a number of time series was cut by merging similar processes: zonal, meridional northern, and meridional southern.

According to results of our analysis, a specific "energy" of short-term disturbance is within 11-92% of the total variance. It is weakly correlated spatially and includes global environmental perturbations, which appear to be caused by ENSO events (Byshev and Lebedev 2000), volcanic activity (Kondrat`ev 1992), etc., and regional fluctuations forced by nonlinear interaction within the climatic system.

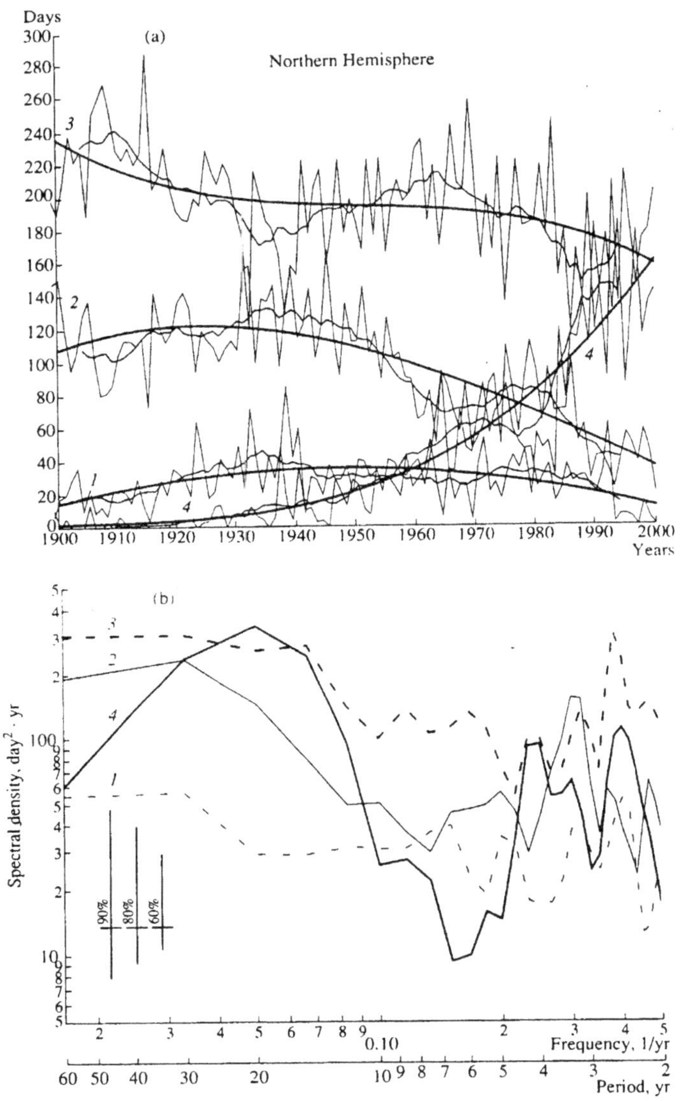

Figure 3. Upper plots (a) show the time series of annual sums (in days) of the four groups of the atmospheric circulation in the Northern Hemisphere. Thin lines: (1) zonal, (2) distortions of zonality, (3) meridional northern, (4) meridional southern. Thick lines: Secular trends based on cube polynomials. Lines of medium thickness: 11-yr running averages. Lower plots (b) show spectra of fluctuations for each group of circulation. Confidence limits of 90, 80, and 60% are shown in the lower left corner.

The time lag in the variation of climatic short-period fluctuations was considered within each accepted group of atmospheric processes. It was found to vary by one order of magnitude during the 20th century, which means the periods of more or less stable climate in each sector were decadal. Statistical probability analysis of monthly sums of ECM with respect to the phases of low (EN) and high (LN) indices of ENSO revealed the response of the atmospheric circulation to El Niño effects. It was significant but its character became qualitatively different by the end of the last century. In fact, in the first half of the century, EN was accompanied by a decrease of the coupled zonal and meridional southern ECMs, while the meridional northern one became intensified. Then, in the end of the century EN (1997-1998) was accompanied by an increase of the meridional southern ECM and a reduction of zonal as well as meridional northern circulation forms.

From 29 % up to 77 % of the climatic variation may belong to the long-term trend, which is an integral feature of atmospheric circulation presented in Fig. 3 (a). One can see that a low-frequency signal probably is the most significant climatic factor. The weight of its input increased sharply in some of the last decades.

Diagnostic investigation of the atmospheric processes clarified that the meridional southern circulation group became more frequent in the second half of the 20th century (Kuvshinova 1987). A similar result is observed in Fig. 3. and seems to determine the state of the present-day climate. An increase of the southern group that appears in the Northern Hemisphere was related to forcing of heat exchange between the Equatorial region and the higher latitudes.

A comparison of the atmospheric circulation peculiarities in particular areas identifies the Equatorial Pacific Ocean as the principal generator of the southern group fluxes in the atmosphere. An increase in number of heavy typhoons and recurrence of ENs, including the strongest (1982-1983 and 1997-1998), indicates increased inter-latitudinal heat flow in the Pacific during the last decades. This process consequently resulted in decreased oceanic heat content, at least in its upper 400 m layer (Liu and Huang 2000). Because the heat store of the Equatorial Pacific water decreased (Byshev 2003), the inter-latitudinal heat exchange that is maintained by this source could reach a turning point and start to decrease as well.

The observation of recurring southern and northern groups of atmospheric processes (Byshev *et al.* 2002) probably shows the absence of balance between them. Thus, one can expect certain compensations in the ocean heat transport system to appear as some kind of change. For example, it is quite possible that a recent displacement of the subtropical gyre in Atlantic (Polonsky 2000) may be attributed to such a process as it obviously demonstrates a feedback effect in the sea-atmosphere climatic system.

4. CONCLUSIONS

The principal result of the present case study seems to denote the possible disturbances of climatic indices on 2-8 yr, 20-60 yr and longer-term time scales. The short-term variation was clearly traced in the Northern Hemisphere, in the atmosphere as well as in the water of the Barents Sea. So, a network of any future appropriate climate monitoring system should be controlled by all revealed climate change scales.

ACKNOWLEDGMENTS

This study was supported by Russian Foundation for Basic Researches, Grant N 02-05-64495.

REFERENCES

Anon. 1987. Large-scale Atmospheric Processes in the Northern Hemisphere and Climatic Extremes in the European Part of the USSR. Data of Meteorological Studies, Moscow, 1987 (13). 180 pp (in Russian).

Anon. 1997. Climate Diagnostic Bulletin. Washington D.C. (1-12).

Anon. 1998. Climate Diagnostic Bulletin, Washington D.C. (1-6).

Adrov NM 1993. *Water Masses Transformation of the Gulf Stream System.* Murmansk Marine Biological Institute, Appatity. 172 pp (in Russian).

Belinskii ON, LV Kirienko and MG Naishuller 1999. Anomalous hydrometeorological phenomena over the Russian Federation in January 1999. *Meteorology and Hydrology* 1999 (4), 112-115 (in Russian).

Birman BA and TG Pozdnyakova 1985. *Climatic characteristics of heat exchange in zones of sea-atmosphere active interaction.* Hydrometcenter USSR. Moscow. 82 pp (in Russian).

Bjerknes J 1966. A possible responce of the atmosphere Hadley circulation anomalies of ocean temperature. *Tellus* 18 (4), 820-829.

Bochkov YA 1982. Water temperature in the 0-200 m layer in the Kola-Meridian in the Barents Sea, 1900-1981. *Proceedings of the Polar Fisheries and Oceanography Institute (PINRO)* 46, 113-122 (in Russian).

Byshev VI 2003. *Synoptical and large-scale variability of ocean and atmosphere.* Nauka Moscow. 344 pp (in Russian).

Byshev VI, LI Galerkin, NL Galerkina, YA Ivanov and LM Fomin 2001. Cooling of the Barents Sea during the El Niño Phenomenon in 1997-1998. *Doklady Earth Sciences* 376 (1), 110-113.

Byshev VI, NK Kononova,, VG Neiman and YA Romanov 2002. Peculiarities of the Climate Dynamics in the Northern Hemisphere in the 20th Century. *Doklady Earth Sciences* 385 (5), 538-535.

Byshev VI, KV Lebedev and MV Matveev 1997. Specialities of modern Northern Hemisphere climate changes. *Izvestiya of the Pacific Research and Fisheries Centre (TINRO-CENTRE)* 122, 16-39 (in Russian).

Byshev VI and MM Lebedev 2000. Probabilistic response of the atmosphere of the Northern Hemisphere to El Niño events. *Oceanology* 40 (5), 631-638.
Byshev VI and VG Neiman. 2000. Response of the Barents Sea to El Niño Events. *Doklady Earth Sciences* 373a (6), 1070-1073.
Dmitriev AA and YP Gudoshnikov 1999. Past, present and future climate of Barents Sea region. *Proceedings of Arctic and Antarctic Research Institute (AARI)*, St-Petersburg, 441, 33-45 (in Russian).
Dzerdzeevskii BL 1968. *Circulation Mechanisms in the Atmosphere of the Northern Hemisphere in the XXth Century*. Institute of Geography of the Academy of Sciences of the USSR. Moscow. 240 pp (in Russian).
Dzerdzeevskii BL and AS Monin 1954. Typical schemes of general atmospheric circulation in the Northern Hemisphere and circulation indices. *Herald of USSR Academy of Sciences, series Geophysics* 1954 (6), 562-574 (in Russian).
Houghton JT, LJ Meira Filho, BA Callander, N Harris, A Kattanberg and K Maskell (eds) 1996. *Climate change 1995*. IPCC (Intergovernmental Panel on Climate Change), Cambridge, UK. 532 pp.
Kondrat'ev KY 1992. *Global Climate*. Nauka St. Peterburg. 359 pp (in Russian).
Kuvshinova KV (Ed.) 1987. *Circulation Mechanisms of Recent Climate Oscillations*. Nauka Moscow. 192 pp (in Russian).
Liu Z and B Huang 2000. Cause of tropical Pacific warming trend. *Geophysical Research Letters* 27 (13), 1935-1938.
Nikolaev YV and GV Alekseev (Eds.) 1989. *Structure and Variability of Large–Scale Oceanographic Processes and Fields in the Norvegian Energy-Active Zone*. Gidrometeoizdat., Leningrad. 128 pp (in Russian).
Polonsky AB 2000. Is warming of intermediate waters of the North Atlantic subtropical circulation really observed? *Marine Hydrology Journal*. 2000 (6), 76-81 (in Russian).
Sidorenko NS 1991. Characteristics of the phenomena of Southern Oscillation - El Niño. USSR *Hydrometeorological Centre Proceedings* 316, 31-44 (in Russian).
Yndestad H 1999. Earth rotation influence on the temperature regime of the Barents Sea. *Journal of Marine Science*. 56, 381-387.

Chapter 9 NATO Advanced Research Workshop

The Circulation of the Nordic Seas

Jan Piechura
Institute of Oceanology, Polish Academy of Sciences, Sopot
E-mail: piechura@iopan.gda.pl

Abstract: The Nordic Seas play significant roles in the Thermo-Haline Circulation of the coupled Atlantic Ocean-Arctic Ocean System. Heat exchange, deep convection and deep-water formation take place there. Barotropic and topographically steered, northward flow on the eastern side of the Nordic Seas carries warm and, salt Atlantic Water to the Arctic Ocean through the Barents Sea and the Fram Strait. On the other side, along the Greenland shelf, cold and fresh water from the Arctic Ocean flows southwards into the Atlantic Ocean. Cold and dense deep-water formed in the Arctic Mediterranean leaves through the deep trenches of the Greenland-Scotland Ridge as overflow to the Atlantic Ocean. Based on publications and the recent data collected during the VEINS project and by IOPAS a general description of the water circulation is given.

Key words: water circulation, transport

1. INTRODUCTION

The Nordic Seas (Greenland, Icelandic and Norwegian), also called the GIN Seas constitute a very important region for the World Ocean circulation and World Climate. The seawater qualities of this region are derived from two main sources: 1) Atlantic Water crossing the Greenland-Scotland Ridge and 2) Polar Water coming from the Arctic Ocean through the Fram Strait.

Exchanges between the Nordic Seas and the North Atlantic in the south, and the Arctic Ocean in the north, both provide the main source for production of the North Atlantic Deep Water (NADW) and supply heat and

S. Skreslet (ed.), Jan Mayen Island in Scientific Focus, 91-99

salt to the Artic Ocean. The exchange occurs through several gaps in the Greenland-Iceland-Scotland Ridge.

As far as the volume flux is concerned, the Iceland-Scotland Gap is the most important one. It carries more then half of the total exchange, i.e. approximately three quarters of the total inflow and one third of the total outflow (Østerhus *et al.* 2001, 2003).

The Fram Strait, that is situated between Svalbard and Greenland is the place where most of the exchange of volume, heat and salt between the Arctic Ocean and the rest of the World's oceans takes place.

Warm, saline Atlantic Water (AW) is carried northwards mainly by the Norwegian Atlantic Current (NWAC) and the West-Spitsbergen Currents (WSC). Their flow along the continental slope represents a major source of heat and salt for the Nordic Seas and the Arctic Ocean.

Most of the AW inflow returns to the Atlantic Ocean as overflow from the Nordic Seas where the processes of forming of intermediate and deep waters are the main forcing mechanism for the Atlantic Inflow.

The AW that is characterised by temperature >3°C and salinity >35 is carried by the WSC that enters the Arctic Ocean through the eastern part of the Fram Strait. On the western side of the Fram Strait the East Greenland Current (EGC) transports colder and fresher (≈ -1°C; <34.4) water southwards. Between WSC and EGC, in the weakly stratified Greenland Sea, the warm and salt AW cools and is modified by mixing and ice forming. It becomes dense enough to convect to great depth, thus forming a variety of deep and intermediate water layers. The pattern of water circulation that distributes volume and heat, starts from the south and proceeds in an anticlockwise direction.

2. ICELAND-SCOTLAND OPENING

The North Atlantic Current (NAC) carries salt and warm water that enters the Nordic Seas in three separate branches (Figs 1 and 2):

- The Faeroe-Shetland Channel: 3.7Sv volume, 140TW of heat and 130kT of salt (Turrell *et al.* 1999, 2003).
- The Iceland-Faeroe opening: 3.3Sv, 110TW, 15kT (Hansen *et al.* 1999).
- The Denmark Strait as a northward extention of the Irminger Current: 1Sv (Kristmannsson 1998).

Adding 1Sv of inflow through the Bering Strait (Roach *et al.* 1995) makes the inflow to the Arctic Mediterranean (Nordic Seas + the Arctic Ocean) a total of 9Sv.

The outflows or overflow of cold deep-water from the Arctic Mediterranean to the Atlantic Ocean takes place through five deep passages:

- Wyville-Thomson Ridge.
- Faeroe Bank Channel.
- Iceland-Faeroe Ridge.
- Denmark Strait.
- Davis Strait.

The Faeroe Bank Channel is the deepest passage across the Greenland-Scotland Ridge.Its deepest part is constantly filled with the cold overflow water which flow into the Atlantic Ocean with average velocities exceeding $1\ m{\cdot}s^{-1}$ in the core. Summing up of the overflow through the Faroe Bank Channel (1.9Sv) across the Iceland- Faeroe Ridge (1.1Sv) and across the Wywille-Thomson Ridge (0.1Sv) gives a total overflow of about 3Sv. A similar number is quoted for the Denmark Strait (Dickson and Brown, 1994), so the remaining 3Sv should come through the Canadian Archipelago and the Davis Strait.

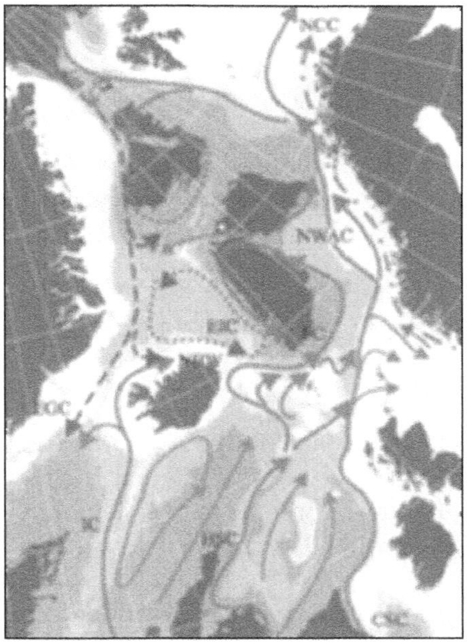

Figure 1. Surface currents in the North Atlantic and the Nordic Seas. CSC: Continental Slope Current. EGC: East Greenland Current. EIC: East Icelandic Current. FC: Faroe Current. HSC. Hatton Slope Current. IC: Irminger Current. NAC: North Atlantic Current. NCC: Norwegian Coastal Current. NIIC: North Icelandic Irminger Current. NWAC: Norwegian Atlantic Current (Modified from Hansen and Østerhus 2000)

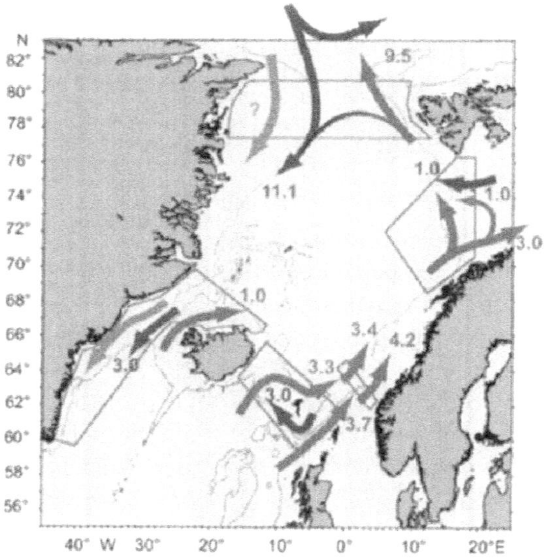

Figure 2. Mass transports (Sv) as measured in the VEINS – period through the major passages between the Arctic and the North Atlantic (VEINS 2000).

The seasonal variations of both inflow and overflow do not exceed 10-20%, and the observations have not lasted long enough to estimate the long-term variability.

3. THE NORWEGIAN ATLANTIC CURRENT (NWAC)

North of the Iceland-Scotland Ridge two branches of Atlantic Water merge into the Norwegian Atlantic Current that is a northern extension of the North Atlantic Current. According to Orvik and Mork (1966) the NWAC flows as an about 30km wide barotropic current trapped over the steepest part of the slope between depth 200 and 900m. The core flow is very stable along isobaths in the north-eastern direction with a mean speed of $40 cm \cdot s^{-1}$ and maxima about $100 cm \cdot s^{-1}$.

The interannual variation in the transport of NWAC ranges from 0 to 12Sv, around the mean value estimated to 5.3Sv. However, the seasonal variability is small.

The latest VEINS data from the Svinøy section outside the North Sea coast of Norway show two branches that are topography trapped. One is a mainly barotropic slope branch and the western branch is related to the Arctic Front with a significant baroclinic part.

The western branch is an about 400m deep and 30-50km wide frontal jet located above the 2000m isobaths. It reaches a maximum speed of 80 $cm \cdot s^{-1}$ and a mean transport of 3.4Sv. The eastern branch is located over the slope between depths of 200 and 800m. The transport is estimated at 4.4 Sv, which make a total of 7.8Sv for transportation of into the Nordic Seas. Closer to the Norwegian coast, on the continental shelf, the Norwegian Coastal Current carries coastal water northwards. It is a mixture of Atlantic water diluted by freshwater from Norway and the Baltic Sea.

Approaching the Barents Sea the NWAC bifurcates. One branch flows eastwards as the North Cape Current. The other is the West-Spitsbergen Current that is directed to the north.

Atlantic Water enters the Barents Sea mainly as a highly variable barotropic flow in two or more distinct cores. One that is 50-70km wide is located between 71°30'and 72°N and the other between 72°45' and 73°N. There is a return flow in between. The transport is highly variable, with average values of 2-3Sv.

4. THE WEST-SPITSBERGEN CURRENT (WSC)

The West-Spitsbergen Current is a complicated, multistream, variable flow with its main branches along the continental slope and along the Mohn's-Knipovich Ridge (Fig.3). The core of Atlantic Water is located as a layer within 100-200m. Barotropic forcing dominates the WSC, but baroclinicity plays a significant role as well. Meso-scale eddies and return flows are permanent features of this current.

The transport of Atlantic Water by the WSC varies broadly but the values of 5-8Sv is most often accepted.

A large portion of the Atlantic Water returns to the west and south as the Return Atlantic Current (RAC). This process starts far to the south, at least at the latitude of 76°N and continues up to the Yermak Plateau.

The West-Spitsbergen Current is the main source of heat and salt supply to the Arctic Ocean and the Greenland Sea. The heat loss to the atmosphere and melting of ice is estimated at about $1000 W \cdot m^{-2} \pm 400 W \cdot m^{-2}$ (Boyd and Asaro 1994). Despite lack of direct contact with the surface the warm core of Atlantic Water cools at about $800 W \cdot m^{-2}$ and this is too much to be caused by diapycnal diffusion alone. Energetic eddies diffuse heat along steeply sloping isopycnals that connect the warm core to the surface, and renew the surface layer several times a day. The local heat loss from the WSC is larger than the local heat gained by air and ice. This would indicate that the heat loss is spread horizontally as well.

The shelf of Svalbard is influenced by the East-Spitsbergen/Sørkapp Current that carries cold and fresh Polar Water from the north-eastern Barents Sea. This current fuses with the West Spitsbergen Coastal Current that also receive freshwater from local glaciers. However, the curents carry only small amounts of water and their participation in the water balance is negligible.

Gravity currents cascading down the slope of Storfjordrenna carry very cold (below -1.5°C) and salt (over 35) water that originates from freezing of seawater and brine rejection from sea-ice on the Storfjord shelf.

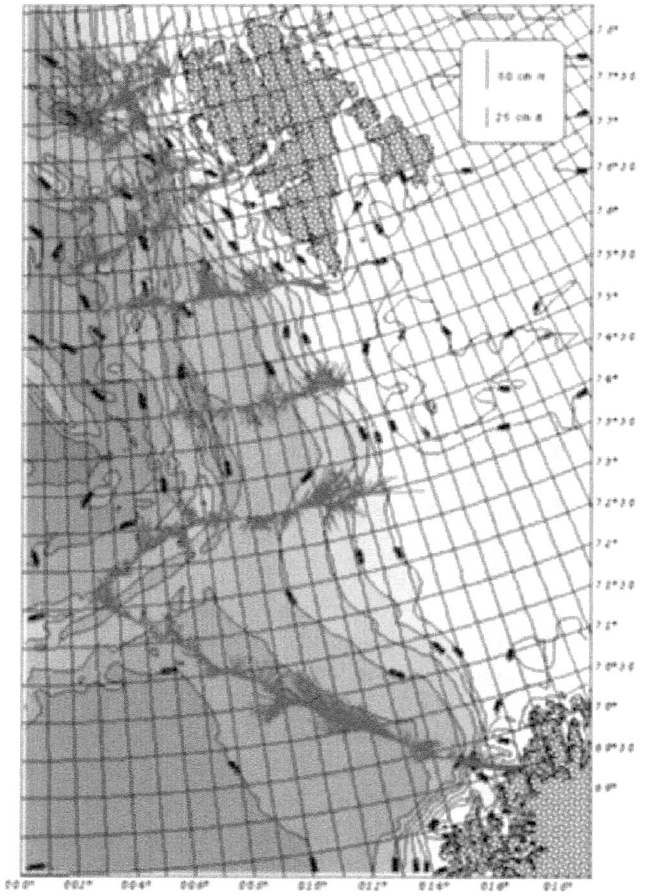

Figure 3. ADCP measured currents at the depth of 100m in June-July 2000 (IOPAS data).

5. THE FRAM STRAIT

Two large currents are dominating in the Fram Strait. The West-Spitsbergen Current flows northwards at the eastern side while the southward flowing East Greenland Current (EGC) flows on the opposite side. According to the VEINS data (VEINS 2000) the WSC obtains maximum speeds over $20 cm \cdot s^{-1}$ in its core over the upper part of the continental slope. The EGC core is located in the upper 200m over the deep part of the slope and flows with a maximum speed of $9\ cm \cdot s^{-1}$. Each of these currents has a horizontal dimension of about 100km. In between them is observed a weak current (about $1 cm \cdot s^{-1}$) of recirculating water. The barotropic component of the WSC is up to $20 cm \cdot s^{-1}$, and is only $5 cm \cdot s^{-1}$ in the EGC. The baroclinic component in both currents is quite similar, and about $5\ cm \cdot s^{-1}$.

Water of Pacific origin has been observed on the East Greenland Shelf (Jones *et al.* 1998) indicating some influence. Maximum recirculation of water was observed just south of the Fram Strait. Average volume transport over two years calculated by averaging monthly mean transport is 10.8 ± 1.6Sv to the north and 11.2 ± 1.7Sv to the south. It is larger than previously given in publications and in better agreement for the EGC than for the WSC because of a stronger baroclinic part relative to a barotropic part in the EGC. The annual mean heat flux by Atlantic water was estimated at 45TW to the North (60W to the N minus 15TW to the S).

Modelling data give a northward mean volume flux around 2.2Sv with the maximum of 4.5-4.2Sv in early spring; while the southerly directed mean flow was calculated at 4-5Sv with a maximum of 7Sv in winter.

VEINS data show a strong month to month variability of north/south fluxes, which can be caused at least partly, by numerous mesoscale circulations: 3 to 150 eddies were crossing the mooring position per month.

6. THE EAST GREENLAND CURRENT (EGC)

The western side of the Nordic Seas is dominated by the East Greenland Current that is a cold, fresh flow of the Polar Water originating in the Arctic Ocean. It flows along the continental slope of the Greenland where large baroclinic components and instabilities produce numerous eddies, particularly along the Arctic Frontal Zone.

. Most of the water carried by the EGC leaves the Nordic Seas through the Denmark Strait, but a part is recirculated to the east as the Jan Mayen Current (JMC) and the East Icelandic Current.The JMC is located north of Jan Mayen Island at the latitude 72-74°N. Its surface water is relatively cold and fresh and readily subject to freezing. The position of JMC in winter is often associated with "Odden", a persistent eastward extension of the drift-ice carried by the East Greenland Ice Stream Current.

Many oceanographers consider the JMC as a meander of the EGC, at least in part. It is also regarded as the south component of the Greenland Sea Gyre.

To the south and west of Iceland warm water is advected with the north flowing Irminger Current (IC), which is a branch of the North Atlantic Current. It meets the cold EGC in the area of the Greenland-Iceland Ridge and splits in two branches. The larger western branch forms a cyclonic eddy in the Irminger Sea. The other branch rounds the north-west penisula of Iceland and flows eastwards off the north coast of Iceland. Further to the north and east of Iceland cold water from the EGS forms the East Icelandic Current that flows eastwards into the Norwegian Sea.

The Denmark Strait overflow is a vigorous topographically steered, near-bottom current. It is 300m thick and reaches a maximum speed up to $100 cm \cdot s^{-1}$ and a mean speed of about $25 cm \cdot s^{-1}$.

7. SUMMARY

At last three features are very important for the Thermo-Haline Circulation and processes associated with the circulation of the Nordic Seas:

- Heat transport to high latitudes by the Atlantic Water carried northwards by the North Atlantic Current, the Norwegian Atlantic Current and the West-Spitsbergen Current.
- Deep convection and formation of dense deep water in the Greenland Basin, which maintains the deep overflow into the Atlantic Ocean.
- Freshwater balance in the Arctic Mediterranean.

Large amounts of heat are transported by Atlantic Water to the Nordic Seas and the Arctic Ocean where it is transferred to the atmosphere and surrounding waters. The heat flux determines the climate over north-western Europe which is 5-10°C warmer than it should be at this latitude.

The northward flowing Atlantic Water is cooled and transformed into high density intermediate and deep water in the process of deep convection. On the other hand, high river runoff and net precipitation deliver fresh water which decreases the density of the surface layer. This decreases the heat loss from the sea surface and results in ice formation which prevents further heat loss to the atmosphere. However, brine rejected from freezing creates dense water which sinks into the deeper layers as convective plumes. A reduction of deep convection in the Greenland Sea has recently been observed. The renewal of the Greenland Sea bottom water now occurs by advection of a bit warmer and less dense deep water from the Arctic Ocean. Less intensive deep convection could influence the Thermo-Haline Circulation and contribute to Climate Change.

These and other observed changes in the circulation of the Arctic Mediterranean are attributed to the changes in the North Atlantic Oscillation (NAO), at least partly.

REFERENCES

Boyd TJ and AD Asaro 1994. Cooling of the West Spitsbergen Current: Wintertime observation west of Svalbard. *Journal of Geophysical Research* 99 (C6), 22591-22618.

Dickson RR and J Brown 1994. The production of North Atlantic Deep Water-sources, rates, and pathways. *Journal of Geophysysical Research* 99 (C6), 12319 -12341.

Hansen B and S Østerhus 2000. North Atlantic-Nordic Seas exchanges. *Progress in Oceanography* 45, 109-208.

Hansen B, KMH Larsen, S Østerhus, WR Turrell, and S Jónsson 1999. The Atlantic Water inflow to the Nordic seas. *International WOCE Newsletter* 35, 33-35.

Hansen B, S Østerhus, H Hátún, R Kristiansen, and KMH Larsen 2003. The Iceland–Faroe inflow of Atlantic water to the Nordic Seas. *Progress in Oceanography* 59 (4), 443-474.

Jones EP, LG Anderson, and JH Swift 1998. Distribution of Atlantic and Pacific waters in the upper Arctic Ocean: implications for circulation. *Geophysical Research Letters* 25, 765-768.

Kristmannsson SS 1998. Flow of Atlantic Water into the northern Icelandic shelf area, 1985- 1989. *ICES Cooperative Research. Report* 225, 124-135.

Orvik KA and M Mork 1996. Atlantic water transport from long term current measurements in the Svinøy section. ICES Annual Science Conference, Reykjavik, Iceland, CM 1996/O:8.

Østerhus S, WR Turrell, B Hansen, P Lundberg and E Buch 2001. Observed transport estimates in the North Atlantic and the Arctic Mediterranean in the Iceland-Scotland region, *Polar Research* 20 (1), 169-175.

Roach AT, K Aagaard, CH Pease, SA Salo, T Weingartner, V Pavlov, and YM Kulako 1995. Direct measurements of transport and water properties through the Bering Strait. Journal of *Geophysical Resesearch 100* (C9), 18443-18457.

Turrell WR, B Hansen, S Hughes, and S Østerhus 2003. Hydrographic variability during the decade of the 1990s in the Northeast Atlantic and Southern Norwegian Sea. *ICES Marine Science Symposia* 219, 111-120.

Turrell WR, B Hansen, S Østerhus, S Hughes, K Ewart, and J Hamilton 1999. Direct observations of inflow to the Nordie Seas through the Faroe Shetland Channel 1994-1997. ICES CM 1999/WL:01. VEINS Final Scientific Report 2000, Contract MAS3-CT96-0070. University of Hamburg.

Chapter 10 NATO Advanced Research Workshop

Jan Mayen Island Ecology

Its Relation to the Arctic Mediterranean Ecosystem

Stig Skreslet
Faculty of Fisheries and Natural Sciences, Bodø Regional university, Bodø, Norway
E-mail: stig.skreslet@hibo.no

Abstract: Jan Mayen Island is remote from all continents, lies downstream to both polar and Atlantic water and its waters feed deepwater into the Atlantic Thermohaline Circulation. Its marine fauna consists of Arctic and boreal species, and species endemic to the sub-arctic transitional zone. Their populations merge in the foodweb of one large ecosystem, the Arctic Mediterranean. Planktonic crustaceans constitute the system's carrying capacity for large populations of marine fish, birds and mammals. Despite nesting sea-birds fertilise the ground, the terrestrial flora and fauna is very poor. Landlocked fish accumulate toxins transferred by contaminated sea-birds. Global warming decreases Jan Mayen's contact with drift-ice biota and may change the foodweb in many respects.

Key words: trophodynamics, island biogeography, tele-connection, pollution, alien species

1. INTRODUCTION

Jan Mayen Island is surrounded by wide oceanic expanses characterized by Arctic as well as Atlantic water (Fig. 1). They form a region that is an integrated part of the Arctic Mediterranean that has been defined by Tchernia (1980) as all marine areas enclosed between the Eurasian and North American continents. The definition also serves the purpose with regard to an ecosystem that is in accordance with the classical principles of Odum (1971). It conserves the flows of energy and matter in a foodweb that contains the dominant populations of producers and consumers.

S. Skreslet (ed.), Jan Mayen Island in Scientific Focus, 101-112

The island is part of the Mid-Atlantic Ridge and composed entirely of volcanic material of tertiary origin (Fitch *et al.* 1965). The rocks easily erode, giving origin to cliffs, boulders and large amounts of loose sand. Places where wind-driven transportation of sand is reduced allow growth of a poor vegetation that is mainly dominated by mosses.

Most of the precipitation falls as snow during winter, but much is blown off the island. During summer the melt-water seeps through the porous ground on top of the permafrost. It seldom appears as surface flow in watercourses but may accumulate in a couple of lakes near sea level. However, except for Nordlaguna, they are shallow and may in some years dry out completely, even Sørlaguna that has the largest surface area.

The geographic position of Jan Mayen, its geology below and above sea-level, and its relationship to the geophysical processes of the atmosphere and the ocean, combines to form a particular set of environmental factors that places it in a unique biogeographical position. The present narrative describes the oceanic, limnetic and terrestrial biota of the Jan Mayen area and sketch how they possibly interact under the influence of their physical and chemical environment. However, it is based on very incomplete information and is meant to emphasise the need for increased scientific studies, rather than express a satisfactory understanding of processes involved.

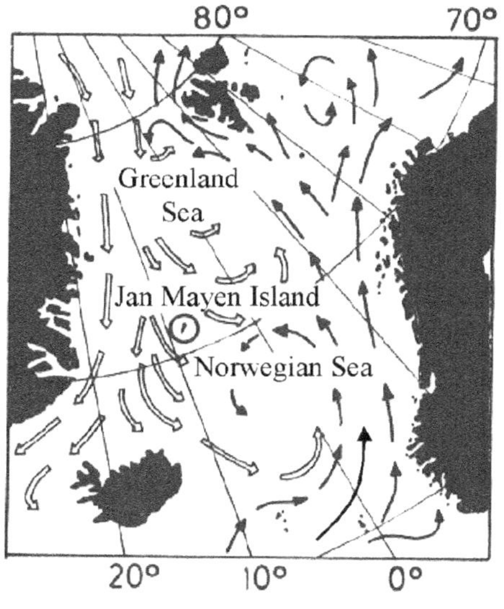

Figure 1. The geographical position of Jan Mayen is down-stream in relation to pollutants from the Arctic Ocean drainage basin, as well as the European continent and shelf. Open arrows: Advection of surface water by the East Greenland Current from the Arctic Ocean to the Nordic Seas. Solid arrows: Advection of Atlantic water.

2. BIOGEOGRAPHY OF THE ECOSYSTEM

The Arctic Mediterranean Ecosystem contains populations of Arctic species that are mostly associated with ice-covered seas. However, some of them have a distribution that extends much farther southwards, and mix with boreal species that extend as far north as to Svalbard (Fig. 2).

The sub-arctic transitional zone where Arctic and boreal species mix, also contains endemic species that are typical for this biogeographical zone (Ekman 1967). The Iceland scallop (*Chlamys islandica*, Mollusca) and the capelin (*Mallotus villosus*, Pisces) are examples of endemic species that are of economical importance. Both are present at Jan Mayen Island. The environmental factors of their population systems are typically influenced by Atlantic as well as by Polar Water.

Jan Mayen Island is situated in the sub-arctic transitional zone and surrounded by the Greenland, Iceland and Norwegian Seas (the Nordic Seas). Its marine fauna is a blend of boreal, Arctic and sub-arctic populations that interact within the ecosystem's food web.

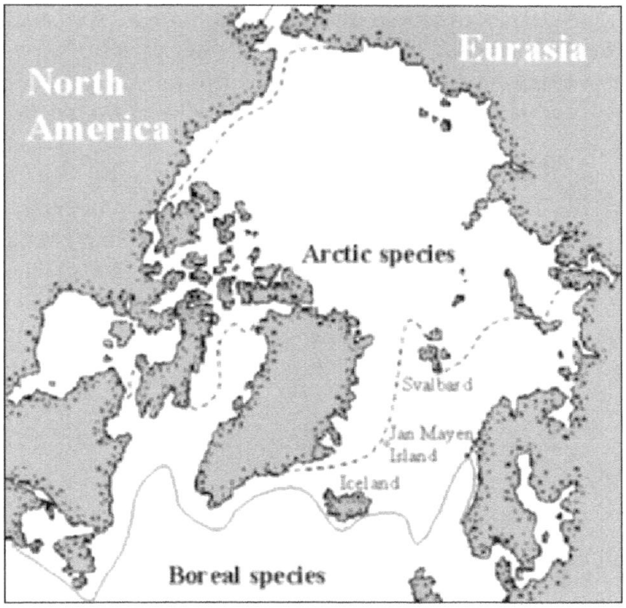

Figure 2. The Arctic Mediterranean with the southern border for Arctic species (solid) and the northern border for boreal species (stippled). The area in between these borders is the sub-arctic transitional zone. Modified from Ekman (1967).

3. THE PELAGIC FOODWEB

Calanus is a genus of planktonic copepods that contains four abundant species in the north-eastern Atlantic. Two of them form populations that play key roles in the production of economically important resources of planktivorous fish. *C. finmarchicus* is endemic in the trans-Atlantic sub-arctic transitional zone where it forms three populations with different genetic constitution. One of the populations inhabits the Nordic and Barents Seas (Bucklin *et al.* 2000). *C. hyperboreus* is a larger Arctic species that perhaps forms one pan-geographic population in the deep oceanic regimes of the Arctic Ocean and the Nordic Seas. The copepods constitute the most productive part of the crustacean zooplankton biomass in the Nordic Seas. However, because much of the produced copepods are rapidly consumed by predators, they are second to the euphausids in terms of average standing biomass (Fig. 3).

When the Soviet Union started investigations on the Norwegian spring spawning herring (*Clupea harengus*) in 1947, this planktivorous fish was the most important predator on *Calanus* spp in the Norwegian Sea. Pavshtiks and Timokhina (1972) observed that *C. hyperboreus* was the dominant copepod in the stomachs of herring caught in the western Norwegian Sea, while *C. finmarchicus* dominated in stomachs from the eastern regions. This is in accordance with newer results (Delpadado *et al.* 2000, Astthorsson and Gislason 2003).

The herring population seeks spawning habitats on the Norwegian continental shelf. There it feeds on reproducing *C. finmarchicus* in its pre-spawning mode. The herring larvae feed on juvenile copepods that are also the main prey for larvae of all commercially important fish that establish population systems in the Nordic and Barents Seas. According to the Soviet investigations, the reproduction of *C. finmarchicus* spread with time across the Nordic Seas, together with other reproducing zooplankton (Fig. 4).

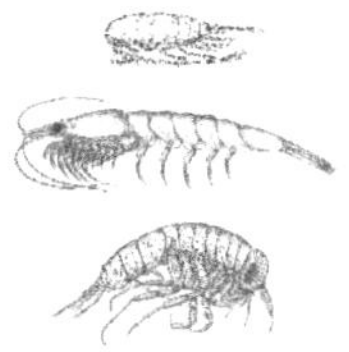

	Production	Biomass
Calanus spp	298	75
Euphausids	242	161
Amphipods	74	49

Figure 3. Estimated annual production and average standing biomass (mill. tons) of zooplankton in the Norwegian Seas and the Barents Sea (Melle and Olsen 2002).

Table 1. Abundance of *Calanus* spp in a north Norwegian fjord in October 1983-2000 correlated with atmospheric pressure indices. NAO: North Atlantic Oscillation. AO: Arctic Oscillation. Data are from Skreslet and Borja (2003).

	NAO			AO		
Species	Season	*r*	*P*	Season	*r*	*p*
Calanus finmarchicus	Mar-Jul	0.717	<0.01	Jul-Sep	0.587	<0.05
Calanus hyperboreus	-	-	>0.05	Jul-Sep	0.832	<0.01

Pavshtiks and Timokhina (1972) observed that *C. hyperboreus* constituted 2-22% of the copepod "production" (=biomass?) in the Norwegian Sea, which shows the dominance of *C. finmarchicus*. They estimated that the "production" of *Calanus* spp varied between 11 and 80 million tons (cf. Fig. 3) in the period 1959-1969, which is an indication of variation in the environmental forcing. One forcing factor was the population of Norwegian spring spawning herring that varied in size and probably exerted inter-annual variation in the predation pressure on Norwegian Sea zooplankton.

The inter-annual variation in zooplankton production is probably not only related to predation. *C. finmarchicus* and *C. hyperboreus* that are produced in the Norwegian Sea during summer accumulate in northern Norwegian fjords during autumn. There, the abundance of *C. finmarchicus* correlates positively with the differences in air pressure between Iceland and the Azores (Table 1). This variable, termed the North Atlantic Oscillation (NAO) reflects geophysical processes in the troposphere. The abundance of *C. hyperboreus* is positively correlated with the Arctic Oscillation (AO) that rather reflects stratospheric processes. Thus, the two species may hold different keys to the understanding of how inter-annual variation in climate interacts with the biological productivity in the Arctic Mediterranean.

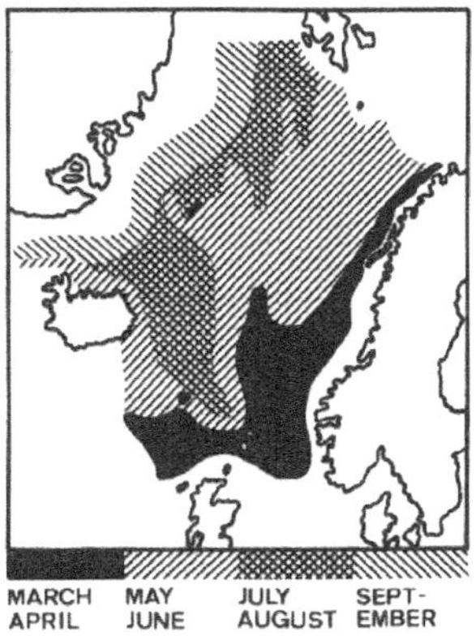

Figure 4. Spreading of the appearance of "spring zooplankton" in the Nordic Seas. Modified after Pavshtiks and Timokhina (1972).

Two fish populations that feed on zooplankton in the Nordic Seas may have profited when the Norwegian spring spawning herring was fished down to a population size that in 1970 was close to extinction. Pavshtiks and Timokhina (1972) argued that this was the case with the blue whiting (*Micromesitius poutassou*) that feed in the Norwegian Sea as juveniles, and as adults between spawning seasons. The population grew to a size that has made it an important resourse for the European fishing industry up to present. The capelin (*Mallotus villosus*) may also have profited. A population that spawns on the eastern Icelandic shelf uses Jan Mayen waters as its adolescent feeding habitat.

After 1970, the Norwegian spring spawning herring was very carefully managed. The growing population only migrated between its wintering habitat in a large northern Norwegian fjord and its spawning and feeding habitats on the Norwegian shelf (Wilhjalmsson 1997). However, from 1990 it migrated into its summer feeding habitats in the Norwegian Sea after spawning (Fig. 5). In 2004 a large part of the population wintered in the Norwegian Sea, which possibly indicates a major change in its migration behaviour. How other fish populations have responded to the return of herring is uncertain.

Fish predators in the Nordic Seas include squids, seabirds, whales and seals. Several hundred years back the Greenland whale (*Balaena mysticetus*) that is endemic in the sub-arctic transitional zone and a specialist on copepod feeding exploited the biomass of *Calanus* spp in Jan Mayen waters. The exploitation of this whale made it nearly extinct and its long absence from the foodweb may have caused an ecological regime shift that has benefited other planktivores (Fig. 6).

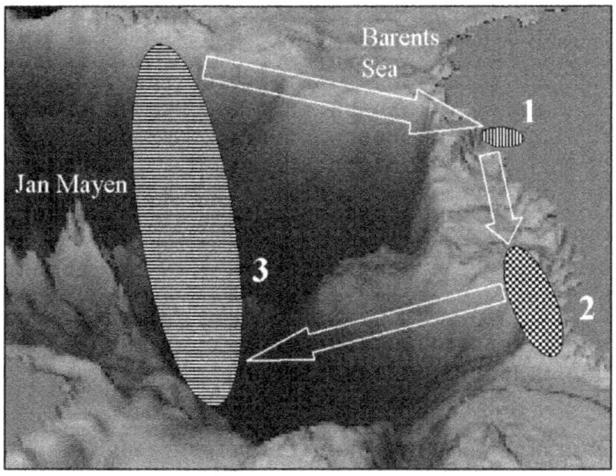

Figure 5. Habitats of sexually mature herring (*Clupea harengus*) 1990-2004. 1) Wintering. 2) Spring spawning. 3) Summer feeding. The herring larvae drift in coastal water from the spawning grounds to nursery habitats in the south western Barents Sea.

Two seal populations are subject to hunting in late winter when they give birth to their juveniles and mate in the drift-ice of the East Greenland Current, often in the proximity of Jan Mayen Island. The juveniles lactate for a few weeks before they are left on their own to find prey. Then, the juveniles of harp seal (*Pagophilus groenlandicus*) feed on zooplankton while the adults feed on epi-pelagic fish, in competition with sea-birds (Fig. 4). Adults of hooded seal (*Cystophora cristata*) dives deeper and may prey on meso-pelagic fish. It may also feed on squid in competition with the bottle-nosed whale (*Hyperoodon ampullatus*) that was previously exploited by Norwegian whalers. Today, whaling at Jan Mayen is restricted to the small minke whale (*Balaenoptera acurostrata*), an opportunistic omnivore that may have profited from the reduction of other planktivorous stocks.

Little is known of the under-ice (sympagic) fauna in the reproduction habitats of harp and hooded seals. However, it is reasonable to suppose that it is not very different from the Svalbard area, where gammarid amphipods and polar cod (*Boreogadus saida*) are important constituents of the foodweb (Loenne and Gulliksen 1989).

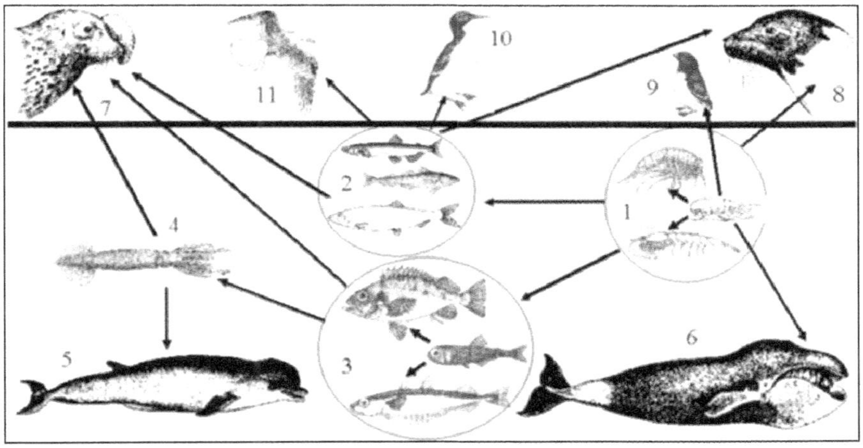

Figure 6. Simplified presentation of the historical trophic energy flow between some of the animal components of the oceanic foodweb of Jan Mayen waters: 1) planktonic crustaceans (*Parathemisto* spp, *Calanus* spp, Euphausiacea), 2) epi-pelagic planktivorous fish (capelin *Mallotus villosus*, polar cod *Boreogadus saida*, herring *Clupea harengus*), 3) meso-pelagic fish (redfish *Sebastes* spp, northern lantern fish *Benthosema glaciale*, blue whiting *Micromesistius poutassou*), 4) squid (*Gonatus fabricii*), 5) bottle-nosed whale (*Hyperoodon ampullatus*), 6) Greenland whale (*Balaena mysticetus*), 7) hooded seal (*Cystophora cristata*), 8) harp seal (*Pagophilus groenlandicus*), 9) little auk (*Alle alle*), 10) Brünnich's guillemot (*Uria lomvia*), 11) glaucus gull (*Larus hyperboreus*).

Some of the cliff-breeding seabirds that feed their juveniles in colonies on Jan Mayen Island during summers probably migrate to the marginal ice zone of the Eat Greenland Current because of its high biological production and accumulation of prey.

4. ISLAND ECOLOGY

Even though Jan Mayen Island is entirely volcanic and its soils therefore contain many essential plant nutrients, its flora mainly consists of moss species (Bryophyta). The vascular flora is poor and its growth is very dependent on nutrients brought onto the island in the form of droppings from seabirds. This is normal in Arctic terrestrial systems where the richest vegetation is found in the surroundings of nesting habitats for seabirds or under their migration routes (Remmert 1980). However, even there Jan Mayen's vegetation is not rich enough to support stable populations of birds that graze or feed on plant seeds. Nor does it support any kind of herbivorous vertebrates like mice or reindeer. Insects that are an important source of proteins for juveniles of birds that migrate to reproduce in the Arctic are very few on Jan Mayen. Only a poor fauna of soil arthropods is present.

The poverty of the terrestrial flora and fauna on Jan Mayen may to some extent be due to the geographical isolation of the island. The large distances to the continents on either side do not favour import of new species by winds. On the other hand, the large amounts of North American and Siberian drift-logs on the shores of Jan Mayen may represent a carrier potential for plant seeds.

According to its remoteness and the small size of Jan Mayen, the general MacArthur-Wilson theory of island biogeography will place the Jan Mayen Island in the lower end of a biodiversity scale. A low rate of immigration and a high rate of extinction will cause equilibrium with few species, both on land and in marine benthic communities.

Human activities connected with whaling, fox trapping, scientific expeditions and activities related to Norwegian public affairs, may have provided bridges for increased immigration of terrestrial plants and invertebrates to Jan Mayen. To the extent that human activities are maintained and expanded, the anthropogenic import of species may continue and even increase, causing a higher biodiversity.

Inaccessible localities may be shown to have a lower biodiversity than the populated areas of the island. Even studies of such relationships may cause changes, if the scientists do not take precautions that prevent transfer of alien species to new habitats.

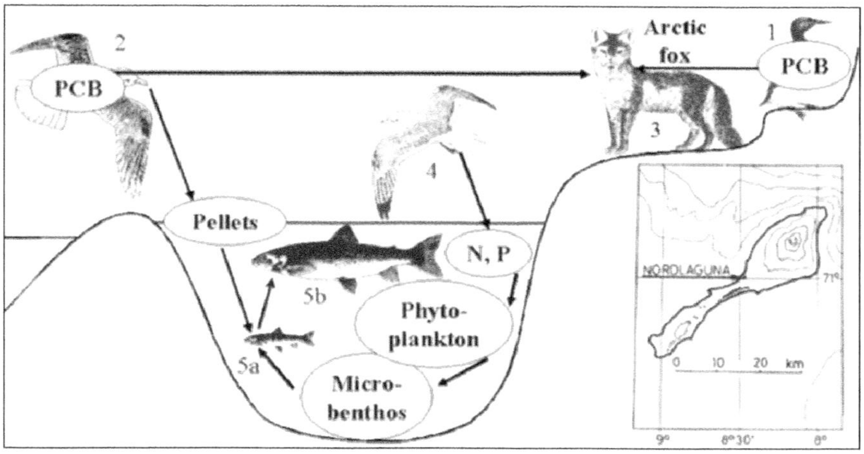

Figure 7. Trophic transfer of matter and energy in the foodweb of Jan Mayen Island. 1) Brünnich's guillemot (*Uria lomvia*), 2) glaucous gull (*Larus hyperboreus*), 3) arctic fox (*Alopex lagopus*), 4) black-legged kittiwake (*Rissa tridactyla*), 5) arctic char (*Salvelinus alpinus*): small spawner (a) and large cannibal (b).

The only mammal resident on the island has been the arctic fox (*Alopex lagopus*). It was known to prey on eggs and juveniles of seabirds and could scavenge on carcasses of fulmar (*Fulmarus glacialis*) that lost control and collided with rocks in strong winds (Richter 1946). It is also likely that the arctic fox benefited in winters when sea ice packed around the island. It might then have located carrion from polar bear (*Thalarctos maritimus*) kills and also placentas during the breeding period of seals. Periods with pack-ice around the island has become briefer and may be one reason for the assumed extinction of arctic fox. However, live trapping was a lucrative business on Jan Mayen because of the high percentage of the blue genotype that was sold to fox farmers in Norway. It may have contributed to the assumed extinction of the population in the latter part of the 20th century.

Arctic char (*Salvelinus alpinus*) is a salmonid fish that populates Nordlaguna, a lake with no outlet to the sea. The population may have originated from anadromous migrants from Greenland or Svalbard. The population probably became land-locked in a period of land-rise 1500 or 4000 years ago (Fitch *et al.* 1965). The char feed on detritus and a very poor fauna of small benthic invertebrates. They grow slowly and mature into small spawners at a low age. Some establish reviers that monopolise their feeding on pellets that are food remains vomited by glaucous gulls (*Larus hyperboreus*). It probably provides the extra energy that allows some char to become cannibals and grow to large size (Skreslet 1973).

Figure 8. A fulmar (*Fulmarus glacialis*) approaches its nesting habitat in the cliffs of Jan Mayen Island. The species may feed on marine zooplankton as well as on carrion from polar bear kills, and dead fish and fish remains from fishing vessels. Its eminent flying capacities may bring on land marine biomass from distant areas. The volcano Beerenberg is a 2277m high landmark on clear days. Photo: ØK Kymre.

Cannibal char in Nordlaguna have livers with high concentration of PCB and DDT (Gabrielsen *et al.* 1997). The source may be pellets and droppings from seabirds, rather than pollutants in precipitation drained from the surroundings.

Planktivorous seabirds sampled on Jan Mayen contain less PCB than seabirds that prey on other seabirds or fish. This indicates trophic accumulation of persistent toxins in the foodweb. Taking into consideration the predation and scavenging on seabirds by arctic fox, a high intake of toxic substances may have contributed to its assumed extinction (Fig. 7).

Trophic transfer by seabirds from distant foraging areas (Fig. 8) may be an important transportation of toxic substances to the island. However, toxins are also present in local marine fish (Gabrielsen *et al.* 1997), which indicates import by currents that transport dissolved toxins or contaminated plankton.

5. BASIN-SCALE ECOLOGY

Jan Mayen Island is tele-connected to human communities in North America by the prevailing hemispheric circulation of air. However, there is no information that toxic aerosols from the continent precipitates on the island. Ecological effects of original airborne pollution may rather be related to Eurasian aerosols that precipitate over the Arctic Ocean where it adds to pollutants contained in freshwater from Russian rivers. Toxic substances that are stored in multi-year ice or bio-accumulated in associated organisms may in theory be transported to Jan Mayen waters by the East Greenland Current. Nations in Western Europe release pollutants into the North Sea by rivers, which adds to the toxins observed in produced water from oil production on the European continental shelf. Both sources contaminate the Atlantic water that is subject to net transport towards Jan Mayen Island (Fig. 1).

Population systems in the Norwegian and Barents Seas are influenced by the inter-annual changes in discharged volumes of freshwater from rivers (Skreslet 1997). Anthropogenic storage of freshwater and out-of season discharge to the coastal seas interfere with turbulent mixing and surface advection in the mixed layer. It probably modifies physical air-sea interaction and interferes with primary production and flow of energy in the marine foodweb.

The anthropogenic Global warming may increase the northward flow of water in the atmosphere and change the seasonal discharge of continental runoff to the Arctic Mediterranean. Altered hydrological exchange of water between atmosphere, continents and oceans may change geophysical processes that forces marine biological production and circulate plankton between different habitats in their population systems. Such changes in the ecosystem may interfere with its production of fisheries resources.

Any major change in the river flow from the continents that border the Arctic Mediterranean will interfere with the stability of the mixed layer in the Nordic Seas, and possibly interact with the generation of Greenland Sea Deep Water. Its advection to the southern hemisphere forces a return flow of surface water from the Pacific to the North Atlantic. This Atlantic Thermo-Haline Circulation is brought into focus whenever the variation in Global climate is discussed (CLIVAR 1998). If the hydrological cycling of water in the World Ocean and the atmosphere will be subject to large anthropogenic change, the entire Arctic Mediterranean Ecosystem may be altered beyond recognition.

A scientific research station on Jan Mayen would be strategically positioned to detect Global climate change, by the monitoring of the island's terrestrial and marine biota, and of processes in their abiotic environment.

REFERENCES

Astthorsson OS and A Gislason 2003. Seasonal variations in abundance, development and vertical distribution of *Calanus finmarchicus*, *C. hyperboreus* and *C. glacialis* in the East Icelandic Current. *Journal of Plankton Research* 25, 843-854.

Bucklin A, OS Astthorsson, A Gislason, LD Allen, SB Smolenack, and PH Wiebe 2000. Population genetic variation of *Calanus finmarchicus* in Icelandic waters; preliminary evidence of genetic differences between Atlantic and Arctic populations. *ICES Journal of Marine Science* 57, 1592-1604.

CLIVAR Scientific Steering Group 1998. CLIVAR Initial Implementation Plan. World Climate Research Programme, Report No. 103. 314 pp.

Depadado P, B Ellertsen, W Melle, and A Dommasnes 2000. Food and feeding conditions of Norwegian spring-spawning herring (*Clupea harengus*) through its feeding migrations. *ICES Journal of Marine Science* 57, 843-857.

Ekman S 1967. *Zoogeography of the Sea*. Sidgwick and Jackson, London. 417 pp.

Fitch FJ, RL Grasty, JA Miller 1965. Potassium-argon ages of rocks from Jan Mayen and an outline of its volcanic history. *Nature* 1349-1351.

Gabrielsen GW, B Brekke, IG Alsos, and JR Hansen 1997. Natur- og kulturmiljøet påJan Mayen. *Norsk Polarinstitutt Meddelelser* 144. 127 pp. (Norwegian).

Lønne OJ and B Gulliksen 1989. Occurrence and ecological importance of of sympagic fauna in Fram Strait, Svalbard area, and Estern Barents Sea. *Papports et Procés-verbaux du Conceil international pour l'Exploration de la Mer* 188, 170-188.

Melle W and RE Olsen 2002. Fórmangel truer veksten i fiskeoppdrett. Kan høsting av plankton løse problemene? *Fisken og Havet Special Publication* 2003 (2), 121-125. (Norwegian, English Summary).

Odum EP 1971. *Fundamentals of Ecology*. Saunders, Philadelphia, PA, 574 pp.

Pavshtiks EA and AF Timokhina 1972. History of investigations on plankton in the Norwegia Sea and the maun results of Soviet investigations. *Proceedings R.S.E.* (3) 73, 267-278.

Remmert H 1980. *Arctic Ecology*. Springer, Berlin. 250 pp.

Richter S 1946. Jan Mayen i krigsårene. *Norsk Geografisk Tidsskrift* 11 (2), 63-86. (Norwegian).

Skreslet S 1973. The ecosystem of the Arctic lake Nordlaguna, Jan Mayen Island III. Ecology of Arctic Char (*Salvelinus alpinus* L.). *Astarte* 6, 43-54.

Skreslet, S. 1997. A conceptual model of the trophodynamical response to river discharge in a large marine ecosystem. *Journal of Marine Systems* 12, 187-198.

Skreslet S and A Borja 2003. Interannual correlation between hemispheric climate and northern Norwegian wintering stocks of two *Calanus* spp. *ICES Marine Science Symposia* 219, 390-392.

Tchernia P 1980. *Descriptive Regional Oceanography*. Pergamon Press, Oxford. 253 pp.

Vilhjalmsson H 1997. Climatic variations and some examples of their effects on the marine ecology of Icelandic and Greenland waters, in particular during the present century. *Rit Fiskideildar* 15, 9-29.

Chapter 11 NATO Advanced Research Workshop

Sea Ice Export as an Indicator of Changing Arctic Ocean Ecology

Igor A Melnikov
P.P. Shirshov Institute of Oceanology, Moscow, Russia
E-mail: migor@online.ru

Abstract: The polar sea ice controls many ecological processes in the Arctic Ocean. Global warming is predicted to shift the ice-edge to the north, decrease the sea ice thickness and surface, and increase ice-free areas. The change may increase the biological productivity and intensify bio-geochemical regeneration processes, especially in the Beaufort Gyre. Less is known about the biological community response to climate in the eastern Greenland Sea, but it may potentially become a key area for monitoring and assessment of ecological change in the Arctic Ocean. Research in the East Greenland ice tongue may establish better estimation and forecasting of ecological effects on higher trophic levels in the North Atlantic marine biological communities including fishes, birds, marine mammals and benthic organisms.

Key words: Arctic, eastern Greenland, ice export, changes, biota, monitoring

1. INTRODUCTION

Polar sea ice may be several meters thick and is an important component of the Global climate system. It controls the fluxes of heat and matter between the atmosphere and the upper polar oceans, and supports unique and tightly coupled biological communities (Melnikov 1989).

Recent climatic changes in the northern hemisphere have led to remarkable environmental changes in the Arctic Ocean:

S. Skreslet (ed.), Jan Mayen Island in Scientific Focus, 113-122

1. A noticeable shrinking of the sea ice cover beginning when the regular satellite observations started in the 1970-ies (Johannessen et al. 1995, 1999, Cavaliere et al. 1997, Vinnikov *et al.* 1999).
2. A rise in temperature and freshening of the Arctic surface waters associated with intensive melting of sea ice (Carmack *et al.* 1995, Serreze and Maslanik 1997, Morison et al. 1998, McPhee *et al.* 1998).

The observed changes predict shifting of the ice-edge to the north, a decrease in the thickness and surface area of sea ice, and an increase of areas with no ice. The scenario suggests an increase of biological productivity, prolongation of the vegetation period, and an intensification of regeneration processes in sea ice systems. However, at present the evidence of impacts of Global climate change on sea ice ecology is sparse or uncertain, though there are fragmentary indications of recent biological changes in the Arctic Ocean. As established now, the biological community response to Global climate change most likely occurs in the regions where the sea ice retreat is rather remarkable, e.g. in the region of the Beaufort Gyre (Melnikov *et al.* 1998, Melnikov and Kolosova 2001, Melnikov *et al.* 2002). Assessment of the present sea ice ecology and modelling of its potential for change are important to estimate and forecast potential changes of the sea ice system and the consequent ecological effects on biodiversity, primary production and carbon fluxes in the Arctic Ocean.

The aim of this paper is to present the Greenland Sea ice tongue as the main region for sea ice export from the Arctic Basin to the North Atlantic Ocean. It is a sensitive indicator of climate and environmental variability in the Arctic Ocean. Jan Mayen Island is situated close to the ice tongue and may therefore be strategically important for monitoring of the sea ice export and conducting of long-term sea ice ecological research.

2. ELEMENTS OF THE SEA ICE BALANCE

In the mid 1970-ies, the Arctic Ocean sea ice area equalled 8.26 million km^2 at the time of maximum development (February) and 6.90 million km^2 at the time of minimum development (September), i.e. 1.36 million km^2 of new ice surface was formed during the winter (Fig. 1).

Two systems of general ice drift are present in the Arctic Ocean:

(1) The Transpolar Drift Stream that contains drift-ice of different age and thickness, varying ice field dimensions and amounts of hummocking ice, etc.

(2) The Beaufort Gyre that is a clockwise gyre in the Canadian Basin (Fig. 2).

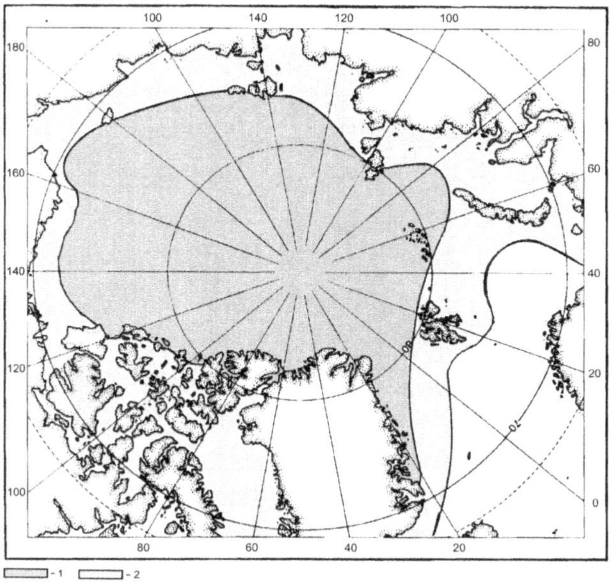

Figure 1. Summer maximum (1) and winter minimum (2) of sea ice distribution in the Arctic Ocean (Atlas of Oceans, 1980).

The residence times of drift-ice in the Transpolar Drift Zone and Beaufort Gyre are different. Sea ice makes the transpolar journey from the Wrangel Island to the Fram Strait within 2.5-3 years, while the ice that circulates in the Beaufort Gyre can remain in the Arctic Basin for many years. The Russian ice station NP-22 drifted for nine years in the Beaufort Gyre, and the US ice station T-3 stayed within the gyre for more than 40 years. Such old ice is thicker than ice located in the transpolar outflow (Zakharov 1981).

From the stable equilibrium of the sea ice cover that existed in the mid 70-ies (Zakharov 1981) follows that new ice should be permanently fed into the Arctic Basin to compensate for the annual loss of older sea ice through the Fram strait. It is logical to assume that the shelf of the Arctic Seas is the source of new sea ice. Gordienko (1958) who studied the regularities of the ice drift in the Arctic Ocean referred to the East Siberian and Chukchi Seas as centers that feed first-year ice into the basin. Sea ice that gets into the Transpolar Drift Zone has a multi-year history by the time it leaves the Arctic Ocean. On an annual basis, most of the seasonal ice that occupies an area of 1.36 million km^2 compensates for the loss of old ice to the Nordic Seas.

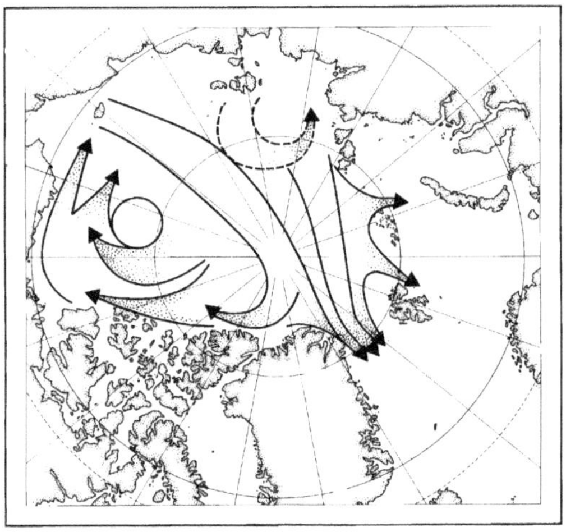

Figure 2. The general circulation pattern of ice drift in the Arctic Ocean as adapted by Hibler (1979) from Gordienko (1958).

The main ice flux of the Transpolar Drift is directed towards the Fram Strait. Several calculations of annual transportation of ice (in million km^2) from the Arctic Ocean are in agreement: 1.25 (Zubov 1945), 0.90 (Volkov and Gudkovich 1967), 1.30 (Hibler 1979), 0.75 - 1.32 (Vinje 1982), 0.90 - 1.30 (Wadhams 1983). Some of the Beaufort Gyre sea ice is fed into the narrow straits between the islands of the Canadian Arctic. However, it comprises only 5-10% of the ice removed by the Eastern Greenland current, and can be neglected (Zubov 1945, Dunbar and Wittman 1963, Mosby 1963, Untersteiner 1963, Shvede 1966, Shneikher 1973). Thus, it can be assumed that the above-mentioned figures represent the total amount of sea ice removed from the Arctic Basin. The values are within the range of 0.75-1.32 million km^2, i.e. approximately 10% of the total sea ice cover in the Arctic Ocean.

3. ENVIRONMENTAL CHANGES

Environmental changes have been observed in the Canadian Basin of the Arctic Ocean by comparing data from the former Soviet ice station North Pole-22 (NP-22) in 1975-81, with data from the SHEBA (Surface Heat Budget in the Arctic Ocean) ice camp in 1997-98, and the "Arctic-2000" expedition with the icebreaker "Ak. Fedorov" (McPhee *et al.* 1998, Morison *et al.* 1998, Melnikov *et al.* 1998, 2000, 2001, 2002).

3.1 Sea Ice Cover

Two features are of primary interest:

1) The ice edge has been displaced northwards by 2-3 degrees of latitude, which is in accordance with satellite information that documents an ice-cover decrease beginning in the 1970-ies (Cavaliere *et al.* 1997).
2) The sea ice thickness has noticeably decreased.

According to recent passive microwave observations by satellites, the Arctic sea ice extent has reached a minimum unprecedented in the last two decades (Serreze *et al.* 2003). In September 2003, it almost equaled the previous year's minimum. The ice edge was well to the north of its median extent in the East Siberian, Chukchi, and Beaufort Seas, and there was a striking lack of sea ice off east Greenland. The absence of ice in the Greenland Sea appears in satellite data recorded for the first time in 2002 (Serreze *et al.* 2003).

In the middle 1970-ies, the mean equilibrium thickness of non-deformed ice in the Canadian Basin comprised about 3m (Busuev 1968, Koerner 1973, Wadhams 1983 *et al.*). Based on data obtained at the NP-22 and AIDJEX ice stations that drifted in the Beaufort Gyre in 1975-1976, the ice thickness after summer melting was 2-3m (Melnikov 1980, McPhee *et al.* 1997). Twenty years later at SHEBA, in October 1997, the mean ice thickness was about 1.5m (Perovich *et al.* 1999). Here, one year later (October 1998), the average thickness of the ice floe where the SHEBA ice camp was located, had decreased to 1m. However, in some parts its thickness was not greater than 0.5-0.8m. Such a noticeable ice thickness decrease for the last two decades could be the result of intense melting of ice due to an increased heat flux.

3.2 Physical and Chemical Components

The present salinity of multi-year ice is almost 50% of the values obtained in the mid 70-ies. Correspondingly, the salinity of first-year ice is now twice as high (Melnikov *et al.* 2002). The concentration of silicate (Si) in both multiyear and first-year ice at the SHEBA station is now very low, being an order of magnitude lower than in previous years, i.e. 3.76 μM and 2.18 μM in multiyear and first-year ice at NP-22 and 0.40 μM and 0.28 μM at SHEBA, respectively. Such a large discrepancy can be caused by a loss of Si with melt-water due to an export from ice to the underlying sea water as a result of active intra-volumetric ice melting in summer at the SHEBA station, while at NP-22 such strong melting was not observed. The loss of Si with melt-water may cause a significant decrease in the number of diatoms (being the main Si-consumers during photosynthesis), which was observed at SHEBA but not at NP-22. The main feature is a two-fold decrease of chlorophyll *a* in multiyear ice at SHEBA and a three-fold increase at NP-22.

3.3 Sea Ice Biota

The species composition of both floral and faunal communities is remarkably changed. The total list of ice algae identified at NP-22 and SHEBA consists of 102 taxa altogether. 84 species (76%) occurred at NP-22 station (1975-81) while only 26 species (23%) occurred at SHEBA (1997-98). The dominance of marine diatoms over the other groups of algae was the most important feature of the sea ice algal community at NP-22. Freshwater algae developed in summer only in melt ponds or in the upper sections of ice. In more recent years, the most important feature of the algal communities of both multiyear and first-year ice is a dominance of freshwater algae (Pyrrophyta and Chlorophyta) over the marine diatom algae (Bacillariophyta). Freshwater algae were in the mid 1970-ies identified only within the upper surface of multiyear ice while they are currently distributed within the whole sea ice interior.

Populations of invertebrates such as Tintinnoidea, Acarina, Nematoda, Turbellaria, Copepoda, and Amphipoda dominated both by number and biomass in the multiyear ice at NP-22 in the mid 1970-ies. They are now rare in collections from SHEBA and the "Arctic-2000" expedition. In ice samples from NP-22, invertebrate animals were recorded in thousands of individuals per m^2 (Melnikov 1989), but in all investigated ice samples from SHEBA only single shells of dead foraminifers, fragments of tintinnids, nematodes, and skins of copepods have been detected.

Observations carried out during the "Arctic-2000" expedition with the icebreaker "Ak.Fedorov" in the Beaufort Gyre have documented that the highest algal biomass (mainly planktonic diatoms) developed within the snow-ice environment (Fig. 3). It is termed the "infiltration" algal community, which is very common in the Antarctic, but previously unknown in the Arctic Ocean.

In previous years, the diatoms developed mainly within the bottom layers of the pack ice and in the ice/water interface where they formed long aggregations attached the under-side of the ice (Melnikov and Bondarchuk 1987). A decrease in sea ice thickness and an increased snow cover determine the recent development of diatoms in the snow/ice interface. The remarkably increased accumulation of snow on the ice surface causes the ice to sink beneath the sea level. That in turn, promotes the development of the infiltration algal community within the snow-ice boundary.

Figure 3. Development of sea ice algae (dark) within the snow-ice boundary of one-year ice. "Arctic-2000" expedition, 82N and 170W, September 2000

The observed phenomenon is especially important with regard to the vegetation's influence on surface temperatures because it affects the albedo of snow-covered surfaces. Experiments have shown a substantial effect during the melting period, which has recently increased in time. The albedo of surfaces covered by snow and ice as well as the extent of melt ponds vary considerably during the vegetation period (Makshtas and Bogorodsky 1996, Ivanov 1999, Andreev and Ivanov 2001). Impact assessments pertaining to Arctic climate change will require high-statistical outputs in order to capture the importance of algal production.

The observed changes in the composition and structure of communities that inhabit sea ice could probably be attributed to an increased melting during the last two decades. Several factors determine these changes, and the most important are:

1. Draining of fresh melt water through the ice strata.
2. Accumulation of freshwater under the ice.
3. Formation of a pronounced pycnocline in 30-35m depth, which reduces the vertical mixing with water below.

These factors modify ecological interactions in the "upper ocean – sea ice" system, changing it from a typically marine to a brackish-water system.

4. EAST GREENLAND SEA ICE EXPORT

The Greenland Sea ice tongue is characterized by outflow of multi-year ice from the Arctic Basin to the North Atlantic, where the ice melts. The ice melting period lasts for several months. The amount of multi-year ice fed into this region is estimated to be 0.75-1.32 million km^2 (Zubov 1945, Volkov and Gudkovich 1967, Hibler 1979, Vinje 1982, Wadhams 1983). Biogeographically, the area has been termed the *sterile region* (Beklemishev 1969). Species expatriation occurs when the sea ice biota degrade because of the melting of the sea ice ecosystem. The biomass is then incorporated into the food web of boreal communities. Calculations show that the reserve of particulate organic carbon (POC) in the Arctic Ocean comprises 8.3 million tons in the spring and 3.7 million tons in the autumn. Accordingly, the reserve of dissolved organic carbon (DOC) equals 124.9 million tons and 56.4 million tons (Melnikov and Pavlov 1978, Melnikov 1981). Bearing in mind that about 10% of the ice is annually removed from the Arctic Basin to the North Atlantic, about 0.4-0.8 million tons of POC and about 5.6-12.5 million tons of DOC are fed into the boreal waters where sea ice melts. This contribution obviously affects the fluxes and total balance of matter in the Nordic Seas (Melnikov 1989, Legendre *et al.* 1992).

Variation in the geographical extension and thickness of sea ice as well as the composition of its biological components may be used to assess Arctic Ocean responses to Global warming. In a context of the International Polar Year (IPY) 2007-2008, the drift-ice region may be considered as a key area for monitoring by ships and satellites. It will then be important to conduct repeated observations by sea ice and water sampling in three transects from Svalbard, Jan Mayen, and Iceland in the direction of the eastern coast of Greenland.

The primary goal of monitoring these transects would be to assess:

1. The effect of recent environmental changes in species composition and structures of drift-ice communities and the ecological fluxes of matter within the system.
2. The reciprocal feedback between the sea ice systems and changing climatic factors by defining the magnitude and variability of:
   a. climatic parameters.
   b. physical processes.
   c. biogeochemical cycling.
   d. biological production processes.

The main focus needs to be on the quantitative and qualitative investigations of all biological components in the Greenland Sea ice tongue. They should include observations on sea ice algae, ice-associated invertebrates, fishes, as well as marine mammals and birds. Key issues include:

1. Satellite sounding of algal production in the sea ice tongue.
2. Ship-board observations of sea ice biota to establish algorithms between satellite and field sounding.
3. Observations in laboratory biota to establish the taxonomy of species that develop within the Greenland Sea ice tongue.

ACKNOWLEDGMENTS

This research was supported by the Russian Foundation for Basic Research (grant #02-05-64357).

REFERENCES

Andreev OM and BV Ivanov 2001. Parameterization of radiation processes in sea ice cover model *Meteorology and Gydrology* 2, 81-88. (in Russian).

Atlas of Oceans 1980. *Arctic Ocean.* Glav. Red. Atlasov Okeanov, VMF USRR, 184 p.

Beklemishev KV 1969. *Ecology and Biogeography of the Open Ocean.* Nauka, Moscow, 291p.

Carmack EC, RW MacDonald, RG Perkin, FA McLaughlin, and RJ Pearson 1995. Evidence for warming of Atlantic water in the southern Canadian Basin of the Arctic Ocean: Results from the Larson-93 expedition. *Geophysical Research Letters* 22, 1061-1064.

Cavaliere DJ, P Gloersen, CL Parkinson, JC Comiso, and HJ Zwally 1997. Observed hemispheric asymmetry in global sea ice changes. *Science* 278, 1104-1106.

Dunbar MJ and W Wittman 1963. Some features of ice movement in the Arctic basin. Pp 90-104 in Proceedings of the Arctic Basin Symposium, October 1962, Arctic Institute Of North America, .

Gordienko PA 1958. Arctic ice drift. In Arctic Sea Ice National Research Council Publication 598, National Academy of Science, Washington, D.C.

Hibler WD 1979. A dynamic thermodynamic sea ice model, *Journal of Physical Oceanography*, 9 (4), 815-846.

Ivanov BV 1999. New data on sea ice albedo in the Laptev and Barents seas. Pp 59-63 in H Kassens, H Bauch, H Eicken, HW Hubberten, M Melles, and J Thiede (eds.) *Land-ocean systems in the Sibirian Arctic.* Springer-Verlag, Berlin.pp.

Johannessen OM, M Miles, and E. Bjorgo 1995. The Arctic's shrinking sea ice *Nature* 376, 126-127.

Johannessen OM, EV Shalina, M Miles 1999. Satellite evidence for an Arctic sea ice cover in transformation. *Science* 286, 1937-1939.

Legendre L, SF Ackley, GS Dieckmann, B Gulliksen, R Horner, T Hoshiai, IA Melnikov, WS Reeburgh, M Spindler, and CW Sullivan 1992. Ecology of sea ice biota. 2 Global significance. *Polar Biology* 12, 429-444.

Makshtas AP and PV Bogorodskiy 1996. A slush-field formation in the Arctic Basin *Meteorology and Gydrology* 1996 (8), 72-80. (Russian, English abstract).

McPhee M, TP Stanton, JH Morison, and DG Martinson1998. Freshening of the upper ocean in the Arctic: is perennial sea ice disappearing? *Geophysical Research Letters* 25, 1729-1732.

Melnikov IA 1981. Importance of drifting ice in the concentrating and transportation of the nutrients and organic matter in pelagic zone of the Central Arctic Basin. Pp 62-63 in *Spatial-temporal structures of the hydrophysical and hydrochemical characteristics of the seas,* All-Union seminar-symposium, Tallin (In Russian).

Melnikov IA 1989. *The Arctic sea ice ecosystem.* P.P. Shirshov Institute of Oceanology, Moscow, 190 pp (In Russian).

Melnikov IA and GL Pavlov 1978. Peculiarities and distribution of organic carbon in waters and ice of the Arctic Basin. *Oceanology* 18 (2), 248-254.

Melnikov IA, and LL Bondarchuk 1987. To the ecology of the mass aggregations of colonial diatom algae under the Arctic drifting sea ice. *Oceanology* 27 (2), 317-321.

Melnikov IA, B Sherr, P Wheeler, and H Welch 1998. Preliminary biological and chemical oceanographic evidence for a long-term warming trend in the Arctic Ocean (current materials of the SHEBA Ice Camp, Beaufort Sea). P 60 in Proceedings of the Arctic Change Workshop, Seattle, University of Washington, June 1997, Report No 8, August 1998.

Melnikov IF and EG Kolosova 2001. The Canada Basin zooplankton in recent environmental changes in the Arctic Ocean. Pp 142-154 in I Semietov (ed.) Proceedings of the Arctic Regional Centre, "Dalnauka", Vladivostok, Vol.3 (2).

Melnikov IAand EG Kolosova, HE Welch, and LS Zhitina 2002. Sea ice biological communities and nutrient dynamics in the Canadian Basin of the Arctic Ocean. *Deep-Sea Research Part 1* 49, 1623-1649.

Morison J, M Steele, and R. Anderson 1998. Hydrography of the upper Arctic Ocean measured from the Nuclear Submarine USS Pargo. *Deep-Sea Research Part 1* 45, 15-38.

Mosby H 1963. Interraction between the Polar basin and peripheral seas, particulary the Atlantic approach. Pp 109-116 in *Proceedings of the Arctic Basin Symposium, October 1962, Arctic Institute of North America.*

Serreze MC and JA Maslanik 1997. Pp p.110-112 in Proceedings from Polar Processes in Global Climate. 13-15 Nov 1996, Cancum, Mexico, American Meteorological Society, Boston.

Serreze MC, JA Maslanik, TA Scambos, F Fetterer, J Stroeve, K Knowles, C Fowler, S Drobot, RG Barry, and TM Harran 2003. A record minimum Arctic Sea ice extent and area in 2002. *Gephysical Resesearch Letters* 30 (3), 1110.

Shneikher AO 1973. North Atlantic iceberges as a indicator of large-scale hydrometeorological processes. *Trudi AANII* 318, 53-68.

Shvede EE 1966. Iceberges of the North-West Atlantic, *Oceanology* 4 (4), 608-614.

Untersteiner, N. 1963. Ice budget of the Arctic ocean. Pp. 219-227 in *Proceedings of the Arctic Basin Symposium, October 1962, Arctic Institute of North America..*

Vinje TE 1982. The drift patten of sea ice in the Arctic with particular reference to the Atlantic approach. Pp. 83-96 in L Rey (Ed.) *Arctic Ocean*, McMillan Press.

Vinnikov KY, A Robok, R Stouffer, J Walsh, C Parkinson, D Cavalieri, J Mitchell, D Garrett, and V Zakharov 1999. Global warming and northern hemisphere sea ice extent. *Science* 286, 1934-1937.

Volkov NA and ZM Gudkovich 1967. General results of studying of the sea ice drift in the Arctic Ocean. *Problemi Arktiki i Antarktiki*, 27.

Wadhams P 1983. Sea ice thickness distribution in Fram Strait. *Nature* 305, 108-111.

Zakharov VF 1981. *Arctic sea ice and modern natural processes.* Gydrometeoizdat, Leningrad, 136 pp.

Zubov NN 1945. *Arctic Ice*, Izd. Glavsevmorputi, 360 pp.

Chapter 12 NATO Advanced Research Workshop

Zooplankton Habitats of the Greenland Sea

An Experimental Laboratory for Studies of Pelagic Ecology

Hans-Jürgen Hirche
Alfred-Wegener-Institut für Polar- und Meeresforschung, Bremerhaven, Germany
E-mail: hhirche@awi-bremerhaven.de

Abstract: The physical oceanography of the Greenland Sea generates three different ecological domains separated by two frontal systems (East Greenland Polar Front, Arctic Front). They are subject to the same light regime which makes the Greenland Sea an ideal place to study physical-biological coupling on various spatial scales. As the three domains represent different marine provinces and climate zones, the effect of climate change on marine ecosystems can be studied very effectively. Examples of physical-biological coupling and phytoplankton-zooplankton relationships in the three domains are shown together with aspects of carbon flux and the contribution of zooplankton. The necessity for seasonal studies in future programs is stressed.

Key words: *Calanus* spp, oceanic, neritic, provinces, Arctic, boreal, biogeography

1. INTRODUCTION

Within the Arctic, the Greenland Sea is unique due to the meridional alignment of three ecological domains that are subject to the same light regime. They contain fauna elements that are predominantly distributed within either of the Boreal or Arctic biogeographical regions that represent latitudinal differences in climate. They overlap in the sub-arctic transitional zone (Ekman 1967) that is extremely sensitive to climate changes such as the recent decrease in ice cover and ice extent in the Arctic Ocean proper (Johannessen *et al.* 2002), and the enhanced Atlantic inflow into the Arctic Ocean (Macdonald 1996).

S. Skreslet (ed.), Jan Mayen Island in Scientific Focus, 123-133

In the Greenland Sea, the major outflow of Polar Water from the Arctic Ocean meets the major inflow of Atlantic water. It is therefore the gateway for exchange processes between high and low latitudes. Its ecological domains are separated by two large frontal systems. The East Greenland Polar Front (EGPF) is the boundary between the East Greenland Current (EGC) and the Greenland Sea Gyre (GSG) in the west, which includes also a marginal ice zone. The other is the Arctic Front (AF) separating the GSG from the North Atlantic Current (NAC) and its continuation, the West-Spitsbergen Current (WSC) to the south and east. The fronts and ice edge systems are ideal study fields for meso-scale physical-biological coupling.

The deep parts of the Greenland Sea are dominated by different species of large copepods of the genus *Calanus*. They undergo large seasonal ontogenetic migrations, transferring a huge amount of carbon to the deep sea. In addition, the deep basins host a rich community of meso-pelagic zooplankton species (Richter 1995). Their roles in the transformation of sedimenting matter are barely studied.

This paper presents bio-physical coupling, phytoplankton-zooplankton interaction and carbon flux in the three domains.

2. SPRING BLOOM DYNAMICS AND PHYTO-PLANKTON-ZOOPLANKTON INTERACTION

The three domains of the Greenland Sea are the arcto-neritic domain on the East Greenland Shelf (EGS), the arcto-oceanic domain of the Greenland Sea Gyre (GSG) and the boreo-oceanic domain of the North Atlantic Current (NAC) and the West-Spitsbergen Current (WSC). The domains have different settings of temperature, nutrients, ice cover, and dominant copepod species of the genus *Calanus* (Table 1), which all affect the dynamics of the biological processes.

2.1 Arcto-Neritic Domain

The arcto-neritic domain includes the East Greenland Shelf (EGS), the marginal ice zone (MIZ) and EGPF. The topography of the EGS is characterised by troughs and banks, which steer the flow and hence affect ice cover and biota (Schneider and Budeus 1994). Both latent and sensible heat polynyas were observed at different times of the year, which have a strong influence on phytoplankton dynamics (e.g. Lara *et al.* 1994, Pésant *et al.* 1998, reviews in *Journal of Geophysical Research* 100, C3, 1995 and *Journal of Marine Systems* 10, 1997).

Table 1. Environmental and biological characteristics of the three ecological domains of the Greenland Sea, and their effects on the plankton.

	Arcto-Neritic	Arcto-Oceanic	Boreo-Oceanic	Effects
T°C	(-1.8) - 0	(-1) - (+1)	2.5 - 6.5	Metabolic rates
NO_3 μmol*	4-6	12	12	New production
SiO_4 μmol*	10-14	6-7	5-6	Phytoplankton community
ice cover	+	+/-	-	Light, stability
Dominant *Calanus* spp	*C.glacialis*	*C.hyperboreus*	*C.finmarchicus*	Life cycle strategy
Weight of adult female (μgC)**	350	2000	80	

*Data from Kattner (pers.comm.)
**Data from Hirche *et al.* 1991

Outside polynyas the spring bloom starts late and is very short. The main zooplankton species is *Calanus glacialis*, a copepod well adapted to the unpredictable food conditions on Arctic shelves, as it is able to reproduce in early spring. It then uses lipids deposited during the previous year, and later switches to freshly ingested food as energy supply for eggs (Hirche and Kattner 1993). It is able to use ice algae to complement its diet (Runge and Ingram 1988) but can starve for a long time (Hirche 1989a) and has a life cycle of several years (Kosobokova 1999).

Although zooplankton biomass is relatively low on the EGS (Hirche and Kwasniewski 1997), copepods used approximately half the primary production in the Northeast Water Polynya and thus represent a major pathway of organic carbon (Daly 1997). However, it seems that most of the egested material is re-mineralised in the water column and does not reach the benthos (Daly 1997). At times large numbers of zooplankton from the GSG or the Return Atlantic Current (RAC) were observed in the troughs on the EGS, but Hirche and Kwasniewski (1997) suggested that they were most likely expatriated and did not grow there successfully.

Marginal ice zones are highly dynamic and productive regions (Smith 1987, Sakshaug and Skjoldal 1989). The MIZ in the Greenland Sea is controlled by the position of the EGPF which separates the EGC from the GSG. Hence, it is relatively invariant in space. This makes it different from other ice edges, which recede in the summer like in the Bering Sea, the Barents Sea etc., where a phytoplankton bloom usually follows the receding ice edge (Alexander and Niebauer 1981, Sakshaug 1987). For the summer MIZ in the Fram Strait Gascard *et al.* (1988) distinguished an inner edge with a stable position from an outer edge, characterized by 1/10 to 4/10 sea ice concentration and composed by small floes. The outer edge is much more variable owing to the action of winds and eddies. Therefore, a combination of ice edge and frontal processes will promote phytoplankton production in the Greenland Sea MIZ. Ice edge upwelling and eddy dynamics may pump nutrients to the surface layer (Niebauer 1982, Niebauer and Smith 1989). Melting of ice at the front provides stability all year round and thus allows algal growth as soon as there is sufficient irradiance. Off-ice winds forcing the ice over the front may extend the productive zone up to 100km eastwards off the front (Hirche *et al.* unpubl.). In cases when oscillation of ice advection combines with ice edge and frontal upwelling, a band of highly productive water will be established all along the MIZ. However, the fate of this production is not yet clear. Increased grazing activity has been observed by Smith *et al.* (1985) and Hirche (1989b), but zooplankton standing stock is mostly low in the MIZ.

2.2 Arcto-Oceanic Domain

The Arcto-Oceanic domain is represented by the GSG. As described above, at the western boundary ice edge several processes strongly affect pelagic production. In addition, an ice tongue ("Isodden") that protrudes from the MIZ advects sea ice north-eastwards from the EGC and may cover most of the GSG in cold years (Johannessen 1986). In warmer years, "Isodden" is separated from the EGC proper by a pronounced tongue of open water ("Nordbukta"). "Isodden" affects the deep convective mixing, but also supports early plankton blooms by melt-water induced stability as found by Legendre *et al.* (1993) on a transect in early June, and observed in sediment traps by Bauerfeind *et al.* (1994). Ramseier *et al.* (1999) estimated that the area of seasonal ice cover, together with an 80km wide band extending along the MIZ, provided about 90% of the total sedimentation.

Silicate depletion may limit diatom growth, which often causes massive blooms of the prymnesiophyte *Phaeocystis pouchetii*. Such blooms were reported from the marginal ice zone as well as from regions remote from the ice (Smith *et al.* 1991, Legendre *et al.* 1993). Smith *et al.* (1991) speculated from production rates of *Phaeocystis* that the Greenland Sea may be a larger sink of atmospheric CO_2 than has been previously thought.

The biomass of herbivorous zooplankton is high in the GSG (Table 2). However, due to permanently low temperatures this biomass is most likely not turned over each year. Instead, Hirche (1997) suggested an at least three-year life cycle for the dominant copepod *Calanus hyperboreus*. Primary production of cells >5µm could more or less provide sufficient carbon to satisfy herbivorous requirements (Table 2). Smaller cells accounted for approximately half of the primary production, indicating size selection by zooplankton.

Table 2. Carbon content of herbivorous copepods, size fractionated primary production (PP) and chlorophyll a, carbon requirements of copepods assuming 10% body carbon d^{-1} (EGS, GSG) and 20% (WSC) on a transect along 74°45´N (phytoplankton data from Legendre et al. 1993, zooplankton data from Hirche *et al.* unpublished).

Domain	Longitude	Zoopl C $g\ m^{-2}$	C-required $g\ m^{-2}$	PP>5 μm $g\ m^{-2}\ d^{-1}$	PP>5μm %	Chla>5μm $mg\ m^{-3}$	Chla>5μm %
EGS	72	7.14	0.71	0.19	54	0.35	48
GSG	79	1.80	0.18	0.24	45	0.75	44
GSG	85	4.20	0.42	0.23	66	0.11	42
GSG	91	2.80	0.28	0.24	62	0.29	37
GSG	98	3.45	0.34	0.19	46	0.20	41
WSC	105	8.20	1.64	0.20	26	0.23	19
WSC	113	5.60	1.12	0.04	9	0.02	7

Little is known of processes on higher trophic levels in the GSG. As commercial fish are absent, chaetognaths, jellyfish and amphipods are probably the main predators of *C. hyperboreus*. It seems therefore that secondary production is mostly under bottom-up control. As most previous sub-arctic observations were limited to short periods and small regions, very little is known on interannual variability of biological production and its fate.

2.3 Boreo-Oceanic Domain

The Boreo-Oceanic domain is separated from the GSG by the AF. It includes the NAC and its northern extension, the WSC. Part of the WSC recirculates in Fram Strait and turns south as the RAC under the EGC along the EGS (Bourke *et al.* 1988), while the other part enters the Arctic Ccean and continues along its rim as the Atlantic inflow (Anderson *et al.* 1989). As most Atlantic floral and faunal plankton components are expatriated in the Arctic Ocean, it is an important carbon sink, the magnitude of which depends on the export of Atlantic water, which shows large seasonal and inter-annual variability (Schmitz 1995, Zhang and Zhang 2001), and on the biological production in the eastern Greenland Sea. The spring bloom in the WSC depends on thermal stability and starts much later than in the GSG. Thus, Legendre *et al.* (1993) observed a beginning diatom bloom in the WSC, when in the GSG already a post bloom situation prevailed. There was a remarkable dominance of small phytoplankton, most likely a result of size selective grazing by a large biomass of the dominant copepod *Calanus finmarchicus* (Table 2). The herbivorous carbon requirements clearly exceeded primary production (Table 2) and resulted in food limitation, as documented by Hirche *et al.* (unpubl.) in egg production experiments with *C. finmarchicus*.

The AF follows the mid-Atlantic ridge from Jan Mayen to Fram Strait (Johannessen 1986). The temperature gradient of 3-6°C makes it a strong bio-geographic barrier (van Aken *et al.* 1991). Numerous eddies and meanders (Trees *et al.* 1992) and associated convergences and divergences should act as nutrient pumps and provide stability, thus allowing much earlier algal growth than in the adjoining Atlantic water. Due to its large extension the frontal zone should considerably contribute to overall production in the Greenland Sea, but not many data are available as yet.

The boreo-oceanic domain is feeding ground of large stocks of commercial fish and therefore of high economic importance. However, the Greenland Sea is outside of international biological monitoring programs such as e.g. the Continuous Plankton Recorder Survey etc. Therefore retrospective studies of climate variability and plankton and zooplankton-fish interaction are not possible as yet.

3. VERTICAL DISTRIBUTION OF ZOOPLANKTON AND CARBON FLUX

3.1 Vertical Structure of the Zooplankton Community

A species inventory for zooplankton in the Greenland Sea was established by Richter (1994). The most important taxonomic groups with regard to biomass are Calanoid copepods (58%) and chaetognaths (14%). Diversity was highest in the 1000 – 1500m layer (Fig. 1). The vertical structure is shaped by a pronounced seasonal vertical migration of the dominant copepods (Hirche 1991, 1997) of the genus *Calanus*, which make up ca. 40% of the total zooplankton biomass.

The species composition is very similar to the Norwegian Sea (Richter 1994) and the Arctic Ocean, especially the Eurasian Basins (Mumm 1993, Kosobokova and Hirche 2000). However, the biomass is much higher in the Greenland Sea (Richter 1994, Hirche, unpubl. data), probably due to suitable feeding conditions and the almost complete absence of large planktivorous fish stocks. Hirche (1997) suggested the Greenland Sea as a main center of *Calanus hyperboreus*, from where it is advected to the south into the Norwegian Sea and to the north into the Arctic Ocean. The same is probably the case with the large meso-pelagic community (Fig. 1b) that may have a role in transformation and sedimentation of particulate organic matter, which is barely studied. Some of the copepod species have specific adaptations to cope with a special food environment, such as the common species *Spinocalanus antarcticus* which has a looped gut (Kosobokova *et al.* 2002).

3.2 Carbon Flux in the Greenland Sea Gyre

The Greenland Sea seems to be a highly productive sea, where a large amount of carbon is drawn down both by biological and physical processes. The area studied most is the GSG (Honjo *et al.* 1987, Bauerfeind *et al.* 1994, Ramseier *et al.* 1999). Hebbeln and Wefer (1991) have shown that the particle flux in the Fram Strait was strongly affected by varied ice regimes. Similarly Ramseier *et al.* (1999) observed a highly variable sedimentation and related particle flux to ice concentration, duration of ice coverage, and distance of the trap location from the ice edge. Elevated export of biologically produced particles was observed in the MIZ as compared to open ocean locations. Multiyear ice released freshwater diatoms and lithogenous material (Kohly 1998). Bauerfeind *et al.* (1994) showed peak diatom fluxes during May-June, whereas by the end of June peak values of faecal pellets were recorded. Due to the time delay of the sedimentation of different components, the authors deduce that the majority of early spring phytoplankton in the Greenland Sea is not controlled by zooplankton grazing. Mass sedimentation of a *Phaeocystis* bloom was reported by Smith et al. (1991) in April. As growth in the MIZ starts about 2 months earlier than in open ocean locations, the herbivorous zooplankton may not have arrived in time at the surface from their over-wintering depths to utilize the algae.

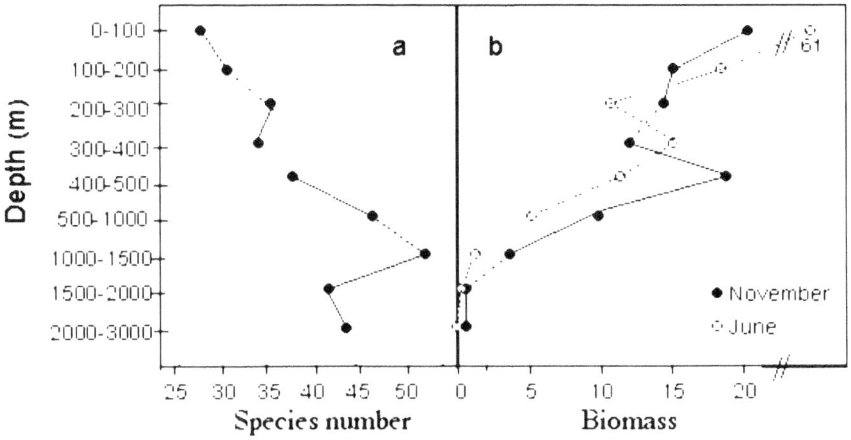

Figure 1. Vertical distribution of zooplankton species (a) and dry mass (b) in mg m^{-3} in the central Greenland Sea. Data from Richter (1994).

Later in the season, sedimentation decreases rapidly with depth, indicating that a major part of the spring phytoplankton biomass is consumed, oxidized and regenerated by the food web during its descent (Noji *et al.* 1996). In summer, after the descent of the large *Calanus* spp. populations, smaller species take over (Hirche, unpubl. data) and the microbial loop may become important. However, grazing and re-mineralization in the euphotic zone, especially the portion of faecal pellets leaving it are poorly studied, as most of the traps were deployed only in deep waters.

Using a 3D model for the carbon cycle and air-sea flux of CO_2 in the Greenland Sea, consisting of three sub-models (hydrodynamics, carbon chemistry, plankton ecology) Slagstad *et al.* (1999) simulated an annual flux of CO_2 from the atmosphere of $53gC{\cdot}m^{-2}$. Of this, $9gC{\cdot}m^{-2}$ is exported by sinking particles, $6gC{\cdot}m^{-2}$ by ontogenetic migration of zooplankton and $38gC{\cdot}m^{-2}$ by advection.

Little is known on the role deep convection plays in the carbon cycle. Ramseier *et al.* (1999) suggest that deep convection per se does not enhance fluxes in a direct way; however, the time available for degradation may decrease dramatically and thus effect the quality of the material arriving at depth.

4. CONCLUSIONS AND PERSPECTIVES

International projects such as the Greenland Sea Project (GSP), European Subpolar Ocean Programme (ESOP 1 and 2), Tracer and Circulation in the Nordic Seas Region (TRACTOR), Variability of Exchanges in the Northern Seas (VEINS) have all focused on the Greenland Sea. They have been based on the fact that inter-annual to decadal climate variability affect the water mass conversion processes of the Northern Seas, the effects of which are propagated by fluxes through the boundaries of the Global Ocean.

Climate variability also affects strongly the pelagic ecosystem and carbon fluxes in the Greenland Sea. The vicinity of ice edge systems, fronts and gyres together with the presence of three hydrographic and bio-geographic domains are aligned on the same latitude at a relatively small spatial scale and under the same light regime. This makes the Greenland Sea an exciting laboratory to study biological-physical coupling on various spatial scales. It is also a hot spot for monitoring climate effects on the ecosystem level. Open questions concern:

- the causes of the apparently high productivity of both phytoplankton and zooplankton and the fate of this production, be it respired or drawn down to deep waters.
- the role that the large meso-pelagic zooplankton community plays in the transformation of particulate matter.

The similarity between the pelagic communities of the Greenland Sea and Arctic Ocean, and their inter-linkage provides the opportunity to compare them under different ice conditions. Furthermore, predicted decreasing ice thickness and shorter duration of ice coverage in the Arctic Ocean may convert it into a system similar to the present day's Greenland Sea.

Deep water formation and export also has a strong effect on the fate of carbon sequestered in the Greenland Sea, both by physical and biological processes. The convective depth determines the duration of carbon removal from the atmosphere, which gives the Greenland Sea a key position as the main gateway for carbon exchange between lower latitudes and the Arctic Ocean. These exchange processes have far-reaching influence on carbon transport.

In the Greenland Sea, there is so far no complete seasonal coverage of biological parameters with sufficient temporal resolution available, not to mention systematic inter-annual observations. Satellite ocean colour is useless due to an almost permanent cloud cover. Hence there is an urgent need for year-round work from land bases and biological moorings equipped with biological sensors, which could be serviced from land base and calibrated with seasonal shipboard observations.

REFERENCES

Aken HM van, D Quadfasel, A Warpakowski 1991. The Arctic front in the Greenland Sea during February 1989: Hydrographic and biological observations. *Journal of Geophysical Research* 96, 4739-4750.

Alexander V and HJ Niebauer 1981. Oceanography of the Eastern Bering Sea ice-edge zone in spring. *Limnology and Oceanography* 26, 111-125.

Anderson LG, EP Jones, KP Koltermann, P Schlosser, JH Swift, and DWR Wallace 1989. The first oceanographic section across the Nansen Basin in the Arctic Ocean. *Deep-Sea Research* 36, 475-482.

Bauerfeind E, Bodungen B von, K Arndt, and W Koeve 1994. Particle flux, and composition of sedimenting matter, in the Greenland Sea. *Journal of Marine Systems* 5, 411-423.

Bourke RH, AM Weigel, and RG Paquette 1988. The westward turning branch of the West Spitsbergen Current. *Journal of Geophysical Research* 93, 14065-14077.

Daly K 1997. Flux of particulate matter through copepods in the Northeast Water Polynya. *Journal of Marine Systems* 10, 319-342.

Ekman S 1967. *Zoogeography of the Sea*. Sidgwick and Jackson, London. 417 pp.

Gascard JC, C Kergomard, PF Jeannin, and M Fily 1988. Diagnostic study of the Fram Strait Marginal Ice Zone during summer from 1983 and 1984 Marginal Ice Zone Experiment. Lagrangian observations. *Journal of Geophysical Research* 93, 3613-3641.

Hebbeln D and G Wefer 2001. Effects of ice coverage and ice-rafted material on sedimentation in the Fram Strait. *Nature* 350, 409-411.

Hirche HJ 1989a. Egg production of the Arctic copepod *Calanus glacialis* - laboratory experiments. *Marine Biology* 103, 311-318.

Hirche HJ 1989b. Spatial distribution of digestive enzyme activities of *Calanus finmarchicus* and *C. hyperboreus* in Fram Strait / Greenland Sea. *Journal of Plankton Research* 11, 431-443.

Hirche HJ 1991. Distribution of dominant calanoid copepod species in the Greenland Sea during late fall. *Polar Biology* 11, 351-362.

Hirche HJ, M Baumann, G Kattner, and R Gradinger 1991. Plankton distribution and the impact of copepod grazing on primary production in Fram Strait, Greenland Sea. *Journal of Marine Systems* 2, 477-494.

Hirche HJ and G Kattner G 1993. Egg production and lipid content of *Calanus glacialis* in spring: indication of a food-dependent and food-independent reproductive mode. *Marine Biology* 117, 615-622.

Hirche HJ 1997. The life cycle of the copepod *Calanus hyperboreus* in the Greenland Sea. *Marine Biology* 124, 607-618.

Hirche HJ and S Kwasniewski 1997. Distribution, reproduction and development of *Calanus* species in the Northeast Water in relation to environmental conditions. *Journal of Marine Systems* 10, 299-317.

Honjo S, SJ Manganini, and G Wefer 1987. Annual particle flux and a winter outburst of sedimentation in the northern Norwegian Sea. *Deep-Sea Research* 35,1223-1234.

Johannessen OM, L Bengtsson, MW. Miles, SI Kuzmina, VA Semenov, GV Alekseev, AP Nagurnyi, VF Zakharov, L Bobylev, LH Pettersson, K Hasselmann, and HP Cattle 2002. Arctic climate change – observed and modeled temperature and sea ice variability. Nansen Environmental and Remote Sensing Center, Bergen. Technical Report No. 218.

Johannessen OM 1986. Brief overview of the physical oceanography. Pp 103-128 in BG Hurdle (ed.) *The Nordic Seas*. Springer Verlag, New York.

Kohly A 1998. Diatom flux and species composition in the Greenland Sea and Norwegian Sea in 1991-1992. *Marine Geology* 145, 293-312.

Kosobokova KN 1999. The reproductive cycle and life history of the Arctic copepod *Calanus glacialis* in the White Sea. *Polar Biology* 22, 254-263.

Kosobokova KN and HJ Hirche 2000. Zooplankton distribution across the Lomonosov Ridge, Arctic Ocean: Species inventory, biomass and vertical structure. *Deep-Sea Research I* 47, 2029-2060.

Kosobokova KN, HJ Hirche, and T Scherzinger 2002. Feeding ecology of *Spinocalanus antarcticus*, a deep-water copeod with a looped gut. *Marine Biology* 141, 503-512.

Lara RJ, G Kattner, U Tillmann, and HJ Hirche 1994. The North East Water polynya (Greenland Sea). II. Mechanisms of nutrient supply and influence on phytoplankton distribution. *Polar Biology* 14, 483-490.

Legendre L, M Gosselin, HJ Hirche, G Kattner, and G Rosenberg 1993. Environmental control and potential fate of size-fractionated phytoplankton production in the Greenland Sea (75°N). *Marine Ecology Progress Series* 98, 207-313.

Macdonald RW 1996. Awakenings in the Arctic. *Nature* 380, 286-287.

Mumm N 1993. Composition and distribution of mesozooplankton in the Nansen Basin, Arctic Ocean. *Polar Biology* 13, 451-46.

Niebauer HJ 1982. Wind and melt driven circulation in a marginal sea ice edge frontal system: a numerical model. *Continental Shelf Research 1*, 49-98.

Niebauer HJ and WO Smith 1989. A mesoscale model of the physical-biological interactions in the Fram Strait marginal ice zone. *Journal of Geophysical Research* 94,16151-16176.

Noji TT, F Rey KY Børsheim, L Miller, HJ Hirche, and J Urban-Rich 1996. Plankton dynamics and sedimentation in the Greenland Sea. Pp 540-554 in P Wadhams, JP Wilkinson, and SCS Wells (eds.) European Subpolar Ocean Programme: Sea Ice - Ocean Interactions Vol. 2. Scott Polar Research Institute, Cambridge.

Pésant S, L Legendre, M Gosselin, C Ashjian, B Booth, L Fortier, HJ Hirche, P Lane, J Michaud, REH Smith, SL Smith, and WO Smith Jr 1998. Pathways of carbon cycling in the euphotic zone: The fate of large-sized phytoplankton in the Northeast water polynya. *Journal of Plankton Research* 20, 1267-1291.

Ramseier RO, C Garrity, E Bauerfeind, and R Peinert 1999. Sea-ice impact on long-term particle flux in the Greenland Sea´s Is Odden-Nordbukta region, 1985-1996. *Journal of Geophysical Research* (C Oceans) 104, 5329-5343.

Richter C 1994. Regional and seasonal variability in the vertical distribution of mesozooplankton in the Greenland Sea. *Report of Polar Research* 154, 90pp.

Richter C 1995. Seasonal changes in the vertical distribution of mesozooplankton in the Greenland Sea (75°N): distribution strategies of calanoid copepods. *ICES Journal of Marine Science* 52, 533-539.

Runge JA and RG Ingram 1988. Under-ice grazing by planktonic, calanoid copepods in relation to a bloom of ice microalgae in southeastern Hudson Bay. *Limnology and Oceanography* 33, 280-286.

Sakshaug E 1987. Arctic planktonic ecosystems. Pp 47-51 in LK Medlin and J Priddle (eds.) *Polar marine diatoms*. British Antarctic Survey, Cambridge.

Sakshaug E and HR Skjoldal 1989. Life at the ice edge. *Ambio* 8, 60-67.

Schmitz WJ 1995. On the interbasin-scale thermohaline circulation. *Review of Geophysics* 33,151-173.

Schneider W and G Budeus 1994. The North East Water polynya (Greenland Sea). I. A physical concept of its generation. *Polar Biology* 14, 1-9.

Slagstad D, K Downing, F Carlotti, and HJ Hirche 1999. Modelling the carbon export and air-sea flux of CO_2 in the Greenland Sea. *Deep-Sea Research* II 46, 1511-1530.

Smith SL, WO Smith Jr, LA Codispoti, and DL Wilson 1985. Biological observations in the marginal ice zone of the East Greenland Sea. *Journal of Marine Research* 43,693-717.

Smith WO Jr, LA Codispoti, DM Nelson, T Manley, EJ Buskey, HJ Niebauer, and GF Cota 1991. Importance of Phaeocystis blooms in the high-latitude ocean carbon cycle. *Nature* 352, 514-516.

Smith WO Jr, MEM Baumann, DL Wilson, and L Aletsee 1987. Phytoplankton biomass and productivity in the marginal ice zone of the Fram Strait during summer 1984. *Journal of Geophysical Research* 92, 6777-6786.

Smith WO Jr 1987. Phytoplankton dynamics in marginal ice zones. *Oceanography and Marine Biology Annual Review* 25,11-38.

Trees CC, J Aiken, HJ Hirche, and SB Groom 1992. Bio-optical variability across the Arctic Front. *Polar Biology* 12, 455-461.

Zhang J and X Zhang 2001. Heat and freshwater budgets and pathways in the arctic mediterranean in a coupled ocean/sea-ice model. *Journal of Oceanography* 57 (2), 207-234.

Chapter 13 NATO Advanced Research Workshop

Continuous Plankton Recorder Observations

Extension of the System into the Nordic Seas

Anthony W Walne and Philip C Reid

Sir Alister Hardy Foundation for Ocean Science, The Laboratory, Plymouth, England

E-mait: anwa@mail.pml.ac.uk

Abstract: The Continuous Plankton Recorder (CPR) Survey has been towed monthly by commercial vessels to monitor the distribution of plankton in the surface waters of the Northeast Atlantic Ocean for over seventy years. In the past, between 1949 and 1981, CPRs were towed to Ocean Weather Station (OWS) Mike in the Norwegian Sea until sampling ceased due to financial constraints. Summary results from this early survey in the Norwegian Sea are outlined. We also discuss options for the resumption of monitoring of the surface plankton of the Norwegian Sea with CPR tows to OWS Mike and new routes around northern Norway and to Svalbard. In a climate change scenario, for assessment of ecosystems and fishery management an urgent need to improve monitoring of the plankton of Nordic Seas is identified.

Key words: time-series, *Calanus finmarchicus*, phytoplankton, Global warming, thermohaline circulation

1. INTRODUCTION

There are few long-term marine biological monitoring programmes in the world. Possibly the longest running, and the only one that operates on an ocean basin scale, is the Continuous Plankton Recorder (CPR) Survey. Since its inception in 1931 this survey has covered about 8 million km of sea using 'ships of opportunity' (SOOP) and approximately 180 000 plankton samples

S. Skreslet (ed.), Jan Mayen Island in Scientific Focus, 135-145

have been analysed. The design of the Recorder itself and the methods of analysis have remained largely unchanged over the past seventy years. An atlas of North Atlantic plankton distributions (CPR Survey Team 2004) summarises information on the occurrence and abundance of 240 species and other taxa from the database. The information provided by the Survey has contributed important information on the response of marine ecosystems to Global Climate Change and the impact of changes in plankton on fish stocks. In addition the data have illustrated the utility of the Recorder as a sampling instrument and the success of the Survey as a concept.

To celebrate the 70th anniversary of the survey a special issue of Progress in Oceanography, 'Achievements of the Continuous Plankton Recorder survey and a vision for its future' has recently been published (Reid *et al.* 2003). This brings together in one place an up-to-date compilation of papers that record the history, technical background, limitations, applications and research findings of the Survey. Examples of other recent findings from the Survey are outlined below. They include evidence in the eastern North Atlantic for a northerly movement of copepods by 10° of latitude in the past 40 years as a possible response to Global warming (Beaugrand *et al.* 2002). It was not possible to determine the full northerly extension of this change as sampling is currently not carried out in the Norwegian Sea. Mechanisms for the decline in salmon returns to home waters linked to plankton and climatic change were outlined by Beaugrand and Reid (2002). Clear evidence for a relationship between cod recruitment and biomass in the North Sea and the abundance, composition and size of plankton was demonstrated by Beaugrand *et al.* (2003). Pronounced changes in phytoplankton on a regional scale were shown by Edwards *et al.* (2002) and evidence outlined for a regime change and other interregional biological responses in the North Atlantic to hydrometeorological forcing by Reid and Beaugrand (2002). These and other publications from the Survey indicate that plankton is able to integrate hydrometeorological variability and act as an indicator of environmental change. Furthermore, the results from these studies and other non-CPR research, indicate that in a Global warming scenario over the next few decades, the Nordic Seas are likely to see substantial ecological changes. At present however, there is limited information available with a large spatial and regular temporal coverage to assess future change in the Nordic Seas and only limited sampling by the CPR on part of two routes that run southwest and southeast from Iceland. Long-term monitoring using the CPR would provide an economical and well tested way of assessing change in these seas. It would be possible to integrate and compare new results from the Nordic Seas with the longer term sampling from the survey to the south. Restarting the T route to Ocean Weather Station Mike in the Norwegian Sea would enable a comparison with an earlier baseline from 1949 to 1981. Such a survey would also provide the necessary spatial coverage in which to place the few coastal transects (in most cases seasonal) and single point ecosystem monitoring sites that exist at present.

This paper describes the Continuous Plankton Recorder, the Survey and analysis techniques and considers past CPR sampling in the Nordic Seas area. It outlines how a synoptic biological monitoring programme could be established in these sub-arctic regions using SOOPs and survey vessels. We consider the Nordic Seas as defined in the Quality Status Report 2000 of the OSPAR Commission (2000) to consist of the Norwegian, Iceland and Greenland Seas extending north of latitude 62°N and the Faroe Islands and from 42°W to 51°E.

2. THE CONTINUOUS PLANKTON RECORDER SURVEY

Merchant vessels (SOOPs) on their standard routes tow CPRs at monthly intervals when practicable, sampling the near-surface plankton at a depth of 5 to 10m. In some cases, and for various reasons such as ice-free conditions, routes are operated at more irregular periods. In addition, weather conditions may cause a vessel to deviate from its usual course. Fig. 1 shows the position of CPR samples analysed from the Nordic Seas in the periods 1948 to 1981 and 1982 to 2002, which highlights the gap in sampling at the present.

The CPR is operated voluntarily by the ship's crew aboard the towing vessel. Thus, no specialist staff accompanies the recorder. The machine is dispatched from the SAHFOS laboratory in Plymouth, U.K. and collected by the SOOP at its port of departure. It is deployed from a small davit and towed between predetermined positions. Officers of the towing vessel complete a log sheet for each tow where they record the time and position of CPR deployment and recovery together with any alterations of course and speed. The goodwill and co-operation of the captain and crew of the tow vessel is essential for the successful operation of each deployment.

The CPR is a metre long, weighs about 85kg and has been manufactured in brass, phosphor bronze and, more recently, stainless steel (Reid *et al.* 2003). Water enters the CPR through a 1.27cm square aperture and passes through an internal cassette that houses the plankton filtering system. The water passes through a continuously advancing band of silk of mesh size 270 μm with a filtering area of 5cm by 10cm. The filtering mesh with the retained plankton is covered by a second band of silk, which together are rolled into a tank of formaldehyde to fix and preserve the plankton.

Figure 1. The distribution of CPR samples in the Nordic Seas area north of 62°N and between 42°W and 51°E. a) 1948-1981: 8073 samples. b) 1982-2002: 2520 samples.

When the CPR is returned to the laboratory the bands of silk are unrolled and cut into samples corresponding to 10 nautical miles (18.5km) of tow. Prior to cutting, the samples are assessed for 'greeness' on a 4 category scale. A three stage enumeration procedure is carried out on each sample, consisting of a high magnification 'phytoplankton' analysis, a lower magnification analysis for smaller zooplankton less than 2mm and finally an analysis for larger 'eyecount' zooplankton greater than 2mm. CPR samples are identified and counted by a specialist team of analysts and a quality control procedure is employed to check for unusual results. Over 450 different species or higher taxa of phytoplankton and zooplankton are routinely identified. Further details can be found in Warner and Hays (1994) and Batten *et al.* (2003).

3. PAST AND CURRENT SAMPLING IN THE NORDIC SEAS AREA

Between 1949 to 1976, with some additional sampling up to 1981, CPRs were towed on the T route from Bergen or from the North Sea to Ocean Weather Station (OWS) Mike at about 66N 3E. In this period over 2700 samples were analysed on 340 tows in the Norwegian Sea north of 62° N.

Starting initially in 1938 and continuously on an almost uninterupted monthly basis since 1948 the V route has operated between Scotland and Iceland. The northern part of this route to the north of 62° N is within the OSPAR definition of the Nordic Seas. In a similar way the northern part of the Z route between Iceland and Newfoundland has sampled part of the Nordic Seas within the same definition. The Z route was operated monthly between 1957 and 1986 and was resumed in 1991. These routes are close to the southern limits of the area and may provide an indication of the boundary conditions of the Nordic Seas.

As a sampler of the information that is available plots of annual and monthly means for Phytoplankton Colour and *Calanus finmarchicus* for these three routes are presented as Fig. 2. Figure 2a shows the mean monthly greenness index averaged annually for samples in the Norwegian Sea (T route) between 1949 and 1981. This Colour index has been used as a proxy for chlorophyll concentration (Reid 1975) and is considered to be indicative of annual primary production (Reid *et al.* 1998). The figure shows a peak in the index between 1950 and 1960 with a decline over the next twenty years.

The results in Fig. 2 are similar to those reported by Reid *et al.* (1998) for their northern North East Atlantic area and Edwards *et al.* (2002) for their Northern oceanic area. This latter area whilst largely outside the Norwegian Seas as defined here may reflect the conditions of the North Atlantic Current upstream of OWS Mike. Reid *et al.* (1998) considered that this area of the Northeast Atlantic Ocean had tended to experience colder surface temperatures between 1948 and 1995. A more recent analysis with the time series extended to 2002 suggests that the decreasing trend of Phytoplankton Colour seen in the northern North East Atlantic area of Reid *et al.* (1998), adjacent to Iceland, has reversed since 1997 and the colour intensity is now increasing (Reid and Beaugrand 2002). Reopening the CPR transect to OWS Mike would confirm if this is also occurring in the Norwegian Sea.

Figure 2b shows the annual mean abundance of the copepod *Calanus finmarchicus* in the Norwegian Sea area for the period between 1949 and 1981. There is no clear trend in the data but it would appear that inter-annual variability was increasing towards the end of the sampling period.

Figures 2c-f shows the annual mean Phytoplankton Colour and *Calanus finmarchicus* abundance for the Nordic Seas area west of 22°W (the CPR survey Z route) and between 22°W and 0°W (the CPR survey V route). The mean Phytoplankton Colour in both these areas (Figs 2c and 2e) shows a distinct upward trend after 1995, which agrees closely with the results of Beaugrand and Reid (2003). The trends in annual mean abundance for *Calanus finmarchicus* (Figs 2d and 2f) are less obvious for both areas.

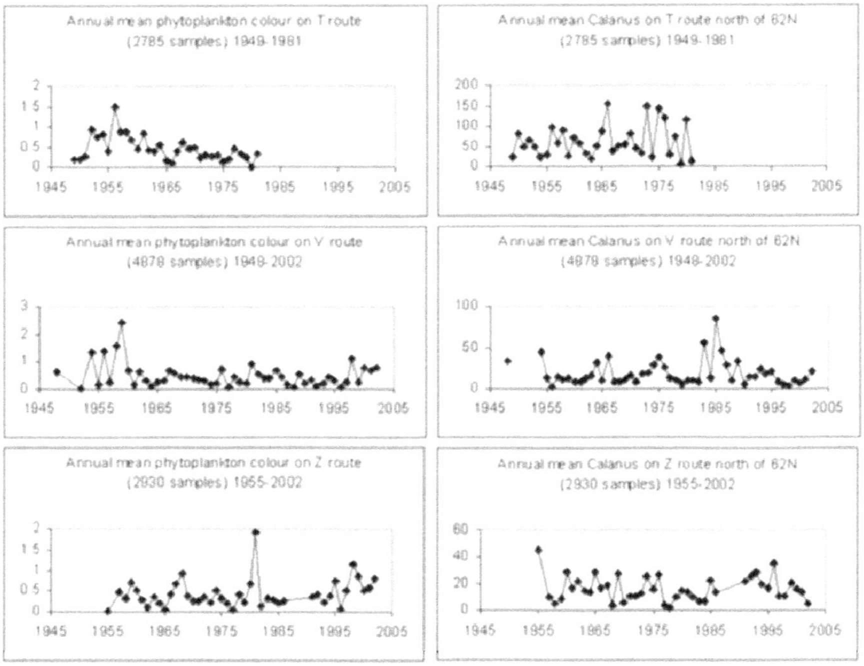

Figure 2. Annual mean CPR results for Phytoplankton Colour Index (Figs 2a, c and e) and *Calanus finmarchicus* abundance (individuals per sample) (Figs 2b, d and f) for the Nordic Seas north of 62°N between 1948 and 2002. Area 1 is east of 0° (2785 samples), area 2 is between 22°W and 0° (4878 samples) and area 3 is west of 22°W (2930 samples).

4. DEVELOPMENT OF A CPR SURVEY IN THE NORDIC SEAS

We outline here possible extensions to the SAHFOS observational system using CPRs and discuss the benefits of establishing such a system in the Nordic Seas area. Ideally a minimal programme should be established in the Norwegian, Barents and Greenland Seas, but restrictions due to ice cover and potential collision with floating ice and limited shipping movement is recognised in the latter case. Continuing operation of the parts of the V and Z routes that already sample these sea areas is assumed. There are several potential opportunities to tow CPRs along standard routes plied by commercial merchant ships in the area as well as use of supply ships and fisheries protection vessels and irregular ferry transects as well as research ships enroute to survey areas.

4.1 Sampling by the CPR in the Norwegian Sea

The T route to Ocean Weather Station Mike (66° N, 01° E) from Utvær that is an island off the coast of Norway was sampled by the CPR survey from 1949 until 1981 when the survey was contracting due to funding constraints. Work in the Northwest Atlantic has shown that valuable comparisons with earlier CPR sampling can be made despite a break in the time series (Johns *et al.* 2001). Resumed sampling in this area of major strategic and economic importance to Norway and other adjacent states would determine whether the Norwegian Sea has undergone changes of a similar scale to those recorded in the Northeast Atlantic to the south. It would also be possible to determine if the Norwegian Sea is responding in a similar way to the areas to the south or is varying independently. The Norwegian Sea is at the northern limit of the distributions of many temperate plankton species and the southern limit (in the eastern Atlantic) for many Arctic species. Changes in the abundance and/or distributions of these species are likely to have an important impact on the ecology of the area and on the carrying capacity of fishery resources as well as indicating wider effects of climate change.

Other options in the Norwegian Sea would be to operate CPRs from supply vessels visiting Jan Mayen Island and the summer ferry between Norway and the Faeroe Islands. These would give long sections across the Norwegian Sea that could be compared to the seasonal coastal transects at fixed sites undertaken by Norway as part of the routine monitoring of the Norwegian shelf.

4.2 Possible CPR Sampling in the Barents Sea

The Survey could be extended into Arctic waters with tows from the Lofoten Islands around North Cape to Kirkenes and from North Norway to Svalbard. Semi-regular supply ships and ferries operate on these routes. The Arctic plankton that characterise these regions is similar in composition to the species found in the Labrador Current. In this latter region the patterns of change seen in the plankton are very different to those in the eastern North Atlantic (Johns *et al.* 2001). Results of sampling in the Barents Sea could provide evidence as to whether different parts of the Arctic pelagic ecosystem are behaving in a common way. A simplistic view of the possible effects of Global warming might suggest that thermophilic species should increase in numbers in the area north of Norway, but a collapse of the Global Thermo-Haline Circulation could be indicated by a reduced northerly flow from the Northeast Atlantic. Changes in the plankton composition indicating important environmental changes have, at times, been evident before a significant signal has been detected in physical variables. In a climate change scenario and with expected reductions in summer ice in the Arctic Ocean large changes are likely in the Barents Sea. Establishing an appropriate baseline monitoring programme to sample the pelagic system of the Barents Sea should thus be a priority.

There is considerable Russian and Norwegian shipping activity in the Barents Sea largely associated with the fishing (e.g. fishery protection vessels) and survey industries that might also provide a platform for CPR tows. Vessels transporting coal from the Russian mines might also be a possibility.

4.3 Possible CPR Sampling in the Greenland Sea

Options are limited for sampling in the Greenland Sea because of the very infrequent shipping in the area and the short ice-free season. Sampling in this region is likely to be opportunistic using research vessels and coastal supply ships from Greenland and Icleand. Because of the interest in deep convection in the Greenland Sea and measurement of outflow from the Fram Strait there are a number of research vessel transects to this area per year that might be utilised.

Table 1. Estimated tow length and number of CPR samples on some proposed extensions of the CPR survey in the Nordic and Arctic Seas.

Route	Distance (nautical miles)	Annual number of CPR samples
Kristiansund to OWS Mike	270	162
Lofoten to Kirkenes	500	300
North Norway to Svalbard	450	180

4.4 Samples Generated and Costs for a Basic Programme

The distance towed and the number of CPR samples that would be generated per annum by establishing routes to the Ocean Weather Station Mike, from the Lofoten Islands around the North Cape and from North Norway to Svalbard is estimated in Table 1. These estimates are based on the usual procedure adopted in the current CPR survey of analysing alternate 10 nautical mile samples. Potential towing vessels have been identified for the T route and for the tow around the North Cape. In the latter case a tow davit would need to be attached to the ship. The sampling estimate for the Svalbard tows are based on eight ice-free months per year. The total cost of running a CPR survey, which includes all aspects from logistics, transport, workshop maintenance, sample analysis, databasing and sample storage and administration is costed at £140 per sample (2004).

Establishing a new tow route would incur some additional costs in the first year. Such costs will depend on whether a new davit is needed and will be unlikely in most instances to exceed £7,000.

5. DISCUSSION

Evidence is accumulating from the plankton, ocean hydrodynamics, greenhouse gases and temperature that the world is undergoing rapid change at the present time at a rate that has not been seen in at least the last 1000 years. The rate of rising temperature is even higher in the Northern Hemisphere and in the Arctic rim. The scientific consensus (IPCC 2001) is that industrial releases of greenhouse gases are fuelling the rise in Global temperatures. Rising temperatures in turn are correlated with observed oceanographic and planktonic ecosystem changes. The extent of the observed changes and especially correlations found between plankton and Northern Hemisphere temperature (Beaugrand and Reid 2002, Beaugrand *et al.* 2002) indicates that areas of the Northeast Atlantic may already be witnessing change over the last 40 years forced by Global warming. Changes in the Nordic Seas are likely to be particularly pronounced because of increased runoff from Arctic rivers and melting permafrost as well as reducing seasonal coverage of sea ice.

It has been well demonstrated that the CPR approach to monitoring planktonic systems at a monthly frequency provides a sensitive approach to evaluating ecosystem change. While a number of Nordic countries undertake occasional or seasonal surveys they do not have the baseline to determine expected future seasonal and interannual changes as detectable with the CPR methodology. In terms of fish stocks timing of the occurrence of plankton has been shown to be crucial to the recruitment and size of stocks (e.g. cod, Beaugrand *et al.* 2003). Expanding the CPR survey into Nordic Seas, with as the first stage the re-occupation of the T route to Weather Station Mike would provide a framework against which past and current national plankton surveys could be integrated into monthly timing studies. With sampling in the Nordic Seas it would be possible to extrapolate patterns of change seen in the wider Atlantic and in particular to determine if the regime shift evident in the North Sea (Reid *et al.* 2001a) and eastern Atlantic (Beaugrand 2004) has had a wider geographical focus. Pulses of warm water in the eastern boundary current at the shelf edge have been shown to be related to planktonic changes in the North Sea (Reid et al., 2001b) and are almost certainly linked to the northerly movement of warm water plankton by about 280km per decade seen to the west of the British Isles (Beaugrand *et al.* 2002). These large scale changes are likely to have had a considerable impact on the Norwegian and possibly Barents Seas. Because plankton sampling at appropriate time and space scales does not exist we do not know the scale of such impacts.

The Nordic Seas are key areas for deep water formation as part of the Global Thermo-Haline Circulation. Some plankton species can provide evidence for change in this water as they migrate down into it to diapause (Reid and Beaugrand 2002). These indicators are potential cheap and easy method of determining changes in deep water formation and movement in the Nordic Seas. This has particular relevance in terms of a potential slowing down of the conveyor and of changing patterns of sea ice cover over the next few decades. In summary overall there is an urgent need to improve systematic monitoring of planktonic ecosystems in the Nordic Seas. On past evidence the CPR approach provides a potential cost-effective way of undertaking such sampling. The CPR survey is already part of the Initial Observing System of the Global Ocean Observing System (GOOS) and the CPR methodology is approved for use in new regional sea.

REFERENCES

Batten SD, R Clark, J Flinkman, G Hays, E John, AWG John, T.Jonas, JA Lindley, DP Stevens, and A Walne 2003. CPR sampling: the technical background, materials and methods, consistency and comparability. *Progress in Oceanography* 58 (2-4), 193-216.

Beaugrand G 2004. The north Sea regime shift: evidence, causes, mechanisms and consequences. *Progress in Oceanography* 60, 245-262.

Beaugrand G, PC Reid, F Ibanez, JA Lindley, and M Edwards 2002. Reorganization of North Atlantic marine copepod biodiversity and climate. *Science* 296, 1692-1694.

Beaugrand G, KM Brander, JA Lindley, S Souissi and PC Reid 2003. Plankton effect on cod recruitment in the North Sea. *Nature* 426, 661-664.

Beaugrand G and PC Reid 2002. Long-term changes in phytoplankton, zooplankton and salmon linked to climate. *Global Change Biology* 9, 801-817.

CPR (Continuous Plankton Recorder) Survey Team 2004. Continuous Plankton Records: Plankton Atlas of the North Atlantic Ocean (1958-1999). II. Biogeographical charts. *Marine Ecology Progress Series* Supplement 2004, 11-75.

Edwards M, G Beaugrand, PC Reid, A Rowden, and M Jones 2002. Ocean climate anomalies and the ecology of the North Sea. *Marine Ecology Progress Series* 239, 1-10.

Hays GC, MR Carr, and AH Taylor 1993. The relationship between Gulf Stream position and copepod abundance derived from the Continuous Plankton Recorder Survey: separating biological signal from sampling noise. *Journal of Plankton Research* 15, 1359-1373.

IPCC 2001. *Climate Change 2001: Synthesis Report*. Cambridge University Press, 398 pp.

Johns DG, M Edwards, and SD Batten. 2001. Arctic boreal plankton species in the North West Atlantic. Canadian *Journal of Fisheries and Aquatic Sciences* 58, 2121-2124.

OSPAR Commission 2000. Quality Status Report 2000, Region I – Arctic Waters. OSPAR Commission, London.

Reid PC 1975. Large scale changes in North Sea phytoplankton. *Nature* 257, 217-219.

Reid PC, M Edwards, HG Hunt and AJ Warner 1998. Phytoplankton change in the North Atlantic. *Nature* 391, 546.

Reid PC and G Beaugrand 2002. Interregional Biological Responses in the North Atlantic to Hydrometeorological Forcing. Pp. 27-48 in K Sherman and HR Skjoldal (eds.) *Large Marine Ecosystems of the North Atlantic. Changing States and Sustainability*. Elsevier Science B.V.

Reid PC, MF Borges, and E Svendsen 2001a. A regime shift in the North Sea circa 1988 linked to changes in the North Sea horse mackerel fishery. *Fisheries Research,* 50 163-171.

Reid PC, NP Holliday, and TJ Smyth 2001b. Pulses in eastern margin current with higher temperatures and North Sea ecosystem changes. *Marine Ecolpgy Progress Series* 215, 283-287.

Reid PC, JBL Mathews, and MA Smith 2003. Achievements of the Continuous Plankton Recorder Survey and a Vision for its Future. *Progress in Oceanography* 58, 115-358.

Reid PC, JM Colebrook, JBL Mathews, J Aiken and Continuous Plankton Recorder Team 2003. The Continuous Plankton Recorder: concepts and history, from Plankton Indicator to undulating recorders. *Progress in Oceanography* 58 (2-4), 117-174.

Warner AJ and GC Hays 1994. Sampling by the Continuous Plankton Recorder survey. *Progress in Oceanography* 34 (2-3), 237-256.

Chapter 14 NATO Advanced Research Workshop

Cyclic Variability of Climate and Fisheries Resources

Relevance for the Nordic Seas

Leonid B Klyashtorin and Alexey A Lyubushin
LB Klyashtorin, Federal Institute for Fisheries and Oceanography (VNIRO), Moscow, Russi
E mail: klyashtorin@mtu-net.ru
AA Lyubushin, Institute of the Physics of the Earth, Moscow, Russia

Abstract: Time series of Global air surface temperature anomaly (Global dT) and Atmospheric Circulation Index (ACI) characterizes a long-term predominant direction of air mass transport, i.e. meridional or zonal (latitudinal). Both were found to be correlated with long-term fluctuations of commercial fish stocks and to exhibit roughly 55-60 years periodicity. It may be proposed that populations of Norwegian spring spawning herring and North East Arctic cod undergo similar climate-linked 55-60 year cyclic fluctuations. A stochastic model was developed to predict the climate-linked stock dynamics of major commercial species for up to 30-40 years ahead. Four climatic indices (Global dT, ACI, "Kola meridian dT", and air surface temperature in the Jan Mayen region) are suggested as indicators of major Arctic fish stock fluctuations.

Key words: climate change, Arctic fisheries, Kola meridian, Atlantic, Pacific, tree rings

1. INTRODUCTION

To begin with, let us define the term climate as weather conditions statistically averaged for a long-term period. The international meteorological conferences (Warsaw 1935, Washington 1957) have recommended using 30 year averaging periods (Monin and Shishkov 2000). The second important dimension of climate is its spatial scale. The minimum area to obtain reliable results from should be about 1–3 mill. km^2. However, in practice researchers use 10-30 year averaging of data over 1 mill. km^2 (Kislov *et al.*2000).

S. Skreslet (ed.), Jan Mayen Island in Scientific Focus, 147-157

The Global Temperature Anomaly (Global dT) is the most commonly used climatic index (Fig. 1). Annual variations of this index are high, and smoothing is necessary to see the multiyear dynamics. In addition, the Global dT has an increasing long-term trend that can be removed to clarify the long-term fluctuations (Sonechkin 1998).

Another important climatic index is the Atmosphere Circulation Index (ACI) or the so-called Vangengeim-Girs Index (Fig. 2). It characterizes the periods of predominant transfer of air masses over the Atlantic-Eurasian region, either meridional (main component directed along a meridian) or zonal (directed along latitudes). ACI does not show the long-term trend, and is out of phase with the dynamics of meridional and zonal transfer types (Klyashtorin 2001).

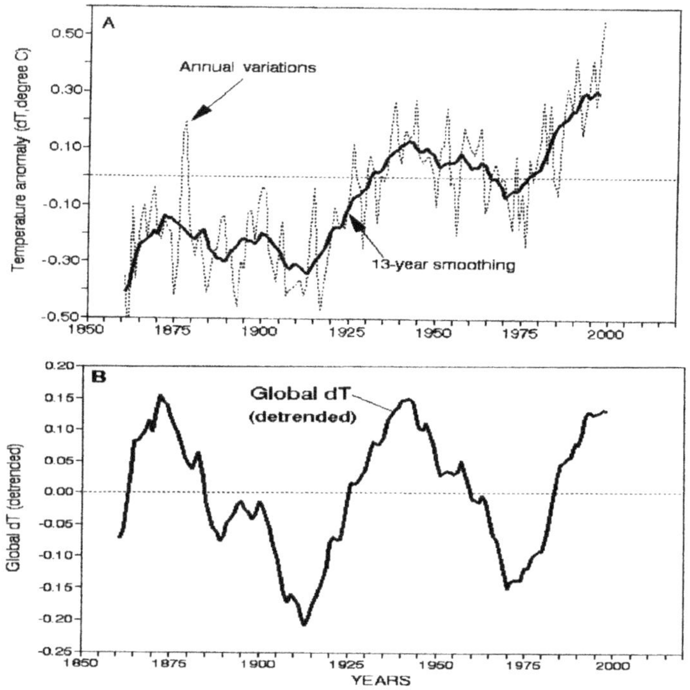

Figure 1. The Global temperature anomaly (Global dT) 1861-2000. A) Thin dashed line: annual dT. Bold line: Global dT (13-year smoothing). B) Global dT after trend removal (Klyashtorin 2001).

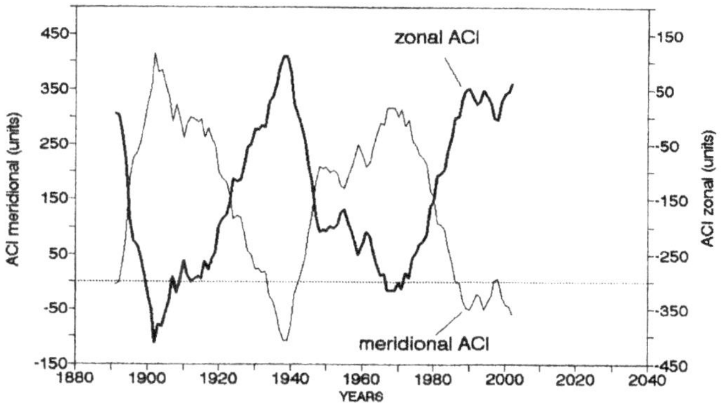

Figure 2. Long-term fluctuation dynamics of meridional (bold) and zonal (thin) ACI (Klyashtorin 2001).

The zonal ACI dynamics is closely correlated to the dynamics of Global dT. Therefore, ACI may be considered as a climatic index of the global scale. Unlike zonal ACI, meridional ACI dynamics is out of phase with the Global dT fluctuations (Fig. 3).

2. FISH CATCH RELATED TO CLIMATE

Both Global dT and ACI dynamics are quasi-regular and exhibit the periodicity of roughly 60 years. Are the climate fluctuations in agreement with the commercial stock dynamics?

The total world catch of marine fish species consists of about 550 species and adds up to roughly 70 mill. t. 12 major commercial species (not more than 2 percent of the total number of commercial fish species) provide 40-50% of the total world marine fish catch and the remaining 538 species (98% of the total species) provide the second half of the world catch.

A number of major commercial species living in temperate waters could be pooled into one group. It includes Peruvian, Japanese, Californian, African and European Sardine that belongs to the herring family, and other species like Pacific salmon, Alaska pollock, and Chilean jack mackerel that belong to other families of fish. The fluctuations in the roughly 60 years catch statistics of fish in this group are similar to the dynamics of Global dT, zonal ACI, and Pacific climatic indices, i.e. the Pacific Decadal Oscillation (PDO) and the Aleutan Low Pressure Index (ALPI) (Klyashtorin 2001).

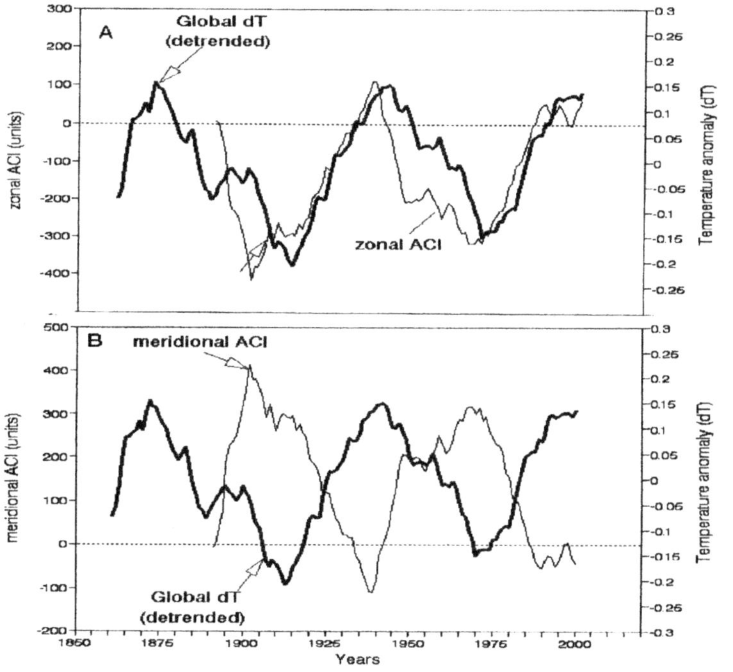

Figure 3. Comparative long-term dynamics of detrended Global dT and Atmospheric-Circulation Index (ACI). A) Zonal. B) Meridional (Klyashtorin 2001).

Then, are the long-term dynamics of sub-arctic populations of herring and cod, *i.e.* the key commercial species of the North East Atlantic, in agreement with Arctic climatic indices? For the Norwegian spring-spawning herring stock there are statistical time series for the last 90 years. The corresponding time series for cod start as late as in the 1940s, i.e. they are by 40 years shorter (Toresen 1999, Anon. 1995)

The estimated spawning stock biomass of Norwegian spring-spawning herring for the last 90 years exhibits a maximum about 1930-1945, a minimum about 1970-1980, and a new ascending trend about 1990 (Fig. 4). It is synchronous with the long-term variation of the total herring stock and to some extent also synchronous with the total population of North-East Arctic cod.

Fig. 5 suggests that the herring population dynamics corresponds to roughly 60 years of variation in zonal ACI, Global dT, and the mean temperature of the 200m water column in the Kola meridian section, i.e. the Kola meridian dT (Yndestad 2002).

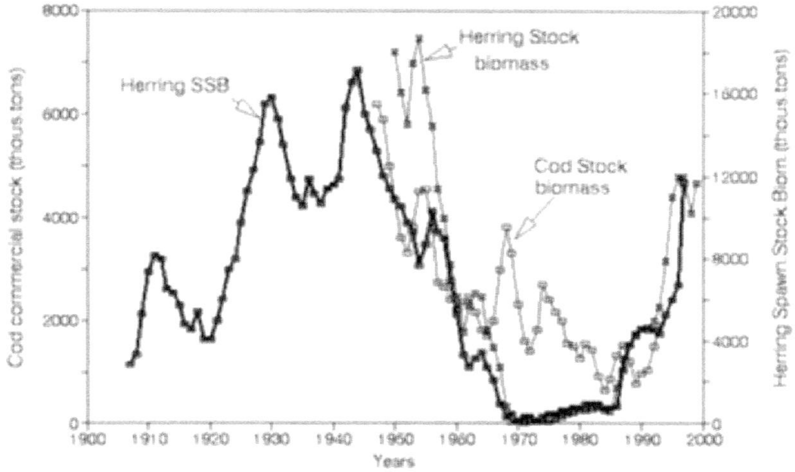

Figure 4. Comparative dynamics of spawning stock biomass (SSB) and the total stock biomass of Norwegian spring-spawning herring, and NE Atlantic cod stock biomass.

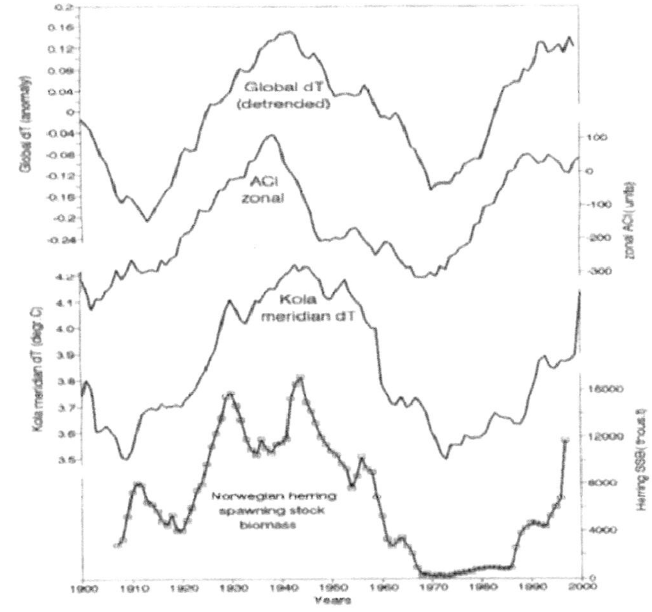

Figure 5. Comparative long term dynamics of climatic indices (Global dT, zonal ACI, Kola meridian dT, and spawning stock biomass of Norwegian spring-spawning herring 1900-2000).

The time series for cod are much shorter. However, the long-term variations of commercial stocks of cod and herring for the last 50 years have occurred almost synchronously (see Fig.4). Thus, it may be proposed that both herring and cod populations exhibit the same quasi-periodic variation with a period of roughly 60 years. The variation is in agreement with the long-term dynamics of key climatic indices, such as ACI, detrended Global dT, and Kola meridian dT.

The preliminary conclusion that coherent climate and fish population cycles of roughly 60 years exist is based on relatively short time series of 100-150 years. Reliable time series of catch statistics are even shorter. To obtain reliable periodicities of climate and fish dynamics on longer time-scales, time series of several hundreds of years and even millennia may provide opportunities. Interesting information on climate variability is found in the smoothed time series of surface air temperature reconstructed from the heavy oxygen (^{18}O) content in the Greenland ice cores (Fig. 6). Other data on long-term temperature dynamics can be derived from long-lived trees like Scottish and North Californian pine that allow analyses of tree rings for about 1500 years.

All time series in Fig. 6 exhibit well-manifested multi-decadal fluctuations, which can easily be quantified.

The sardine and anchovy biomass in the Californian up-welling region for the last 1700 years undergo rather regular short-scale fluctuations (Fig.7). Interestingly, the dramatic outbursts and collapses of sardine and anchovy populations for nearly two millennia obviously depended on natural factors only and were not influenced by fishery.

The so-called Temporal Spectrum Analysis (TSA) analysis was applied to both climatic and fish population dynamics for quantitative estimation of the visible periodicity. We were focused on the quasi-cyclic natural processes with a period from 20 to 100 years, characterizing the climatic periodicity. The results of spectral analyses are (Table 1):

- The dominant periodicity of temperature fluctuations calculated from the Greenland ice cores (the most reliable data) is 54 years. The periodicity calculated from the tree rings of Scottish and Californian pines are 60 and 76 years, respectively.
- The spectral analysis applied to the time series of Californian pine tree rings (the last 8 thousand years) results in the dominant periodicity of 55 years.
- The spectral analysis applied to the anchovy and sardine time series for the last 1700 years results in three dominant periods: 57, 76, and 72 years (Baumgartner *et al.*1992).
- The spectral analyses applied to the instrumental time series of ACI and Global dT for the last 100 and 140 years (modern data measured

instrumentally) return the 50-years and 55-years periodicity, respectively.

- The average dominant periodicity calculated form all the data is about 60 years, but the periodicity calculated from the Greenland ice cores (55 years) seems to be the most reliable.

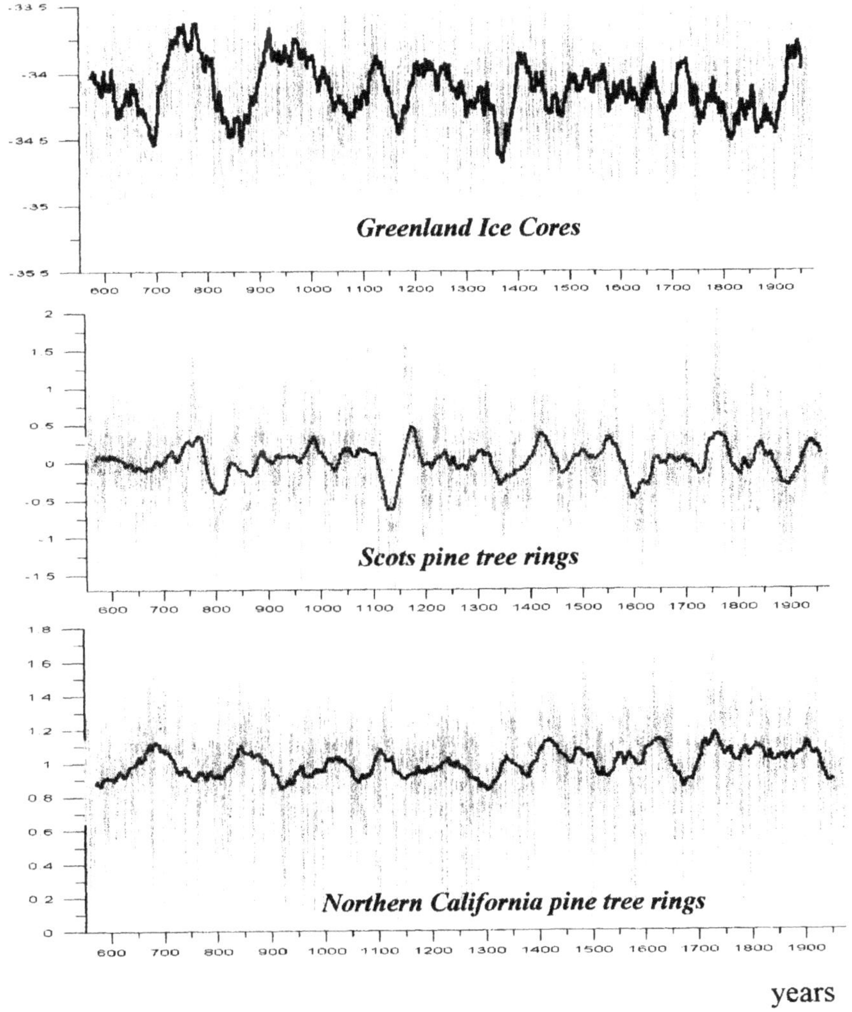

Figure 6. Fluctuations of air surface temperature for the last 1500 years reconstructed by analysis of O^{18} in ice cores and pine tree growth. Thin line: Annual temperatures. Bold line: 13-years smoothing temperatures. (Note: Temperature reconstructed by Greenland ice cores expressed as ppm O^{18}).

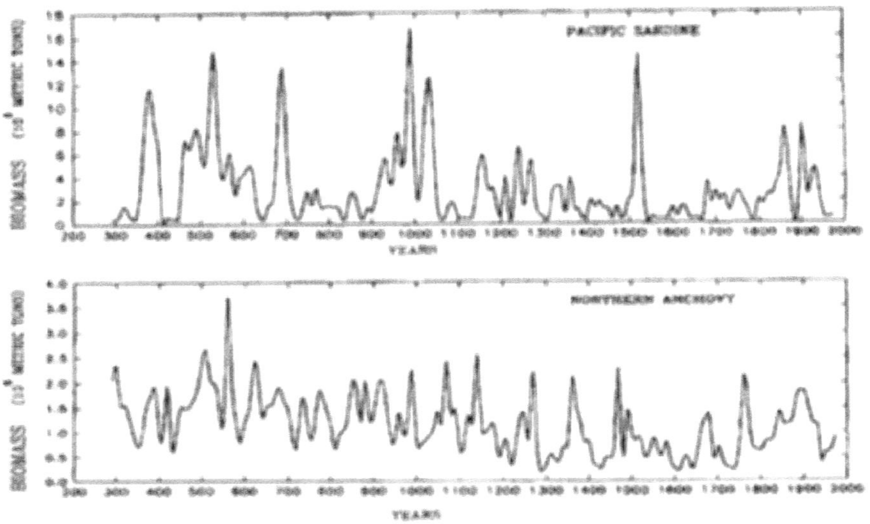

Figure 7. Fluctuations of Californian sardine and anchovy biomasses for the last 1700 years reconstructed by sediment core analysis (from Baumgartner *et al.* 1992).

The above results characterize the integrated spectrum of the most manifested (dominant) periodicity for each of the time series as a whole. However, it is important to know how these dominant periods change with the time.

The results obtained from our spectral analyses may be summarized as follows. During the last 500-1000 years, the dominant periodicity of air temperature fluctuations varied in the range of 55 to 70 years. The intensity of this periodicity has increased monotonously for the last millennium and reached the maximum by the late 20th century (Klyashtorin and Lyubushin 2003). This fact is an extra evidence for our earlier hypothesis that the 50-70 year cycles of climate and biota dominates in the present times.

3. MODELLING OF FISH PRODUCTION FORCED BY CLIMATE

Proceeding from all the results presented above, a simple stochastic model was developed to predict the climate-linked catch dynamics of 12 major commercial species and particularly the major North Atlantic and Arctic commercial fish species.

Table 1. Dominant periodicity of the investigated time series within the frequency range of 20-100 years.

Title of time series	Time series (years)	Primary (dominant) peaks (years)	Secondary peaks (years)
Greenland Ice cores	1422 (552-1973)	54	32
Arctic pine tree rings	1480 (500-1980)	60	32
Californian pine tree rings	1500 (479-1979)	76	32
Californian pine tree rings	8000 (-6000-1979)	55	20-35
Bottom sediments (Sardine)	1700 (270-1970)	57 and 76	33
Bottom sediments (Anchovy)	1700 (270-1970)	57	72 and 99
Global temperature anomaly (dT)	140 (1861-2000)	55	18
Atmospheric Circulation Index (ACI)		50	19

The main result of the model is that in the near future the climate and the dynamics of major commercial stocks will roughly continue their 60-year cycles, which they followed rather accurately during the last two millennia.

The model has predicted the future climate-linked dynamics of the Norwegian spring spawning herring stock (Fig. 8). According to the prediction the gross production of Norwegian herring may have reached its maximum at the beginning of the first decade of the new century and would after that decrease gradually. The long-term cod stock variation will possibly be similar to the herring dynamics.

4. REGIONAL CLIMATE IN THE NORDIC SEAS

The surface air temperature in Iceland, Svalbard and Jan Mayen (Fig. 9) show the same pattern of fluctuation throughout the 20th century. The data do not demonstrate any effect of climate change related to Global warming for the last 75 years. There are similar results from time series of air temperature in Norway, Northern Sweden and Finland (http://www.smhi.se/hfa_coord/nordklim/nkds.htm).

The mean temperature of the upper 200m layer measured for almost 100 years at the "Kola meridian" (north of the Kola peninsula) also does not demonstrate a long-term ascending trend, *i.e.* it does not tally with the Global warming concept (see Fig. 9). Note also that ACI measured in the North Atlantic–Eurasian region for more than 100 years shows no long-term increase (Figs 2 and 3).

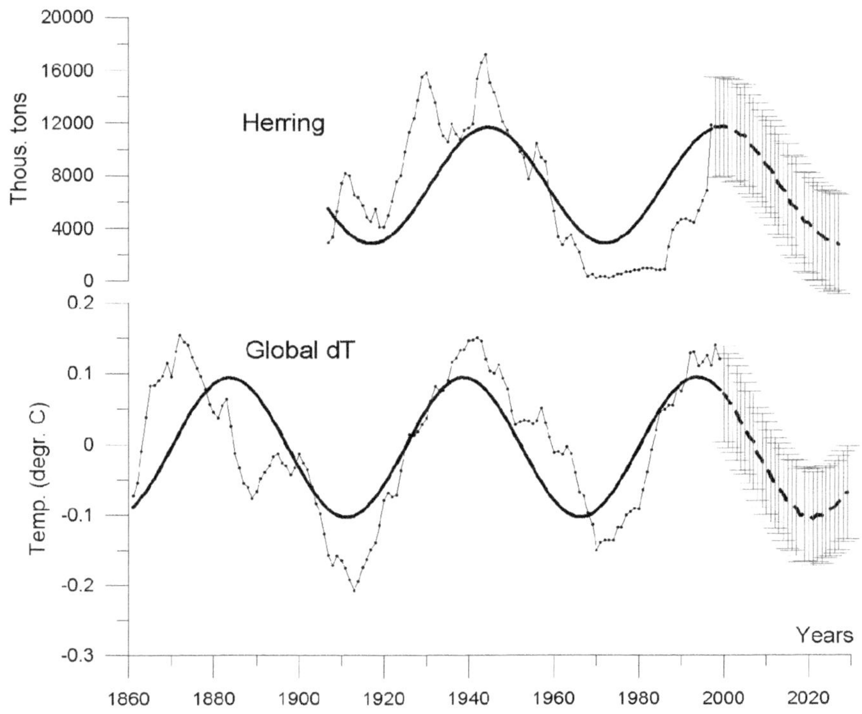

Figure 8. Comparative cyclic variations in detrended Global dT, Norwegian spring-spawning herring stock biomass (thin lines), model-generated 60 years climate and herring fluctuations (bold line) and proposed trends into the 21th century (dashed line with error bars).

The long-term dynamics of air temperature in the Arctic region of the NE Atlantic for the last 100 years (Fig. 9) is in good agreement with the dynamics of the Kola Meridian dT and de-trended Global dT.

Taking into account the different observations, the Nordic Seas may not respond to Global climate change as expected from observations in other parts of the Global system. That may have some consequences for the exploitation and management of region's living marine resources. However, the lack of adequate data from the western parts of the Nordic Seas is a disadvantage that prevents closer investigations and prognoses. The geographic position of Jan-Mayen in the center of the Atlantic sector of the Arctic makes it a particularly important place for regional monitoring of fluctuations in climate and biota.

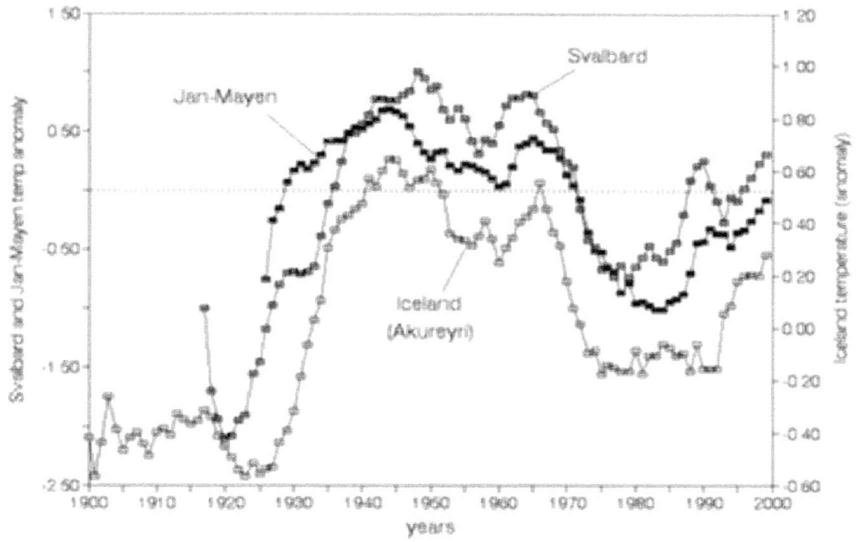

Figure 9. Comparative long-term dynamics of air temperature anomaly at the Svalbard, Iceland and Jan Mayen 1900–2000 (15-year smoothing).

REFERENCES

Anon. 1995. Report of the Arctic Fisheries Working Group. ICES C.M.1995/Assess:3.

Baumgartner TR, A Soutar, and V Ferreira-Bartrina 1992. Reconstruction of the history of Pacific sardine and Northern Pacific anchovy populations over the past two millennia from sediments Santa Barbara basin. CalCOFI Report 33, 24-40.

Kislov A, A Krenke, L Kitaev, N Shuvaeva and E Volodina 2000. Reproduction of temperature, precipitation and snow cover by the model in the framework of AMIP II experiment. *Izvestia Atmospheric and Oceanic Physics V* 36, 409-424.

Klyashtorin LB 2001. Climate change and long-term fluctuations of commercial catches. FAO Fisheries Technical Paper No 410, 86pp.

Klyashtorin LB and AA Lyubushin 2003. On the coherence between dynamics of the world fuel consumption and global temperature anomaly. *Energy & Environment* 14, 773-782.

Monin AS 2000. Climate kak problema phisiki (Climate as a problem in physics). *Uspekhi phizicheskikh nauk* (Achievements in the Physical Sciences) 170 (4), 419-445. (in Russian).

Sonechkin DM 1998. Climate dynamics as a non-linear Brownian motion. *International Journal of Bifurcation and* Chaos 8(4): 799-803.

Toresen R 1999. Variasjoner i mengde av norsk vårgytende sild i det tjuende århundre og sannsynlige miljømessige påvirkninger. *Fisken og Havet*, Special Publication 1999 (1), 125-127. (in Norwegian).

Yndestad H 2002. The code of long-term fluctuations of Norwegian Spring Spawning herring. ICES C.M. 2002/Q:02.

Chapter 15 NATO Advanced Research Workshop

The Marine Sublittoral Fauna of Jan Mayen Island

Zoogeography and Succession on "New" Lava Grounds

Bjørn Gulliksen, Frank Beuchel, Torleiv Brattegard, and Rune Palerud
B Gulliksen, Norwegian College of Fishery Science, University of Tromsø, Norway
E-mail: bjoerng@nfh.uit.no
F Beuchel, Norwegian College of Fishery Science, University of Tromsø, Norway
T Brattegard, Department of Biology, University of Bergen, Norway
R Palerud, Akvaplan.niva AS, Tromsø, Norway

Abstract: A volcanic eruption at Jan Mayen in 1970 created "new" submarine lava grounds. This incident initiated spatial and temporal studies of the marine sublittoral fauna in 1972, 1978, 1994 and 1999. The spatial inventories indicate that Jan Mayen Island in zoogeographical terms has not a unique and endemic benthic fauna. The fauna is generally an impoverished fauna of that found at the Norwegian coast and Svalbard. The studies of colonisation and succession of the "new" grounds indicate that the time needed to reach a successional stage (climax-stage) similar to that recorded on similar substrates at the same depth (deeper than about 15m) elsewhere at the island is a process which needs at least 30 years, and probably much longer time.

Key words: Jan Mayen Island, sub-littoral fauna, zoogeography, succession

1. INTRODUCTION

A volcanic eruption at Jan Mayen in 1970 created "new" submarine lava grounds (Gjelsvik 1971). Lava flowed into the sea and created a "new" coastal platform about 1km wide and 4km long (Gulliksen 1974a). This incident initiated spatial and temporal studies of the marine sublittoral fauna in 1972, 1978, 1994 and 1999 (Gulliksen *et al.* 1980, Gulliksen unpubl.).

S. Skreslet (ed.), Jan Mayen Island in Scientific Focus, 159-171

This presentation reviews the main results from these investigations. Information related to the spatial studies is primarily focused on comparisons of the sub-littoral benthic fauna at Jan Mayen with those of Svalbard, Bear Island and the Norwegian coast. The temporal studies focus on the status on colonization and succession of sublittoral organisms on the "new" submarine grounds at Jan Mayen created in 1970.

The aims of this work are to present an updated version of the biogeographical status of the marine fauna of Jan Mayen Island, and also to give a rough estimate of the time needed before the "new" grounds have a fauna similar to that found at similar habitats on the "old" grounds.

2. MATERIAL AND METHODS

The investigations in 1972 were carried out by the University of Trondheim in collaboration with The Norwegian Polar Institute using the rescue vessel "Sjøfareren". The investigations in 1978 were carried out by the University of Tromsø using the research vessel "Johan Ruud", and the investigations in 1994 and 1999 were joint ventures between the University Centre of Svalbard (UNIS) and the University of Tromsø with the research vessel "Jan Mayen".

2.1 Spatial Studies (Zoogeography)

The comparisons between the benthic faunas of Jan Mayen with those of Svalbard, Bear Island and at the Norwegian coast are mainly based on data published in two documents (Brattegard and Holthe 1997, Gulliksen *et al.* 1999). The authors regularly update the species lists in these publications with newly published and unpublished information regarding nomenclature and records, and these updated versions are used when faunas are compared.

An overview of the collected material from Jan Mayen in 1972 is given in Gulliksen (1974). Parts of the collected material from 1972 were sent to taxonomic specialists who later published more detailed accounts of Mollusca (Sneli and Steinnes 1975, Sneli 1977), Porifera (Tendal 1983), Echinodermata (Skjæveland 1973) and Ascidiacea (Gulliksen 1974b).

2.2 Temporal Studies (Succession on "New" Grounds)

Quantitative studies did not take place on the "new" grounds in 1972. Samples from "old" grounds were collected with triangular dredges and handpicking of organisms while SCUBA diving. Investigations at the "new" grounds 1972 were restricted to observations and handpicking of organisms while SCUBA diving due to the scarcity of organisms.

Quantitative sampling of both "old" and "new" grounds was carried out in 1978. The benthic samples were collected by SCUBA diving using a suction sampler working on the "air lift" principle (Barnett and Hardy 1967, Hiscock and Hoare 1973). 43 samples (sample size = $1/8m^2$) were taken from the "new" grounds and 18 samples from the "old" grounds covering an area of $7.625m^2$. More details about the sampling are given in Gulliksen et al. (1980).

The sampling and treatment of the samples in 1978 were quite time-consuming and laborious. For example, two divers were usually not able to obtain more than two samples during one dive. Weather and currents also limited efficient use of the suction sampler. The sampling therefore took four experienced divers about three weeks of intense diving and sampling. Less boat-time and fewer divers were available in 1994. It was therefore decided to use quantitative underwater photography, which is faster and less laborious. An experience from 1978 was also that a sample size of $1/8\ m^2$ probably was too small for reliable estimates of "colonial" organisms such as poriferans and ascidians. The sample size was therefore increased to $1/4\ m^2$. A major drawback of underwater photography compared to suction sampling is, however, that smaller macrofauna such as small and inconspicuous amphipods and polychaetes can be overlooked. The fauna recognizable on underwater photographs is therefore now commonly called "megafauna" (Gage and Tyler 1991, Piepenburg and Schmid 1997). A general advantage of image analysis is that the technique is non-destructive, and samples (photographs) can be re-examined if necessary.

A result from 1978 was also that the fauna on "new" grounds shallower than about 15m depth had reached a stage similar to that found on the "old" grounds. However, the fauna on deeper "new" grounds had not yet reached the same succession stage as "old" grounds. The quantitative sampling in 1994 was therefore limited to the depth interval 15–32m depth. A total of 59 samples (photographs) allocated at three localities from the "old" grounds and 32 photographs (two localities) from the "new" grounds covering a total area of $23.5m^2$ were analysed. The photographs from 1994 were analysed by image analysis techniques, using Adobe Photoshop with the IPTK- plug-in ("Fovea pro") for verification of morpho-metric measurements.

In 1978, the abundance of colonial species and most poriferans were given relative abundance values according to their weight in each sample. The relative abundance values were ascribed as follows: Species with a cumulated specimen weight less than 5 % of the total sample weight were given abundance value 1, and species weighing 5-25 % of the total sample were given abundance value 2, while species with more than 25 % of the total sample weight were given abundance value 3. As regards the sponge

Sycon, the vase-shaped "bodies" of this genus were counted as if they were solitary organisms.

In 1994, the technique of image analysis enabled us to calculate the cover of such organisms. The cover of sponges and colonial organisms were therefore measured in percentage of the total area analysed (91 * 1/4 m^2) while the abundance of solitary organisms were calculated as number of individuals per unit area.

The investigations on the "new" grounds in 1999 were restricted to qualitative observations using SCUBA diving and the ROV "Aglantha". The ROV enabled observations from greater depths than possible using conventional SCUBA-equipment.

3. RESULTS AND DISCUSSION

3.1 Spatial Studies (Zoogeography)

A total of 607 marine invertebrates and fish species are recorded in the coastal waters of Jan Mayen Island (Table 1). No endemic species are recorded, and all species have been recorded somewhere else in North Atlantic waters. Comparable numbers elsewhere are 3678 spp at the Norwegian coast, 1673 spp at Svalbard, and 445 spp at Bear Island.

Table 1. Benthic and hyperbenthic* macro-invertebrates and demersal fish in four regions. Compiled and updated by Torleiv Brattegard (October 2003).

	Norway	Bear Island	Svalbard	Jan Mayen
Porifera	270	18	141	24
Hydrozoa	174	42	84	31
Scyphozoa (Stauromedusae)	5	2	6	1
Anthozoa	114	18	41	10
Nemertini	76	1	10	1
Kamptozoa (Entoprocta)	27	0	4	0
Priapulida	4	0	3	2
Echiurida	5	0	1	0
Sipuncula	16	2	5	2
Annelida	656	37	259	75
Pycnogonida	42	15	26	11

Table 1 (continued)

	Norway	Bear Island	Svalbard	Jan Mayen
Ostracoda	144	5	46	1
Copepoda Calanoida*	28	0	1	0
Cirripedia	32	6	12	1
Leptostraca	4	1	1	1
Mysidacea	46	2	16	12
Cumacea	56	6	31	11
Tanaidacea	30	1	7	8
Amphipoda	401	51	265	63
Isopoda	115	10	34	17
Decapoda	101	15	23	10
Caudofoveata	7	1	3	1
Solenogastres	22	0	2	0
Polyplacophora	13	3	5	3
Prosobranchia	229	42	140	73
Heterobranchia	186	2	37	11
Cephalopoda	10	0	4	3
Bivalvia	207	24	86	41
Scaphopoda	10	0	3	1
Phoronida	4	0	0	0
Brachiopoda	9	5	7	3
Bryozoa	272	29	192	99
Chaetognatha (benthic)	1	1	0	0
Hemichordata	5	0	0	0
Crinoidea	6	2	4	2
Asteroidea	45	24	31	16
Ophiuroidea	36	11	16	12
Echinoidea	16	3	4	4
Holothuroidea	31	9	16	7
Ascidiacea	77	26	50	17
Cephalochordata	1	0	0	0
Petromyzoniformes	3	0	0	0
Myxiniformes	1	0	0	0
Chondrichthyes	19	5	4	3
Osteichthyes	122	26	53	30
Sum	3678	445	1673	607

The differences in number between the areas are due to factors such as:

- Areas compared have different size, and some types of habitats present at the Norwegian coast and Svalbard are lacking at Jan Mayen Island and Bear Island. Examples of such habitats are fjords with cold-water basins, soft bottom habitats and habitats influenced by heavy sedimentation of terrigenous inorganic particles. This "lack of habitats" reduces the numbers of taxa at Jan Mayen and Bear Island.

- The Norwegian coast has generally warmer waters compared to the three other areas. Some taxa such as Decapoda and Cirripedia do have fewer species in cold waters, and thus higher numbers at the Norwegian coast.
- Some taxa, such as Nemertini and Heterobranchia, are less studied at Svalbard, Jan Mayen and Bear Island compared to the Norwegian coast, and the numbers for such groups are thus higher for the Norwegian coast where investigations of most taxa have been more thorough.
- The low species number at Jan Mayen compared to Svalbard and the Norwegian coast is due to its smaller size and isolated position. Recruitment and settlement of some benthic species is thus dependant on invasion of pelagic larvae from continental shelves far away. Such larval invasions are stochastic events, which may lead to periodical absence or disappearance of species recorded in earlier investigations. An example of such a species is the barnacle *Balanus balanus*.

3.2 Temporal Studies (Succession on "New" Grounds)

The "new" grounds were dominated by rocky substrates, and techniques adapted to those types of substrates have been applied (observations while SCUBA diving, a diver-operated suction sampler, underwater photography, video-recordings using ROV).

3.2.1 1972

Observations while SCUBA diving (Gulliksen 1974) did reveal little epifauna on the "new" grounds. Scattered and small individuals of two sedentary taxa, namely hydrozoans and bryozoans, were observed. Motile animals present on the "new" grounds were primarily amphipods (*Gammarellus homari*, *Gammarus setosus*, *Ischyrocerus* sp.) and nudibranchs (*Dendronotus* sp.). Five species of benthic algae were recorded (Gulliksen op. cit.).

3.2.2 1978

Studies in 1978, eight years after the volcanic eruption, suggested that faunal assemblages on the "new" grounds below 15m depth had not yet reached the same succession stage as the "old" grounds (Gulliksen *et al.* 1980). Above 15m depth, however, where more severe physical conditions (strong water movements and scouring or grinding by sand and ice) limit the community complexity, the "new" grounds where quite similar to those on the "old" grounds at the same depth.

The "new" grounds had in 1978 a lower diversity (Gulliksen *et al.* 1980), higher biomass and a dominance of filter feeders compared to "old" grounds with quite similar substrate. The space on "new" grounds was highly monopolized by the bivalve *Hiatella arctica*, which contributed more than 80 % of the total biomass (Fig. 1).

The "old" grounds did not reveal the same single-species domination regarding biomass as the "new" grounds (Fig. 1). The biomass of *Hiatella arctica* was about 14 % of the biomass recorded at the "new" grounds and contributed to 20 % of the total biomass. Other dominants regarding biomass were the sponge *Halichondria* sp. (26 %), the actinian *Urticina felina* (L.) (20 %) and the sea urchin *Strongylocentrotus droebachiensis* (Müller) (20 %). These three are all organisms with long maximum life spans (> 40 years). *Strongylocentrotus droebachiensis* and the sea anemone *Urticina felina* are both predators. Two other species among the nine species with the highest values of biomass, the brittle star *Ophiopholis aculeata* and the soft coral *Gersemia rubiformis* are also species with long life spans.

The abundance data (Fig. 2) revealed the presence of a high number of amphipods (i.e. *Ischyrocerus anguipes*, *Caprella septentrionalis*, *Metopa bruzelii*) and polychaete species (i.e. *Spio filicornis*) both on the "old" and "new" land, but no clear trends regarding differences between "old" and "new" grounds were evident regarding these taxa. However, the gastropods *Margarites groenlandicus* and *Margarites helicinus* were numerically more dominant at the "new" grounds compared to the "old grounds". The lack of predators on the "new" grounds may explain the high abundances of these snails.

3.2.3 1994 and 1999

The change of sampling methods from 1978 (Fig. 2) to 1994 (Fig. 3), and 1999 should be taken in account when comparing data. In addition, no biomass data were collected in the two latter years.

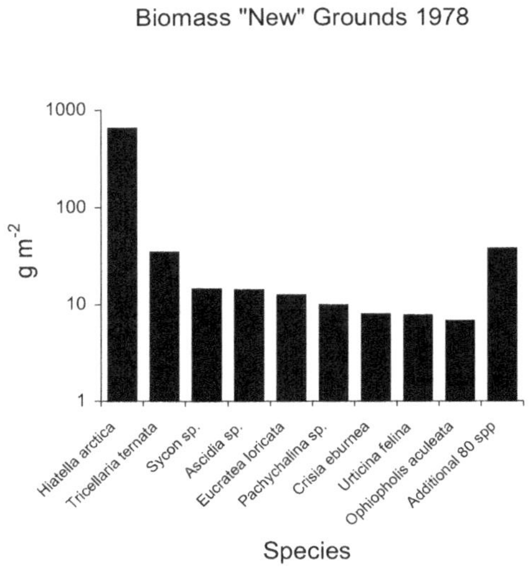

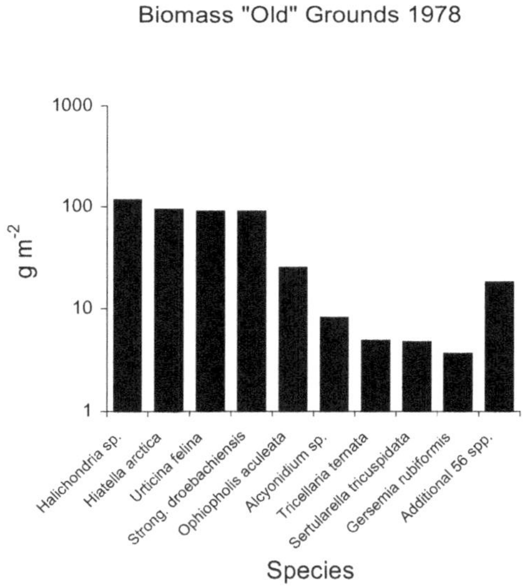

Figure 1. Biomass of benthic organisms (alcohol weights in g m^{-2}) from "new" and "old" grounds in 1978.

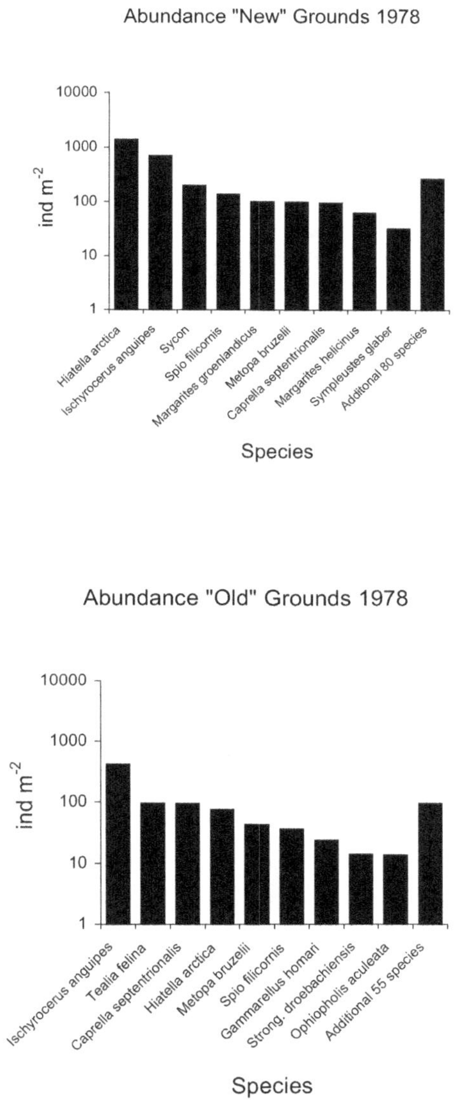

Figure 2. Abundance of solitary and colonial organisms in numbers$\cdot m^{-2}$ in 1978. Colonial organisms were given relative abundance values according to their weights in each sample. Sponges and colonial organisms were counted, as they were solitary individuals.

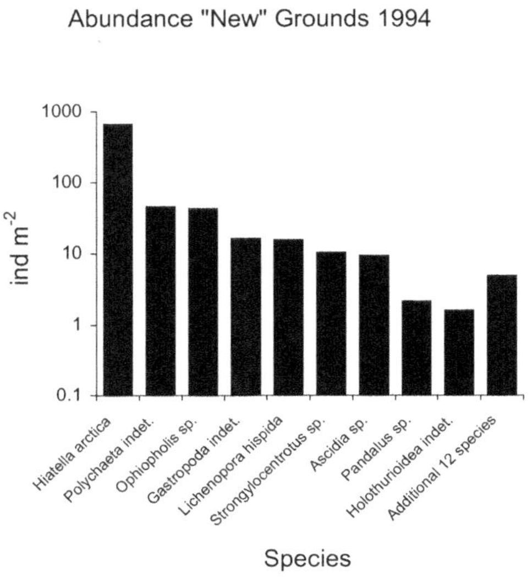

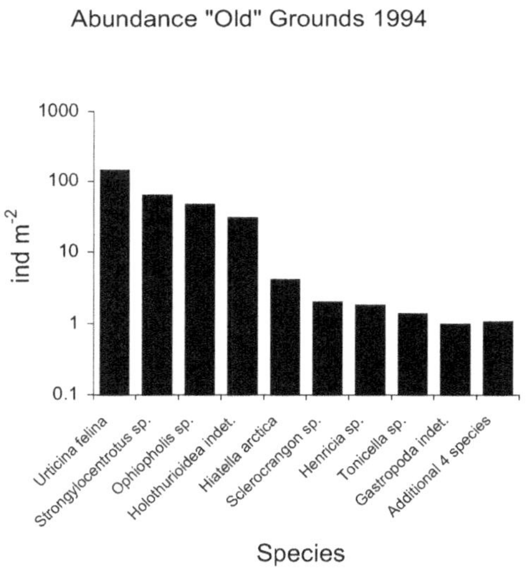

Figure 3. Abundance of solitary and colonial organisms in numbers·m^{-2} in 1994. Data from image analyses. Only solitary organisms were regarded.

The photographs from 1994 revealed that the faunal assemblages below 15m depth on the "new" grounds still were quite different from those recorded at the "old" grounds (Fig. 4), even when the smaller amphipods and polychaetes, which are difficult to observe and count on photographs and video, are disregarded. The bivalve *Hiatella arctica* was still a dominant species on rocky bottom below 15m depth on the "new" grounds (Fig. 3).

Even if long-lived motile and colonial organisms, which are conspicuous members of the climax-assemblages on "old" grounds, occurred in higher numbers in 1994 and 1999 compared to 1978 on the "new" grounds, there were still large differences between "old" and "new" grounds (Beuchel and Gulliksen 2003).

Digital image analysis was used on a set of underwater pictures obtained in 1994 from "old" and "new" grounds. The resulting MDS-plots (MDS=Multi-Dimensional Scaling) (Fig. 4) confirm the differences between "old" and "new" grounds regarding both colonial and solitary benthic organisms (Beuchel and Gulliksen 2003). The scattered distribution on "new" ground places indicates that the colonisation process is heterogeneous and probably irregular.

Observation at deeper waters (50 – 100m depth) on the "new" land using the ROV equipped with a video-camera indicated also here a dominance of *Hiatella arctica*, but the distribution was quite patchy. There were large areas completely covered with the mussels while others hardly had any.

4. CONCLUSIONS

In zoogeographical terms, Jan Mayen Island does not have a unique and endemic faunal composition of benthic organisms. The fauna is generally an impoverished fauna of that found at the Norwegian coast and Svalbard. However, studies of the species composition at Jan Mayen may reveal important knowledge regarding life strategies of selected benthic organisms, and their abilities to invade and colonize distant islands.

The temporal studies of colonization and succession indicate that the time needed to reach a climax stage of the "new" grounds similar to that recorded on "old grounds" (relevant for depths >15m) is a process which needs at least 30 years, and probably much longer time.

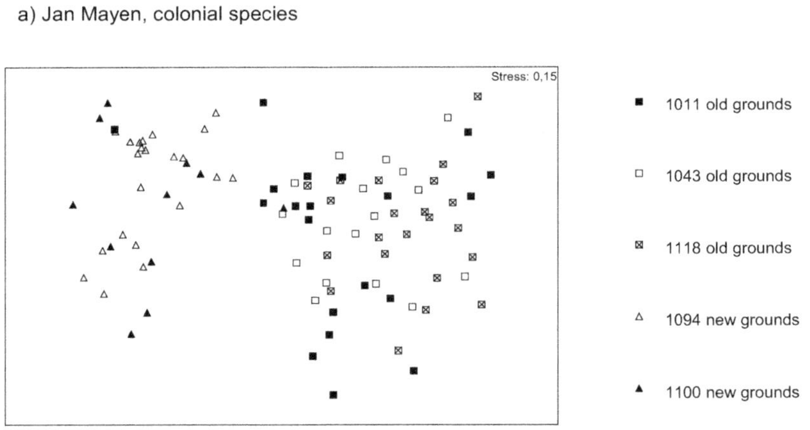

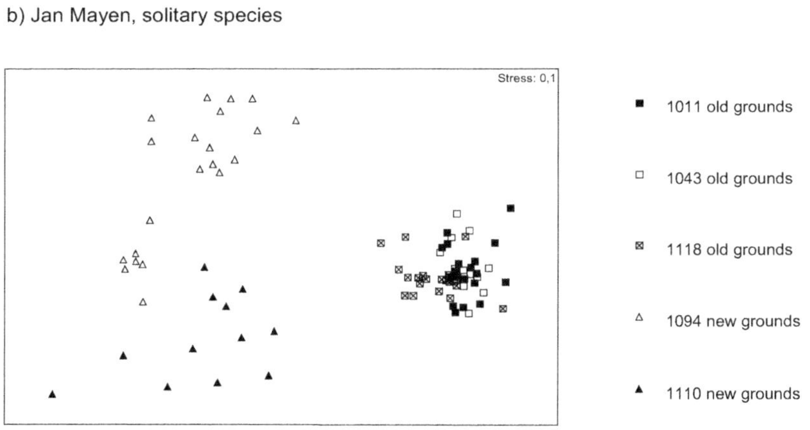

Figure 4. MDS-plots. Data (log-transformed) from 1994 comparing colonial (a) and solitary (b) organisms on "old" and "new" grounds. Each point represents one picture (all together 91 pictures over five localities). Data from colonial organisms are based on %- cover of picture-area, solitary organisms on numbers * m^{-2} (The numbers in the legend refers to the ship's station-numbers of the sampled localities).

REFERENCES

Barnett, PRO and BLS Hardy 1967. A diver-operated quantitative bottom sampler for sand macrofauna. *Helgoländer wissenshaftliche Meeresuntersuchungen* 15, 390-398

Beuchel F and B Gulliksen 2003. Monitoring of rocky-bottom macrobenthic communities on locations at Svalbard and Jan Mayen using digital image analysis. Conference paper, ICES-ASC 2003, 4pp.

Brattegard T and T Holthe (eds.) 1997. Distribution of marine, benthic macroorganisms in Norway. Norwegian Directorate for Nature Management. Research Report for DN 1997-1. 409 pp.

Gage JD and PA Tyler 1991. *Deep-Sea Biology: A Natural History of Organisms at the Deep Sea Floor.* Cambridge University Press, Cambridge. 504 pp.

Gjelsvik T 1971. Volcano on Jan Mayen alive again. *Nature* London 228, 352.

Gulliksen B 1974a. Marine investigations at Jan Mayen in 1972. *Miscellanea. Det Kgl. Norske Videnskabers Selskab. Museet* 19, 1-46.

Gulliksen B 1974b. Ascidians of Jan Mayen Island. *Astarte* 7, 25-28.

Gulliksen B, T Haug and OK Sandnes 1980. Benthic macrofauna on new and old grounds at Jan Mayen. *Sarsia* 65, 137-148.

Gulliksen B, R Palerud, T Brattegard and JA Sneli 1999. Distribution of marine benthic macro-organisms at Svalbard (including Bear Island) and Jan Mayen. Norwegian Directorate for Nature Management. Research Report for DN 1999-4. 148 pp.

Hiscock, K and R Hoare 1973. A portable suction sampler for rock epibiota. *Helgoländer wissenschaftliche Meeresuntersuntersuchungen* 25, 35-38.

Piepenburg D and MK Schmid 1997. A photographic survey of the epibenthic megafauna of the Arctic Laptev shelf: distribution, abundance, and estimates of biomass and organic carbon demand. *Marine Ecology Progress Series* 147. 63-75.

Skjæveland SH 1973. Echinoderms of Jan Mayen Island. *Astarte* 6, 69-74.

Sneli JA and A Steinnes 1975. Marine Mollusca of Jan Mayen Island. *Astarte* 8, 7-16.

Sneli JA 1977. Jan Mayen molluscs. *Malacologia* 16 (1), 257-260.

Tendal OS 1983. Sponges of Jan Mayen Island. *Astarte* 12 (1979), 53-55.

Chapter 16 NATO Advanced Research Workshop

The Marine Fauna of Arctic Islands as Bioindicators

Jan M Węsławski
Institute of Oceanology, Sopot, Poland
E-mail: weslaw@iopan.gda.pl

Abstract: European Arctic and sub-arctic islands represent a variety of physical and biological characteristics. For the species that inhabit the inter-tidal zone, the open sea expanses act as effective barriers for dispersal. The four scientific challenges that comes with the Arctic islands littoral studies are discussed in this paper: 1) the island biogeography theory, 2) larval drift and dispersal, 3) bio-indication of climate change and water masses displacement, 4) bio-monitoring of climate change the biological consequences. The science of marine ecology may benefit from studying the little known, isolated islands in the Arctic.

Key words: islands biogeography, marine ecology, dispersal, littoral

1. INTRODUCTION

European Arctic and sub-arctic islands present a variety of physical and biological characteristics, from the small, remote Bjørnøya, the extensive archipelagos of Franz Josef Land and Svalbard, and the ocean-exposed Jan Mayen, to the shelf and brackish waters around Kolguyev. For the shallow water fauna, especially for organisms which inhabit the inter-tidal zone, open sea expanses act as effective barriers against dispersal, and this is why an island's littoral fauna may serve as an indicator of long-term processes. The four scientific challenges presented by Arctic island littoral studies discussed in this paper are as follows:

S. Skreslet (ed.), Jan Mayen Island in Scientific Focus, 173-180

- Island biogeography theory that is of significant importance to modern ecology and biodiversity studies, still used infrequently with marine benthos.
- Larval drift and dispersal. Islands act as stepping stones for the expanding geographical ranges of marine benthos, and larval plankton studies are the focus of research issues.
- Bio-indication of climate change and water mass displacement. The number of littoral animals and seaweeds are confined to specific hydrological and climatic conditions, so they are useful indicators of change in physical conditions integrated over long time scales.
- Bio-monitoring of the biological consequences of climate change by food web studies, palaeo-oceanography and comparative research in similar biota. Much can be learned about the effects of changes on the functioning of coastal ecosystems.

2. ISLAND BIOGEOGRAPHY

Island biogeography theory was developed in the 1960s by McArthur and Wilson (1967) during experimental and field research on the insect fauna from small sandy islands. The basic observation was that small islands have fewer species than large islands, and that species richness on them also depends on the distance from the source area (mainland). Furthermore, observations were made on the arrival rates of new species to the island and the rates of extinction on the island. These basic facts have been applied and developed in a great number of papers and scientific books, especially in the field of conservation biology (the marine example is Johanesson and Johanesson 1995). The principle of the minimum effective area needed to sustain a stable population and habitat fragmentation comes from this last application. The theory has important implications for evolutionary research and studies on the colonization and reclamation of deserted areas. Still, the great majority of studies are based on the terrestrial fauna or flora of islands. There is very little interest in the island biogeography theory among marine biologists. Few recent exceptions are papers by Clarke and Warwick (1999), Morton and Briton (2000), Benedetti-Cecchi *et al.* (2003). The reason for this absence is simple. Islands need to be separated by an effective barrier. In the case of terrestrial organisms it is water. However, there is a group of marine organisms that has similar difficulties in crossing the open sea barrier. Species that belong to the intertidal and littoral fauna and flora need solid ground in shallow water, good aeration, sufficient light and water movement for survival. When washed offshore, theses species barely survive and are unable to reproduce or grow.

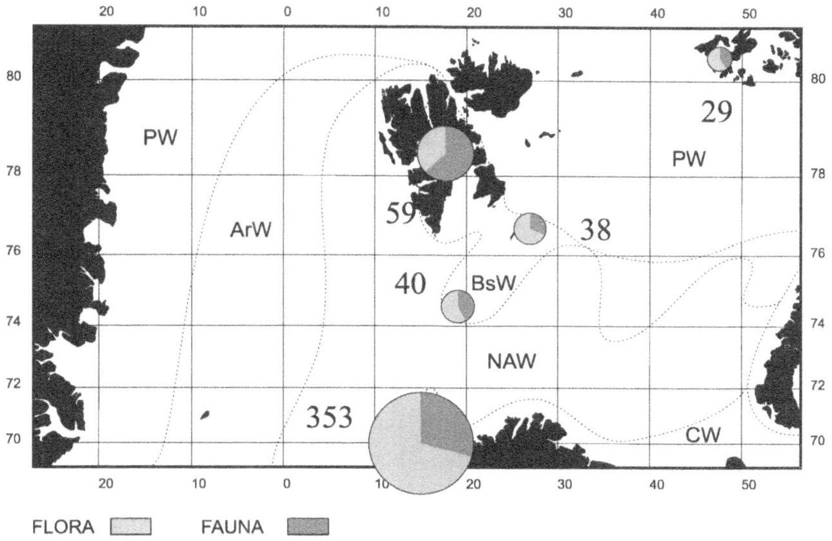

Figure 1. Littoral fauna and flora of northern European and Arctic islands. Compilation from Golikov and Averintzev (1977), Gulliksen (1979), Gulliksen *et al.* (1980), Ingolfsson (1996), Munda (1991), Sneli (1968), author's unpublished data.

Traditionally, the Arctic intertidal zone is regarded as very poor since it is exposed to a harsh climate, ice scouring and extreme variability in the sea water salinity during the melting season (Madsen 1936). However, closer examination indicates that even in the high Arctic, the littoral contains several specific animal and plant species (Fig. 1). Low temperature and salinity are not the major inhibitors of opportunistic species occurrence, and mechanical erosion by ice does not have an impact in sheltered localities (Hopner-Petersen 1962, Weslawski *et al.* 1993, 1997, Gutt 2001).

The majority of Arctic inter-tidal zones have been free of ice cover for the past 8-10 thousand years (Hjort 1995). This means that the ecosystem there is very young. It can be assumed that the observed scarcity of Arctic littoral is an effect of the slow colonization of remote Arctic islands, with smaller islands having fewer species (Fig. 1). Hence, the arrival rate of new species from the source area is of great interest as is the turnover of species. The paradox is that species with better dispersal ability (those with pelagic larvae) are typically the first to both appear and disappear from the island. Their offspring cannot return to the home shore. One of the few described examples is the disappearance of the barnacle *Balanus balanus* from Jan Mayen (Gulliksen 1974, and pers. comm.). This fact is most likely responsible for the occurrence of the viviparous periwinkle *Littorina saxatilis* and the absence of the pelagic-breeding *Littorina* species in the vicinity of Spitsbergen (*L. saxatilis* has brood protection and do not lose offspring by currents, which e.g. *L. littorea* does, because they spawn pelagic eggs and have planktonic larvae). The more remote the island, the more interesting the process of its shore fauna formation is. Jan Mayen presents the ideal example of a small isolated island far away from the Boreal and Arctic source areas.

3. LARVAL DRIFT AND DISPERSAL

This problem is closely related to the previous one, as the islands act as stepping stones for the expanding geographical ranges of marine benthos through larval plankton (mero-plankton). Again, knowledge of this issue is very limited, and substantial data only exist from the northern Norwegian Sea and Barents Sea (Mileikovskij 1968). He observed that pelagic larvae of different taxa of benthic animals had unequal capacities for dispersal (Fig. 2). Larval drift by advection of seawater is closely coupled with predominant currents. Thus, the inter-annual variability of the North Atlantic Current plays a very important role. The study of larval drift and survival of rafters can provide information about the natural processes of faunal dispersion and species turnover (Bhaud 2000, Hobday *et al.* 2000, Weber *et al.* 2000). The Jan Mayen Island is positioned in a very interesting place, where drifting propagules from different source areas may occur. The indirect evidence for this is the studies on the driftwood dispersal in the Arctic, where Jan Mayen, have been found as recipient of timber from the NW Siberia (Johansen 1998).

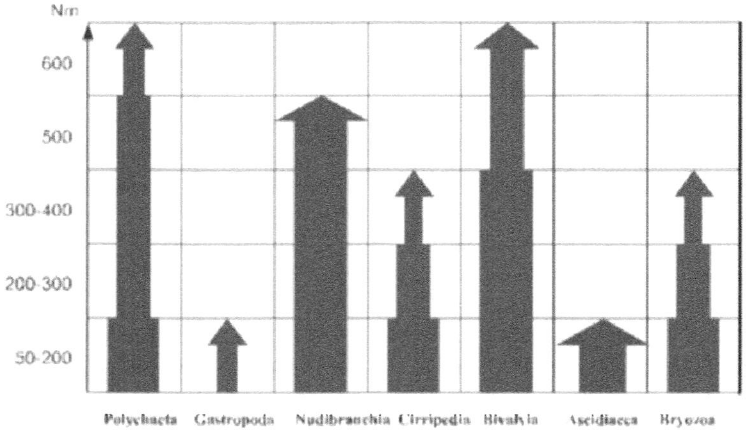

Figure 2. Distances in the drift of different larva (Modified from Mileikovski 1968).

4. BIO-INDICATION OF CLIMATE CHANGE AND WATER MASS DISPLACEMENT

The number of littoral animals and seaweeds is confined to specific hydrological and climatic conditions that interact with dispersal of marine species. The presence of species indicates not only present conditions. If their life cycle is not too long (only a few years), stable presence shows that favorable conditions for breeding have prevailed over a longer period. Thus, multiyear marine invertebrates may serve as useful indicators of physical conditions integrated over long time scales (Blacker 1957). From palaeontological excavations, we know that the common blue mussel *(Mytilus edulis)* was very abundant on Svalbard during a warm period of up to 8 thousands years ago (Hjort 1995). Then it disappeared from the archipelago. Its reappearance may provide us with a clear signal of Global warming effect. The finding of live *M. edulis* on Bjørnoya in 1994 may be a first sign of the warming phenomenon, and is an example of the usefulness of islands research (Weslawski *et al.* 1997).

Quite often there are sequences of sibling species that have evolved on the border between the Arctic and boreal provinces. If they occur together (sympatry), the colder and more brackish part of the area of occurrence is inhabited by Arctic species, while Atlantic species inhabit the warmer, more saline part. A good example of this is the distribution of such a pair of species in the Svalbard littoral (Fig. 3).

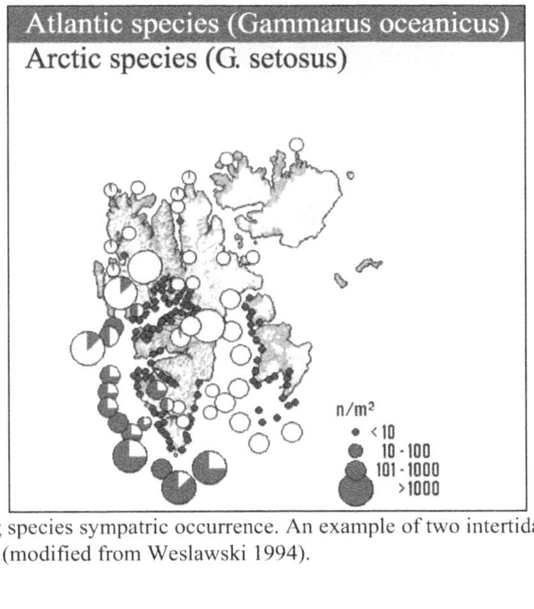

Figure 3. Sibling species sympatric occurrence. An example of two intertidal *Gammarus* species (modified from Weslawski 1994).

Short-lived pelagic species like the sibling *Calanus* forms of the boreal *C. finmarchicus* and the Arctic *C. glacialis* are equally good indicators of water mass displacement (Karnovsky *et al.* 2003). The area of Jan Mayen, undergoes strong inter-annual changes and may be subject to gradual Global warming, which makes the island a very attractive site for monitoring occurrence of bioindicators.

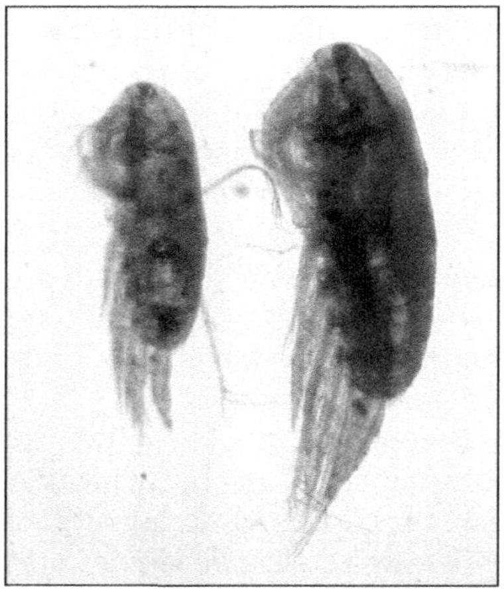

Figure 4. Two sibling species of pelagic copepods with great difference in caloric value and availability to top predators. Left: *Calanus finmarchicus*. Right: *Calanus glacialis*.

5. MONITORING THE BIOLOGICAL CONSEQUENCES OF CLIMATE CHANGE

By analyzing year to year pelagic food web changes, much can be learned about major biological shifts in the system. Pelagic organisms in the Jan Mayen area are supplied by two major sources that are the Atlantic and Arctic water masses. Both can occur at varied intensities and frequencies. In general, Atlantic plankton species are small and lean, very abundant and distributed below the mixed layer or in its deep parts. The Arctic plankton species are bigger (Fig. 4) and fatter, less abundant and more surface oriented (Weslawski and Kwasniewski 1990).

Inter-annual shifts from Atlantic to an Arctic food web may have profound effects on the direction of the energy flow, the dominance changing from pelagic to benthic (Petersen 1984). Furthermore, the ratio of crustacean to gelatinous biomass can also change from year to year, thus changing the nutritional quality of the food for planktivores. The top predators that feed on macro-zooplankton function as very efficient samplers, which permit checking the abundance and quality of the food base (Karnovsky *et al.* 2003).

Populations of seabirds nesting on islands are of special interest in summer, because during feeding of chicks the adult birds may forage intensively in the sea near the nesting habitat. Sampling the food brought by adults to chicks need not be destructive as the birds are sampled alive and released. The method permits relatively large statistical material. Colonies of auks and gulls on Jan Mayen are very well fit for such research.

REFERENCES

Benedetti–Cecchi L, E Maggi, I Bertosci, S Vaselli, F Micheli GC Osio, and F Cinelli 2003. Variation in rocky shore assemblages in the northwestern Mediterranean:contrasts between islands and the mainland. *Journal of Experimental Marine Biology and Ecology* 293, 193-215.

Bhaud MR 2000. Some examples of the contribution of planktonic larval stages to the biology and ecology of polychaetes. *Bulletin of Marine Science* 67, 345-358.

Blacker RW 1957. Benthic animals as indicators of hydrological conditions and climatic changes in Svalbard waters. *Fishery Investigations* 2, 1-49.

Clarke KR and RM Warwick 1999. The taxonomic distinctness measure of biodiversity: weighting of step lengths between hierarchical levels. *Marine Ecology Progress Series* 184, 21-29.

Golikov AN, Averintzev VG 1977 Benthic biocenoses of Franz Josef Land. Iss. Fauny Moriej 14, 5-54 (in Russian).

Gulliksen B 1974. Marine investigations at Jan Mayen in 1972. *Miscellanea. Det Kgl. Norske Videnskabers Selskab. Museet* 19, 1-46 (Trond-heim).

Gulliksen B 1979. Shallow water benthic fauna from Bear Island. *Astarte* 12, 5-12.

Gulliksen B, T Haug, and OK Sandnes 1980. Benthic macrofauna of new and old lava grounds at Jan Mayen. *Sarsia* 65, 137-148.

Gutt J 2001. On the direct impact of ice on marine benthic communities, a review. *Polar Biology* 24, 553-564.

Hjort C 1995. Radiocarbon dated common mussel *Mytilus edulis* from eastern Svalbard and the Holocene marine climatic optimum. *Polar Research* 14, 239-243.

Hobday AJ 2000. Persistence and transport of fauna on drifting kelp (Mactrocystis pyrifera (L) rafts in the southern California Bight. *Journal of Experimental Marine Biology and Ecology* 253, 75-96.

Ingolfsson A 1996. The distribution of intertidal macrofauna on the coasts of Iceland in relation to temperature. *Sarsia* 81, 29-44.

Johannesson K and B Johanneson 1995 Dispersal and population expansion in a direct developing marine snail (*Littorina saxatilis*) following a severe population bottleneck. *Hydrobiologia* 309, 173-180.

Johansen S 1998. The origin and age of driftwood on Jan Mayen. *Polar Research* 17, 125-146.

Karnovsky NJ, S Kwasniewski, JM Weslawski, W Walkusz, and A Beszczynska-Moller 2003 Foraging behavior of little auks in a heterogenous environment. *Marine Ecology Progress Series* 253, 289-303.

MacArthur R and OE Wilson 1967. *The Theory of Island Biogeography*. Princeton University Press, 203 pp.

Madsen H 1936. Iinvestigations on the shore fauna of East Greenland with a survey of the shores of other Arctic regions. *Meddelelser om Grønland* 100, 8, 1-79.

Mileikovsky SA 1968 Distribution of pelagic larvae of bottom invertebrates of the Norwegian and Barents Seas. *Marine Biology* 1, 164-167.

Morton B and JC Briton 2000. Origins of the Azorean intertidal biota: the significance of introduced species, survivors of chance events. *Arquipelago. Life and Marine Sciences.* Supplement 2, 29- 51, Ponta Delgada.

Munda IM 1991. Shoreline ecology in Iceland with special emphasis on the benthic algal vegetation. *Ecosystems of the World* 24 (5), 69-81.

Petersen GH 1962 The distribution of Balanus balanoides (L.) and Littorina saxatilis, Mencke in Northern West Greenland. *Meddelelser om Grønland* 159 (9), 1-39.

Petersen GH 1984. Energy flow in comparable aquatic ecosystems from different climatic zones. *Rapports et Procès-Verbaux des Réunions du Conseil International pour l'Exploration de la Mer* 183, 119- 125

Sneli JA 1968. The intertidal distribution of polychaetes and mollusks on a muddy shore in Nord–Møre, Norway. *Sarsia* 31, 63- 68

Weber LI, RG Hartnoll, and JP Thorpe 2000. Genetic divergence and larval dispersal in two spider crabs (Crustacea: Decapoda). *Hydrobiologia* 420, 211-219

Węsławski JM 1994. Gammarus (Crustacea, Amphipoda) from Svalbard and Franz Josef Land. Distribution and density. *Sarsia* 79, 145- 150.

Węsławski JM and S Kwaśniewski 1990. The consequences of climate fluctuations for the food web in Svalbard coastal waters. Pp. 281-295 in M Barnes and RN Gibson (eds.) *Proceedings 24th European Marine Biology Symposium.* Aberdeen University Press.

Węsławski JM, J Wiktor, M Zajączkowski, and S Swerpel 1993. Intertidal zone of Svalbard 1. Macroorganism distributiona nd biomass. *Polar Biology* 13, 73- 79

Węsławski JM, M Zajączkowski, J Wiktor, and M Szymelfenig 1997. Intertidal zone of Svalbard 3. Littoral of a subarctic, oceanic island; Bjrnøya. *Polar Biology* 18, 45- 52.

Chapter 17 NATO Advanced Research Workshop

Seabird Research and Monitoring on Jan Mayen

Geir W Gabrielsen and Hallvard Strøm
Norwegian Polar Institute, Tromsø, Norway
E-mail: geir@npolar.no

Abstract: This paper presents seabird research and monitoring at Jan Mayen. We present an overview of the population status of seabirds, migratory birds, and the contaminant levels in different seabird species at Jan Mayen. By examining data on the breeding population of major seabird species, together with data on energy expenditure and diet, we estimated the food consumption by these populations and the fluxes of energy and nutrients from marine prey through the seabirds. With regard to future seabird monitoring it is recommended that an initial survey and comprehensive census are carried out on Jan Mayen.

Key words: seabirds, monitoring, persistent organic pollutants, energy expenditure, diet, food consumption

1. INTRODUCTION

The first publication on the birdlife of Jan Mayen was by Fisher and Pelzeln in 1886. Since then few ornithological studies have been carried out on the island, mainly during the first half of the 1900s (Deichmann 1904, Musters 1930, Schaanning 1933, Bird and Bird 1935, Olrog 1938, Seligman and Willcox 1940, Richter 1946, Marshall 1952, Holgersen 1953, Langseth 1965, Dullerud 1973). Since the 1980s, van Franeker and Camphuysenhave provided new information, based on their fieldwork on the island in 1983-88 (van Franeker *et al.* 1986, 1998). The Norwegian Polar Institute has otherwise collected data on the birdlife of Jan Mayen for many years. These data, which are based on observations by visitors and crewmembers at the

S. Skreslet (ed.), Jan Mayen Island in Scientific Focus, 181-194

station on Jan Mayen, have been kept on file and updated regularly. This summary of the current information on the breeding and non-breeding bird species at Jan Mayen, is based mainly on the observations by van Franeker *et al.* (1986, 1998).

The presence of persistent organic pollutants (POPs) in the Arctic, especially in the marine environment, has been documented in the literature since the 1970s. In seabirds, high tissue levels of POPs were reported in glaucous gulls (*Larus hyperboreus*) from Bear Island in the early 1970s (Bogan and Bourne 1972) and since the beginning of the 1990s, a considerable amount of work has been done on POPs in Arctic seabirds (AMAP 1998, 2004). High concentrations have been found in some seabird species from Svalbard (Daelemans *et al.* 1992, Gabrielsen *et al.* 1995) and in 1995 the Norwegian Polar Institute studied the contaminant levels in several seabird species at Jan Mayen (Gabrielsen *et al.* 1997a, see below).

There has been an increasing awareness of the roles of seabirds in marine ecosystems (e.g. Croxall and Prince 1982, Schneider *et al.* 1986) and some recent estimates have shown that seabirds may have a considerable impact on the prey populations in some areas (Cairns *et al.* 1991, Mehlum and Gabrielsen 1995). Barrett *et al.* (2001) estimated the annual consumption of prey by seabirds living in the Norwegian and Barents Seas based on numbers of birds present, their energy expenditure (and hence food demand) and the composition of their diet. As part of this study we report on the energy expenditure and food consumption of seabirds from Jan Mayen.

We conclude this paper with recommendations for future monitoring of seabirds on Jan Mayen.

2. BREEDING BIRD SPECIES ON JAN MAYEN

A total of 98 bird species has been recorded at Jan Mayen (van Franeker *et al.* 1998), of which 27 have been reported to breed on the island or were likely to do so in small numbers. The majority of the breeding species are marine or related to the marine environment. It is estimated that a total of 270 000 pairs of seabirds breed on Jan Mayen (Gabrielsen *et al.* 1997b). According to van Franeker *et al.* (1998) the four most common species are the northern fulmar (*Fulmarus glacialis*), black-legged kittiwake (*Rissa tridactyla*), Brünnich's guillemot (*Uria lomvia*), and the little auk (*Alle alle*). The breeding distributions of northern fulmars, black-legged kittiwakes, and Brünnich's guillemots on Jan Mayen are shown in Fig. 1.

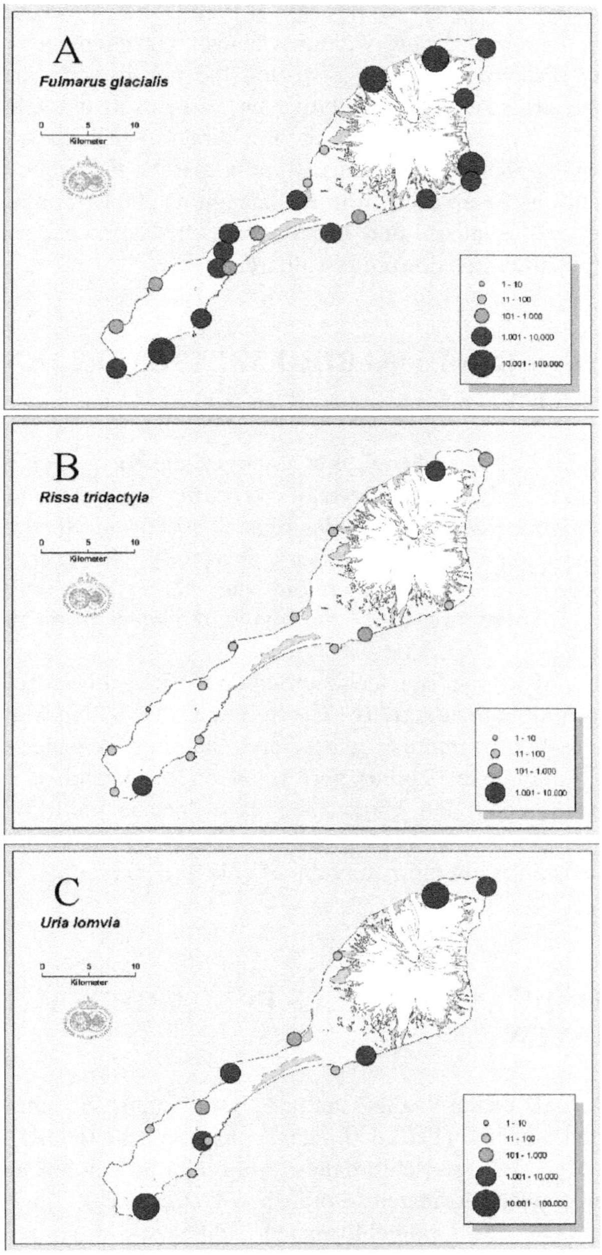

Figure 1. Maps showing the most important seabird colonies at Jan Mayen. A) Northern fulmar. B) Black-legged kittiwake. C) Brünnich's guillemot.

Few terrestrial birds breed on Jan Mayen. The commonest are the meadow pipit (*Anthus pratensis)*, the wheatear (*Oenanthe oenanthe),* the snow bunting (*Plectrophenax nivalis),* and the purple sandpiper *(Calidris maritima)*. The terrestrial birds are, however, also closely associated with the marine system as they feed regularly in the vicinity of seabird colonies.

Jan Mayen supports 28 species, approximately the same number of breeding species as Svalbard (Strøm and Bangjord 2004). The most striking difference between Svalbard and Jan Mayen is that no geese species breed on Jan Mayen while three do so on Svalbard.

3. NON-BREEDING BIRD SPECIES ON JAN MAYEN

Some species use Jan Mayen as a stopover site for resting and feeding during their annual migration between the breeding areas in the north and the wintering areas in the south. Examples of such species are the ringed plover (*Charadrius hiaticula)* and dunlin (*Calidris alpina).* However, most of the 98 species recorded on Jan Mayen are rare vagrants recorded only once or on few occasions. These are birds that have deviated from their normal migration route.

The number of vagrant species recorded on the island is closely related to the level of ornithological activity. The avifaunal list of Jan Mayen holds a remarkable number of regular guests and vagrants considering the low number of ornithological studies performed on the island. Being a small island in a vast ocean, Jan Mayen acts as a "rescue platform" for long-distance travellers. It is, however, likely that most of the vagrants become trapped on the island, and ultimately die of cold and starvation (van Franeker *et al.* 1998).

4. POLLUTANT LEVELS IN SEABIRDS FROM JAN MAYEN

Environmental contaminants include both industrial chemicals, i.e. polychlorinated biphenyl (PCBs) and hexachlorobenzene (HCB), chlorinated pesticides, e.g. DDTs, chlordanes, hexachlorocyclohexane (HCH), aldrin/dieldrin, polychlorinated boranes (Toxaphenes), and industrial byproducts such as polychlorinated dibenzo-p-dioxins (PCDDs), polychlorinated dibenzofurans (PCDFs), polybrominated biphenyls (PBBs), and polybrominated diphenyl ethers (PBDEs) (Muir *et al.* 1992). Most of these contaminants were developed and put into production more than 50 years ago and large quantities were produced at the end of the 1960s and the start of the 1970s (Blus 1995). New substances like PBBs and PBDEs are still being produced globally at an estimated 300 000 tonnes per year (Sellstrøm and Jansson 1995, de Wit *et al.* 2004). These emerging POPs (persistent organic pollutants) may, without proper restriction, cause the same environmental problems as the old POPs (i.e. PCBs and DDTs).

Most POPs found in the Arctic are transported from distant industrial and agricultural sources by atmospheric and oceanic currents, and river discharge, of which the most important transport route is atmospheric current (AMAP 1998 MacDonald *et al.* 2000). POPs accumulate in lipids and biomagnify throughout food chains (Muir et al. 1992, Borgå *et al.* 2001). Consequently the highest POP levels in Arctic marine food chains have been reported in apex predators such as the glaucous gull (Gabrielsen *et al.* 1995).

During the summer of 1995, five Little auks, six black-legged Kittiwakes, six northern fulmars, six glaucous gulls, three great skuas (*Catharacta skua*), one great black-backed gull (*Larus marinus*) and one black guillemot (*Cepphus grylle*) were collected at Jan Mayen. The birds were weighed and frozen until chemical analysis (for pesticides and PCBs) at the Norwegian School of Veterinary Science and the National Veterinary Institute in Oslo, Norway (Bernhoft and Skaare 1994, Gabrielsen *et al.* 1997a).

The POP levels in tissues from seabirds at Jan Mayen are primarily related to trophic position and secondarily to migration pattern. The highest POP concentrations were found in the great skuas, great black-backed gull, and glaucous gulls (Fig. 2). These are scavenger species preying on other seabird species, and occasionally on carcasses of marine mammals. The glaucous gull migrates to the northern Atlantic region in the winter, and the other two to SW Europe (Gabrielsen *et al.* 1997b, Bakken et al. 2003). At Jan Mayen, little auks had the lowest POP concentrations, followed by black-legged kittiwakes, northern fulmars, and black guillemots (Fig. 2). The little auk is a migrating species and feed at lower trophic levels (e.g. copepods). Black-legged kittiwakes also feed at similar trophic level, but migrate long distances exposing themselves to higher levels of PCB in more southern habitats (Hobson *et al.* 2002, Hop *et al.* 2002). Black guillemots are mainly inshore feeders, also feeding mainly on zooplankton and fish, but they do not migrate.

Levels of POP on Jan Mayen varied among seabird species and the chemical type. The results generally corroborate with earlier studies of seabirds and mammals, showing that levels of POPs in seabirds are highest in the Russian part of the Barents Sea and lowest in the North American Arctic (except sum HCH, which is higher in the North American Arctic)(de

Wit *et al.* 2004). Sum HCH concentrations were highest in the Canadian seabirds which may reflect a closer proximity to Asia. Sum DDT concentrations in black-legged kittiwakes and black guillemot from Jan Mayen were similar to levels found in the Canadian Arctic and in the Barents Sea (Fisk *et al.* 2001a, Borgå *et al.* 2004). There was much more variation in sum PCB concentrations with lowest levels being found in little auks. The levels of PCBs in little auks from Jan Mayen were the same as found in little auks from the Canadian Arctic and Svalbard (Fisk et al. 2001a; Borgå et al. 2004). In black-legged kittiwakes the levels of sum PCB were lower at Jan Mayen than in the eastern Barents Sea (Savinova *et al.*, unpublished), but similar to those found in the Canadian Arctic (Fisk *et al.* 2001a). For black guillemots the highest sum PCB concentrations were found at Jan Mayen. Concentrations of sum PCB and sum DDT in glaucous gulls were higher in the Barents Sea and Jan Mayen than in Baffin Bay, with maximum concentrations at Franz Josef Land. Of all seabird species analysed from Jan Mayen, the highest sum PCB and sum DDT were found in great skuas. Since there are few POP data reported for great skuas, it is difficult to make comparisons. However, the sum PCB and sum DDT levels found are high enough to induce biological effects (for more information see de Wit *et al.* 2004).

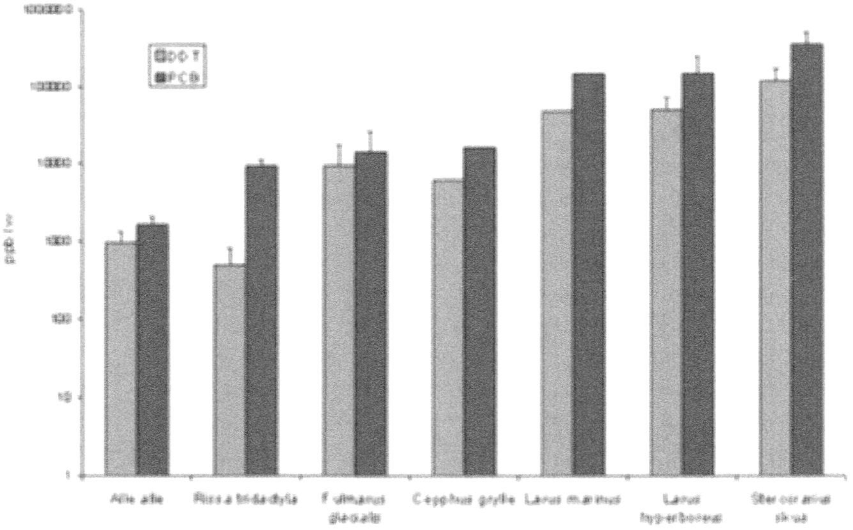

Figure 1. Pollutant levels in different seabird species from Jan Mayen. Concentrations of sum PCB and sum DDT are in $\mu g \cdot kg^{-1}$ lipid weight (or ppb).

Table 1. Pollutant levels in different seabird species and arctic char from Jan Mayen. Concentrations of different substances are presented as μg·kg^{-1} lipid weight (or ppb).

Name (number of samples)	HCB	HCH	Chlordanes	DDT	Mirex	PCB
Little auk (5)	169.0	39,4	324,1	921,9	30,9	1607,7
Alle alle	± 50,5	± 13,5	± 137,7	± 382,1	± 8,9	± 445,7
BL kittiwake (6)	239,3	19.0	328,4	487,4	205,7	9079
Rissa tridactyla	± 75,4	± 7	± 87,3	± 308,1	± 62,3	± 2008,2
N fulmar (6)	749,9	11,3	3512,5	9198,5	207,7	13420,6
Fulmarus glacialis	± 327,2	± 5,2	± 1072,9	± 7966	± 134,4	± 11992,8
Black guillemot (1)						
Cepphus grylle	174,7	47,4	401,8	5816,3	163,3	15267,9
Black-backed gull						
Larus marinus (1)	2834,3	382,3	7892,4	45639,5	2543,6	139026,2
Glaucous gull (6)	2355,9	134,9	8040,5	47193	2532	143248,4
Larus hyperboreus	± 255,9	± 98,3	± 3803	± 20421,8	± 1511	± 78371,3
Great skua (3)	4965,4	364,2	22491,7	113197,3	5485,9	332036
Catharacta skua	± 892,8	± 116,8	± 12906,9	± 41448	± 1719,2	± 122547,1
Arctic char (5)	53,2	70,3	310,8	1970,6	74,1	4569,3
Salvelinus alpinus	± 7,2	± 15	± 52,7	± 448,6	± 22,4	± 1090,2

Based on chemical analysis of the diet of pelagic organisms from the Barents Sea (Borgå *et al.* 2001, 2002, 2004) and in the area of Jan Mayen (Gabrielsen *et al.* 1997a) the low POP levels found at lower trophic level could not explain the high levels found in top predators such as the glaucous gull and the great skua. The migration pattern of the great black-backed gull, glaucous gull, and great skua to more polluted areas of the North Atlantic during the winter may explain the high POP levels found at Jan Mayen.

On Jan Mayen, as for most Arctic seabirds, PCB is the most prevalent contaminant followed by DDT and chlordanes (Barrett et al. 1996; Borgå et al. 2001). The relative amounts are mainly related to the trophic positions of the species (Borgå *et al.* 2001). Higher trophic-level seabirds have generally larger proportions of PCBs as percentage of the total POPs. This is due to the greater bio-magnification potential of PCBs and the ability of seabirds to metabolize other POPs, such as alfa HCH (Moisey *et al.* 2001) and some chlordane components (Fisk *et al.* 2001b).

When comparing PCB congeners among species, there seems to be a good agreement between seabirds from lower and higher trophic levels. Little Auks have higher amount of lower chlorinated PCBs than birds from higher trophic levels. Furthermore, when comparing the PCB congeners of similar species (black-legged kittiwakes and glaucous gulls) from Jan Mayen and Russia (Savinova *et al.*, unpublished), we found that the most persistent congeners were found in birds from Jan Mayen while the most volatile compounds were found in the birds from the Pechora area. This may indicate that Jan Mayen is further away from industrial areas than the Russian locality.

5. TRANSFER OF POLLUTANTS TO FISH

In Ellasjøen, a lake at Bear Island, fish have shown to have exceptionally high levels of POPs (Skotvold *et al.* 1999, Evenseth *et al.* 2004). One explanation for the high levels found in Ellasjøen was that thousands of seabirds breed close to the lake and use the lake as a resting area. The guano contains elevated levels of POPs that (especially PCBs) is deposited directly into the lake or in the catchment area of the lake. In another lake (Øyangen) on Bjørnøya, with no nearby seabird colonies, levels of POPs were several times lower than in Ellasjøen.

At the lake Nordlaguna on Jan Mayen, high levels of sum PCB and sum DDT were also found in arctic char (*Salvelinus alpinus*) (Table 1). Observations of seabirds at Jan Mayen showed that seabirds used Nordlaguna as a resting area after feeding excursions at sea. As for Ellasjøen it is suggested that seabirds serve as a biological pathway carrying contaminants from the marine to the freshwater environment. These findings showed that seabirds can serve as a biological pathway carrying contaminants from the marine to the terrestrial and freshwater environment.

6. FOOD CONSUMPTION BY SEABIRDS FROM JAN MAYEN

In 2001, Barrett *et al.* (2001) presented a model for food consumption by seabirds in Norwegian waters. This model estimated the annual consumption of prey by seabirds living in the Norwegian and Barents seas (ICES Areas I, IIa and IIb), including species from Jan Mayen (Area IIa). In this paper we report on the energy expenditure and food consumption of seabirds from Jan Mayen.

This study includes all seabird species that obtain their food items from the sea and which stay in the Norwegian Sea during a large part of the year. Many of the marine seabird species breeding on Jan Mayen migrate out of the Norwegian Sea during the winter. We estimated on average that the seabird population are present in the area for 250 days of the year (see Table

2). The calculations were based on three species-specific parameters: the numbers of birds in each region at any time of the year, their diet composition, and their energy expenditure (and hence their food demand).

6.1 Number of Seabirds

Population numbers of birds breeding at Jan Mayen were obtained from the database at the Norwegian Polar Institute. The food consumption was modelled for the breeding populations on the assumption that they remained within their normal feeding range and hence did not leave the region during the breeding season. The breeding population of seabirds at Jan Mayen was approximately 270 000 pairs. The main species are northern fulmars (100 000 pairs), little auks (100 000 pairs) and Brunnich's guillemots (50 000 pairs). The population of black-legged kittiwakes and Atlantic puffin were estimated to 10 000 and 5 000 pairs, respectively. The population of other seabird species (common guillemots (*Uria aalge*), black guillemots, great black-backed gulls, glaucous gulls, arctic terns (*Sterna paradisea*), common eider (*Somateria mollissima*), arctic skua (*Stercorarius parasiticus*), great skua (*Catharacta skua*) were estimated to total 5 000 pairs. The number of non-breeding adults, chicks and immature was estimated to be half to the number of breeders.

6.2 Energy Expenditure

Basal metabolic rates (BMR) were estimated using Ellis and Gabrielsen's (2002) allometric equations for Procellariiformes (BMR = $2.763m^{0.726}$, mass in g) and for Charadriiformes (BMR = $2.162m^{0.803}$, m in g). Field metabolic rates (FMR) of breeding birds within the breeding season were estimated for Procellariiformes and Charadriiformes using Ellis and Gabrielsen's (2002) allometric equation (FMR = $22.06m^{0.594}$ and FMR = $11.49m^{0.718}$ respectively, m in g). The length of the breeding season was set as the incubation period plus fledging period (in days, as given in Cramp and Simmons, 1977) plus 20 days. For the non-breeding part of the breeding population (chicks, immature individuals and deferred breeders) during the breeding season and for all birds outside the breeding season, the FMR was set as 2.5 x BMR (Gales and Green, 1990, Gabrielsen GW unpubl. data). The energy expenditure for breeding seabirds was thus calculated separately within the breeding season (using the largest FMR values) and outside the breeding season, and for non-breeding birds throughout the period they occupied the area in question using the lower FMR values. These were then combined to give the annual overall energy-expenditure.

6.3 Diet

To calculate the overall food consumption, we used information on the diet composition from different seabird species at Jan Mayen. Most diet studies have, however, been restricted to a few breeding colonies during the summer only, and little is known about diet outside this period. At Jan Mayen the main prey species for seabirds are capelin (*Mallotus villosus),* polar cod (*Boreogadus saida*), herring (*Clupea harengus*), amphipods (*Parathemisto libellula* and *Gammarus wilkitzkii*) and euphausiids (*Thysanoessa inermis*).

In our calculation of the prey consumption by seabirds we used a mean energy density values ($4.5 kJ \cdot g^{-1}$ fresh mass) reported in the literature and found during our own studies (Gabrielsen unpubl.). Assimilation efficiencies of 0.75 for all seabirds were taken from Brekke and Gabrielsen (1994).

6.4 Food Consumption

The highest food consumption rate by seabirds is reached during the chick-rearing period. When seabirds have nestlings, they have to provision the chicks while feeding themselves. Estimates of energy expenditure and food consumption for individual seabird species are presented in Table 2.

All seabirds at Jan Mayen have a daily consumption of 219 tonnes per day or approximately 63,500 tonnes per year. Of the total consumption in the Jan Mayen area the northern fulmars and Brünnich's guillemots are the main consumers (49 and 37%, respectively). Little auks and northern fulmars feed mainly on invertebrates whereas Brünnichs guillemots and black-legged kittiwakes feed on different fish species.

Compared to prey consumption by seabirds from the Norwegian Coast and the Norwegian Sea (ICES Area IIa), the seabirds from Jan Mayen consume 9.3% of the total consumption in the area (Barrett *et al.* 2001).

7. SEABIRD RESEARCH AND MONITORING AT JAN MAYEN

We suggest a comprehensive survey and census of the most important seabird breeding colonies to be carried out on the island, and a monitoring programme for the northern fulmar, black-legged kittiwake and Brünnich's guillemot populations to be established. This should be an integrated part of the seabird monitoring programme at Svalbard, and the monitoring plots should be counted every second year. The inland bird fauna also needs to be surveyed. With regard to monitoring of pollution it is recommended that eggs are collected from selected seabird species every fifth year in order to examine spatial and temporal trends of POP levels.

Table 2. Estimated total annual prey consumption by seabirds at Jan Mayen. Adol.: Non-breeding adolescent birds. FMR: Field metabolic rates.

		Population				FMR (kJ d^{-1})		Consumption (tons d^{-1})		
Species	Body mass g	Brood pairs 10^{-3}	Total ind. 10^{-3}	Period Present	Days	Breeders	Adol.	Breeders	Adol.	Total
Fulmarus glacialis	820	100	300	Jan-Dec	120	1187	901	70.2	26.7	96.9
Alle alle	160	100	300	Apr-Sep	80	439	318	26.0	9.4	35.4
Cepphus grylle	1030	50	150	Jan-Dec	90	1673	1419	49.6	21.0	70.6
Rissa tridactyla	410	10	30	Mar-Oct	90	864	677	5.1	2.0	7.1
Fratercula arctica	480	5	15	Mar-Oct	110	967	769	2.9	1.1	4.0
Other species	700	5	15	Mar-Oct	110	1268	1041	3.8	1.5	5.3
Sum			810							219.3

REFERENCES

AMAP (Arctic Monitoring and Assessment Programme) 1998. AMAP assessment report: Arctic pollution issues. Oslo, Norway. P.xii+859.

AMAP, 2004. AMAP Assessment 2002: Persistent Organic Pollutants in the Arctic. Arctic Monitoring and Assessment Programme (AMAP), Oslo, Norway. 310 pp.

Bakken V, O Runde and E Tjørve 2003. *Norsk Ringmerkingsatlas*. Vol.1. Stavanger Museum, Stavanger (in Norwegian).

Barrett, RT, JU Skaare, and GW Gabrielsen 1996. Recent changes in levels of persistent organochlorines and mercury in eggs of seabirds from the Barents Sea. *Environtal Pollution* 92, 13-18.

Barrett RT, T Anker-Nilssen, GW Gabrielsen, and G Chapdelaine 2001. Food consumption by seabirds in Norwegian waters. *ICES Journal of Marine Science* 59, 43-57.

Bernhoft A and JU Skaare 1994. Levels of selected individual polychlorinated biphenyls in different tissues of harbour seals (*Phoca vitulina*) from the southern coast of Norway. *Environmental Pollution* 86, 99-107.

Bird CG and EG Bird 1935. The birds of Jan Mayen Island. *Ibis* 77, 837-855.

Blus LJ 1995. Organochlorine pesticides. Pp 275-300 in DJ Hoffman *et al.* (eds.) *Handbook of Ecotoxicology*. Lewis Publishers, Boca Raton.

Bogan JA and WRP Bourne 1972. Organochlorine levels in Atlantic seabirds. *Nature* 240, 358.

Borgå K, GW Gabrielsen, and JU Skaare 2001. Biomagnification of organochlorines along a Barents Sea food chain. *Environmental Pollution* 113, 187-198.

Borgå K, B Gulliksen, GW Gabrielsen, and JU Skaare 2002. Size related bioaccumulation and between year variation of organochlorines in ice-associated amphipods from the Arctic Ocean. *Chemosphere* 46, 1383-1392.

Borgå K, H Wolkers, JU Skaare, H Hop, DCG Muir, and GW Gabrielsen 2004. Bioaccumulation of PCBs in Arctic seabirds; influence of dietary exposure and congener biotransformation. *Environmental Pollution* (Submitted).

Brekke B, and GW Gabrielsen 1994. Assimilation efficiency of adult kittiwakes and Brünnich's guillemots fed capelin and arctic cod. *Polar Biology* 14, 279-284.

Cairns DK, G Chapdelaine, and WA Montevecchi, 1991. Prey exploitation by seabirds in the Gulf of St. Lawrence. *Canadian Special Publication of Fisheries and Aquatic Science* 113, 277-291.

Cramp S and KEL Simmons (eds) 1977. *The birds of the Western Palearctic I.* Vol I. Oxford University Press. 722 pp.

Croxall JP and PA Prince 1982. A preliminary assessment of the impact of seabirds on marine resources at South Georgia. Le Comite National Francais des Recherches Antarctiques 51, 501-509.

Daelemans FF, F Mehlum and PJC Schepens 1992. Polychlorinated biphenyls in 2 species of Arctic seabirds from the Svalbard area. *Bulletin of Environmental Contamination and Toxicology* 48, 828-834.

Deichmann H 1904. Birds of East Greenland. *Meddelelser om Grønland* 29, 141-156.

de Wit CA, A Fisk, K Hobbs, D Muir, GW Gabrielsen, R Kallenborn, M Krahn, R Norstrom, and J Skaare 2004. Persistent Organic Pollutants. Pp 78-201 in SJ Wilson, JL Murray, HP Huntington (eds.) AMAP II Assessment Report. Arctic Pollution Issues, Arctic Monitoring and Assessment Program, Oslo, Norway.

Dullerud, O. 1973. Observasjoner fra Jan Mayen 1971/72. *Fauna* 26, 70-71.

Ellis HI and GW Gabrielsen 2002. Energetics of free ranging seabirds. Pp 359-407 in EA Schreiber and J Burger (eds.) *Biology of marine birds* CRC Press, Boca Raton.

Evenset A, GN Christensen, T Skotvold, E Fjeld, M Schlabach, E Wartena, and D Gregor 2004. A comparison of organic contaminants in two high arctic lake ecosystems, Bjørnøya (Bear Island), Norway. *The Science of the Total Environment* 318, 125-141.

Fischer F and A von Pelzeln 1886. Vögel von Jan Mayen. *Mitteilungen der Ornitologishe Verein*, Wien, 10.

Fisk AT, J Moisey, KA Hobson, NJ Karnovsky, and RJ Norstrom 2001a. Chlordane components and metabolites in seven species of Arctic seabirds from the Northwater Polynya: relationships with stable isotopes of nitrogen and enantiomeric fractions of chiral components. *Environmental Pollution* 113, 225-238.

Fisk AT, KA Hobson, and RJ Norstrom 2001b. Influence of chemical and biological factors on trophic transfer of persistent organic pollutants in the Northwater Polynya marine food web. *Environmental Science and Technology* 35, 732-738.

Gabrielsen GW, JU Skaare, A Polder, and V Bakken 1995. Chlorinated hydrocarbons in glaucous gulls (*Larus hyperboreus*) in the southern part of Svalbard. *The Science of the Total Environment* 160/161, 337-346.

Gabrielsen GW, IG Alsos, and B Brekke 1997a. Undersøkelser av PCB i jord, fisk og sjøfugl I området rundt avfallsfyllingen på Jan Mayen (FBT lokalitet 1805 001). *Norsk Polarinstitutt Rapportserie* 104, 1-29 (in Norwegian).

Gabrielsen GW, B Brekke, IG Alsos, and JR Hansen 1997b. Natur- og kulturmiljøet på Jan Mayen. *Norsk Polarinstitutt Meddelelser* 144, 1-127 (in Norwegian).

Gales RP and B Green 1990. The annual energetics cycle of Little Penguins *Eudyptula minor*. *Ecology* 71, 2297-2312.

Hobson KA, G Gilchrist, and K Falk 2002. Isotopic investigations of seabirds of the North Water Polynya: contrasting trophic relationships between the eastern and western sectors. *Condor* 104, 1-11.

Holgersen H 1953. On a collection of birds from Jan Mayen. *Astarte* 5, 1-9.

Hop H, K Borgå, GW Gabrielsen, L Kleivane, and JU Skaare 2002. Food web magnification of persistent organic pollutants in poikilotherms and homeotherms from the Barents Sea. *Environmental Science and Technology* 36, 2589-2597.

Langseth R 1965. Låvesvale på Jan Mayen. *Sterna* 6, 312 (in Norwegian).

MacDonald RW, LA Barrie, TF Bidleman, ML Diamond, DJ Gregor, RG Semkin, WMJ Strachan, YF Li, F Wania, M Alaee, LB Alexeeva, SM Backus, R Bailey, JM Bewers, C Gobeil, CJ Halsall, T Harner, JT Hoff, LMM Jantunen, WM Lockahrdt, D Machay, DCG Muir, J Pudykiewicz, KJ Reimer, JN Smith, GA Stern, WH Schroeder, R Wagemann, and MB Yunker 2000. Contaminants in the Canadian Arctic: 5 years of progress in understanding sources, occurence and pathways. *The Science of the Total Environment* 254: 93-234.

Marshall AJ 1952. Non-breeding among arctic birds. *Ibis* 94, 310-333.

Mehlum F and GW Gabrielsen 1995. Energy expenditure and food consumption by seabird populations in the Barents Sea region. Pp 457-470 *in* by HR Skjoldal, C Hopkins, KE Erikstad, and HP Leinaas (eds) *Ecology of fjords and coastal waters*. Elsevier Science, Amsterdam.

Moisey J, TT Fisk, KA Hobson, and RJ Norstrom 2001. Hexachlorocyclohexane (HCH) isomers and chiral signatures of a-HCH in the Arctic marine food web of the Northwater Polynya. *Environmental Science and Technology* 35, 1920-1927

Muir DCG, R Wagemann, BT Hargrave, DJ Thomas, DB Peakall, and RJ Norstrom 1992. Arctic marine ecosystem contamination. *The Science of the Total Environment* 122, 75-134.

Musters JLC 1930. Fuglefaunaen på Jan Mayen. *Norsk Ornitologisk Tidsskrift* III (11-129), 216-219 (in Norwegian).

Olrog CC 1938. Jan Mayens fåglar. *Fauna och flora* 33, 94 (in Swedish).

Richter S 1946. Jan Mayen i krigsårene. Norges Svalbard- og Ishavsundersøkelser Meddelelse 66, 1- 24 pp (in Norwegian).

Schaanning HTL 1933. A contribution to the bird fauna of Jan Mayen. Skrifter om Svalbard og Ishavet 49, 25-39.

Schneider DC, GL Hunt Jr, and NM Harrison 1986. Mass and energy transfer to seabirds in the southeastern Bering Sea. Continental Shelf Research 5, 241-257.

Seligman OR and JM Willcox 1940. Some observations on the birds of Jan Mayen. Ibis 82, 464-479.

Sellestrøm U and B Jansson 1995. Analysis of tetrabromobisphenol A in a product and environmental-samples. Chemosphere 31 3085-3092.

Skotvold T, EMM Vartena, G Christiansen, and M Schlaback 1999. Organochlorine contaminants in biota and sediments from lakes on Bear Island. Akvaplan-niva report no 510.1443.1. 47 pp.

Strøm H and G Bangjord 2004. The bird and mammal fauna of Svalbard. Pp 123-137 in P Prestrud, H Strøm and H Goldman (eds) A catalogue of the terrestrial and marine animals of Svalbard. Norwegian Polar Institute Skrifter 201.

van Franeker JA, K Camphuysen, and F Mehlum 1986. Status over Jan Mayens fugler. *Vår Fuglefauna* 1986.

van Franeker JA, CJ Camphuysen, and F Mehlum 1998. The birds of Jan Mayen. Circumpolar Journal 13, 28-43.

Chapter 18 NATO Advanced Research Workshop

Dispersal of Plants in Remote Arctic-Alpine Habitats

Modes and Time-Scales for Colonization

Volodymyr Kyyak
Institute of the Carpathians Ecology, L'viv, Ukraine
E-mail: natalia@mail.lviv.ua

Abstract: Modes and time-scales for colonization and dispersal of plants in remote Arctic islands have been analysed as well as these have been studied in the highlands of the Ukrainian Carpathians. The rates of plant colonization in Arctic and alpine areas are very slow and tangible changes take place after tens and hundreds of years. The ways of dispersal of diaspores for tens, hundreds and thousands kilometers has not yet been properly studied.

Key words: plant, diaspore, dispersal modes, colonization, Arctic, alpine

1. REGION AND OBJECTS

The alpine zone of the Ukrainian Carpathian Mountains (Fig. 1) is found at the height above 1700-1800m over the sea surface. Sub-nival coenoses occur at the height of 1900 to 2000m on abundant in snow northern slopes.

Low temperatures, high air humidity and dense clouds, strong winds and sharply diverse environmental factors in adjacent areas seem to be typical for the climate. In winter, areas with deep snow alternate with snow-less frozen earth.

Pleistocene icing in its movement to the south did not reach the Carpathians. Meanwhile, several migrations of Arctic and alpine plants have taken place, with latitudinal spreading to the Arctic and longitudinal spreading to the highest mountain ridges during the periods of warming.

S. Skreslet (ed.), Jan Mayen Island in Scientific Focus, 195-206

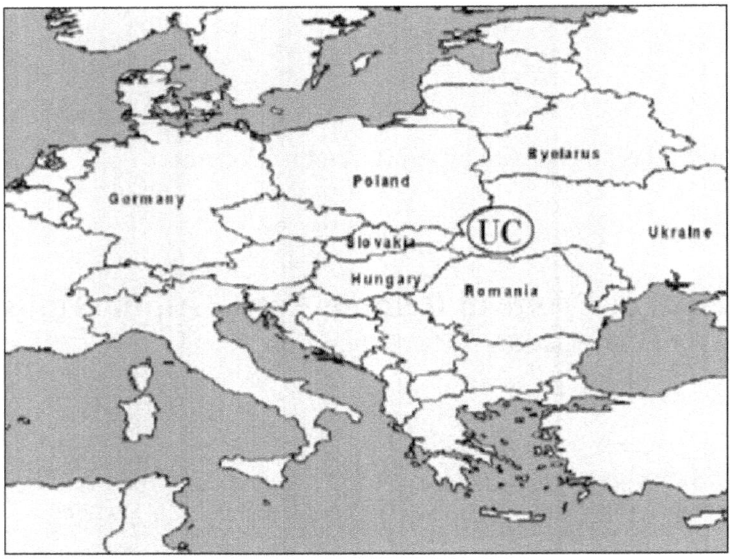

Figure 1. The research area in the Ukrainian Carpathians (UC).

As the climatic conditions of the alpine zone in Central Europe and high latitude tundra show many similarities, the flora also has considerable resemblance. If we compare the flora of Jan Mayen and the Carpathians, it turns out that out of 75 species of vascular plants that the Jan Mayen area has (Gabrielsen *et al.* 1997, Lid 1964, 1941), 30 occur in the Carpathians and 25 in the Ukrainian Carpathians (Table 1), with anemochores dominating by the mode of dispersal.

2. DISPERSAL MODES OF PLANT DIASPORES AND SPEED FOR COLONIZATION IN REMOTE HABITATS

The seeds can be carried over large distances by winds, birds, animals or humans, in the high mountains as well as by sea to distant islands.

There are no permanent synanthropic plants in Jan Mayen. But definitely, there is a potential danger of other plants spreading as the anthropogenic influence intensifies.

Wind is the main factor behind the dispersal of bryophytes. The distances to which spores are carried away depend on their size. The bulk of spores land very closely to the mother plant, but we must take into account their huge numbers, ranging between 50 thousand to 600 thousand for small spores and between 5 thousand and 10 thousand for large spores.

For liverworts the number of spores can be both small and high, reaching up to 5 million. Due to their large numbers, even a small percentage of spores spreading to other areas are significant in quantity. As tests have shown, spores have a great interspecific variability in their resistance to cold, ultraviolet radiation and drying, which happens when spores are carried by air at high altitudes of the Earth's atmosphere.

The correlation between survival under these conditions and the range of a certain species has been established (van Tooren 1989).

Unlike the spores of mosses, the seeds of flowering plants are much heavier and are carried by wind for lesser distances. According to Zhyliaev (2001), with the speed of wind at $5 m \cdot sec^{-1}$ the seeds of *Festuca rubra, Deschampsia caespitosa* are carried away for not more than 100m and not more than 1 km even by hurricane winds. There is a lack of data today as to the distances at which the seeds of anemochore species can be carried away by hurricanes, storms or ascending air flows. Nothing is also known about the role of such dispersal and which species tend to form air plankton.

Birds, basically those who swallow whole fruits and seeds, carry them for large distances. In the stomachs of Galliformes, particularly Tetraonidae, fruits and seeds are partially damaged, but some escape digestion and are carried away for large distances (Nechayev 2001).

In the Arctic, the only two groups of typical phytophagous are Tetraonidae and Anserinae. Red-breasted goose (*Rufibrenta rupicollis*), for instance, feeds on 28 plants species from 9 families, notably *Festuca rubra, Poa arctica, Polygonum viviparum, Carex bigelowii* etc. (Rosenfeld and Volkov 2001).

In the Carpathian highlands seeds are actively spread by black-cock birds, notably, *Prunella collaris,* living on mountain tops, rocks and screes in Eurasia (Chervona 1994).

Birds' digestion is very intense and they defecate in about 30 minutes, rarely in 60 to 90 minutes after feed. That means that maximum distances at which seeds are carried can reach tens of kilometers. Only when seeds act as millstones in birds' stomachs can they remain long there, many hours or days, and can be carried away for large distances. But the number of such seeds is small and only some plant species can spread that way.

With some plants, epiornithochory, or dispersal of seeds on feathers or feet with stuck soil, is very important for some plant species. Seeds of plants growing in swampy areas, particularly Juncaceae, can be spread for hundreds and thousands kilometers (Inozemtsev 1987).

Unfortunately, not much is known about the scale of seed dispersal for herbaceous species. There is a lot of examples, when plant species without any special adaptation for dispersal are able to distribute for large distances, even to remote islands and mountain territories (Brochmann *et al.* 2003). However, the mechanisms of their distribution are not studied enough.

Table 1. Vascular plants known to Jan Mayen (Norway) and Carpathian highlands and modes of their dispersal.

Species	Whole Carpathian*	Ukrainian Carpathian*	Dispersal modes
	Seedless vascular plants		
Cystopteris fragilis			anemochore
	Monocotyledones		
Calamagrostis stricta			?
Carex bigelowii			?
Carex lachenalii		*C. lagopina*	?
Deschampsia caespitosa			anemochore
Festuca rubra			anemochore
Luzula spicata			anemochore
Poa alpina ssp. vivipara	*P. alpina*	*P. alpina*	anemochore
Poa glauca		*P. balfourii*	anemochore
	Dicotyledones		
Alchemilla glomerulans			epizoochore
Arabis alpina			anemochore
Bistorta vivipara (Polygonum viviparum)			endozoochore
Cerastium alpinum	*C. alpinum ssp. alpinum*	*C. alpinum ssp. lanatum*	hydrochore
Cerastium cerastoides			anemochore, hydrochore
Empetrum nigrum ssp. hermaphroditum			zoochore
Epilobium anagallidifolium			hydrochore
Loiseleuria procumbens			anemochore
Matricaria inodora	*Matricaria perforata*	*Tripleurospermum inodorum*	anthropo-chore
Omalotheca supina			anemochore
Oxyria digyna			anemochore
Potentilla crantzii			anemochore
Ranunculus acris ssp. acris			anemochore
Ranunculus glacialis		—	?
Rumex acetosa			anemochore
Salix herbacea			anemochore
Saxifraga cernua		—	?
Saxifraga oppositifolia			anemochore
Veronica alpina ssp.alpina	*V. alpina*	*V. alpina*	anemochore, hydrochore

* According to Tasienkevytch (1998).

Hydrochorous seeds are successfully carried away by sea water for long distances. In this respect, it would be interesting to study how common is the indirect dispersal of seeds on other drifting objects (pieces of wood etc.).

Of the seeds carried to distant habitats only a small fraction survives until the mature plant is formed. The rate of colonization is especially low in high latitudes. The example of the volcanic Surtsey island that formed in 1963 40km off Iceland and 5km off the nearest small island covered with vegetation is very telling. The first vascular plants appeared on it only after 2 years. After 10 years Surtsey had only 11 species of vascular plants and 62 species of mosses, with the bulk of seeds arriving by sea.

The low rate of primary succession in the Arctic was shown in the research on the development of vegetation after former glacier on the Treskelen-Treskelodden peninsula on Svalbard (Birkenmajer 1960, Kuc 1964). The glacier began to recede in 1890. At first, the rate of primary succession was very slow. Then, for 50 years there was no change in the number of species of vascular plants which totalled 15 (Pirożnikow and Górniak 1992). The only parameter which gradually increases was their coverage.

3. DISPERSAL AND COLONIZATION OF PLANTS IN ALPINE HABITATS IN THE CARPATHIAN MOUNTAINS

According to the ecological and phytocoenological characteristics the rocky communities of the alpine zone as well as for the top and near-the-top sites of the highest mountain ranges of the Ukrainian Carpathians – Chornohora, Svydivets and Marmarosh Alps are especially close to the Arctic tundras of Jan Mayen. Because of weathering local primary and secondary successions take place there. For the screes under the rocks, on "mobile" fine-earth sites and on rocky slopes in places with deep snow the cyclic succession is characteristic, where the processes of colonization, elimination and re-colonization alter each other. At these sites it is possible to observe on small distances (of tens to hundreds meters) the continuum of changes in environmental factors and vegetation from primary or secondary succession to established alpine coenoses. Comparison of changes in environmental factors and vegetation in the series: subnival sites – alpine grass sites – alpine grass – small shrub sites shows much analogy to the change of tundras in the series: Arctic deserts – Arctic tundras – small-shrub tundras.

At the same profiles the changes of colonist species characteristic for the one and another phase are clearly observed. The species distribution of different life forms and the phytocoenotic relation correlates with the gradient of the following factors:

- availability, depth and structure of the soil.
- duration of the vegetation period.

- snow and wind conditions.
- availability of phytoenvironment (micro-phytoclimate).

This profile largely illustrates the time scales of colonization and distribution of species in the alpine zone of the Carpathians.

In what follows the plant species characteristic of different stages of soil development are examined:

1. The primary stage of soil development is characterized by the participation of algae, lichens, mosses and fungi. The vascular plants are absent. In the high mountains and high latitudes this stage is especially long.
2. The stage of very limited soil resources, when the depth of the soil layer is scanty and mostly irregularly distributed, particularly in the rock crevices. The species of the genera *Poa* (*P. alpina, P. granitica*), *Silene* (*S. dubia, S. zawadskii*), *Saxifraga* (*S. carpatica, S. maculata, S. luteo-viridis*), *Cerastium* (*C. alpinum, C. cerastoides*), are most characteristic for the high-mountain zone of the Carpathians while *Allium sibiricum* and *Lloydia serotina* are characteristic of the humid sites.
3. The stage of fine-earth soil with irregular depth numbering a few centimeters and frequent rocky exposure. The species of a wider spectrum of life forms and of patient life strategy: *Rhodiola rosea, Festuca ametystina, Ranunculus thora, Primula halleri, Leontopodium alpinum, Veronica alpina, Oxyria digyna, Loiseleuria procumbens, Salix herbacea* occur here.
4. The stage of relatively formed soil. The species of different life forms and strategies are characteristic. On dry habitats grow *Polygonum bistorta, Ranunculus acris, Rumex acetosa, Festuca rubra*, while *Cardamine pratensis, Geum rivale, Deschampsia caespitose* grow on humid sites

In the Arctic the snow cover and the humidity of the soil surface layer are more important factors influencing the vegetation development than soil structure (Matveyeva 1994). Snow in winter helps to form (preserve) thicker vegetation cover, but excessive humidity in summer causes its degradation. The influence of the type of rocks on the vegetation composition and allocation is weak. The main limitative factor of plant development, especially for flowering plants, is the minimum of heat.

The main feature of vegetation in Arctic tundras is its sparse cover. Adaptive abilities of flowering plants are insufficient to stand the harmful influence of unfavorable physical factors and to form the close vegetation cover (Chernov and Matveyeva 1997, Elvebakk 1997). The poorest is the flora of the swamp habitats, where the lichens disappear, the number of flowering plant species reduces to 8-10 and the moss cover includes one or a few dominants.

The characteristic feature of vertical vegetation structure is reduction of the layering. Only some reproductive shoots of grasses reach the maximum height (20cm). Bryologists consider, that constant small disturbance prevents any moss species from competition with other species at the stage of secondary succession, and forecast the decrease of the number of species colonizing such plots (van Tooren and During 1988).

Obviously, on one hand the regular small disturbances slow down the speed of colonization of disturbed habitats, but they also restrict the number of moss species as well as vascular plants, which are active participants in colonization of such habitats. In different communities of the alpine zone of the Carpathians this is caused, first of all, by wind erosion on flattened, convex and top sites, as well as by water erosion on the slopes.

The moss cover reduces the germination of many species of vascular plants but promotes the growth of the seedlings if they survive the primary stage of development (van Tooren 1989). This phenomenon is explained by the authors differently. Decrease of the seed germination of many species may be connected with the light reduction, spectral change of the light or with allelopathic relationship. Possibly, seedling survival favored by the moss cover is connected with the biological ability of mosses to keep moisture as compared with bare soil, or with emission of the nutrients from decomposed moss cover.

The deep snow plays a special role. It has mainly considerable direct mechanical impact or indirect one – by shortening of the vegetative period on the sites with long-lasting snow cover. The pressure of very thick snow is so strong, that even the dwarf shrubs (*Pinus mugo, Alnus viridis*) are damaged. Thus, if in the subalpine zone the snow depth exceeds 3 m then patches of extrazonal grass and grass-shrub communities characteristic for the alpine zone appear. Heavy snow on the slopes causes the soil shifts. Consequently, on the steep 30°-40° slopes, where the snow depth even on horizontal sites makes 3-6m the speed of secondary succession is very low.

The specific dynamics of the vegetation spread on disturbed sites is possible to display according to the following correlated processes:

1. change of configuration of disturbed sites, namely their elongation downhill because of the soil shifts, water erosion and deepening of the disturbed sites;
2. vegetation spread on the disturbed sites from their margins;
3. formation of the elongated deepened gutters .

In the cause of quick (spring) snow melting there are sites with high floristic diversity, numbering dozens of species. At these sites different phytocoenoses are occur – herbaceous and low-shrub. But at the areas with deep long-lasting snow, which melts during the second half of summer, the species composition declines to a few chionophilous species (*Ligusticum mutellina, Soldanella hungarica, Luzula spadicea, Poa alpina* etc.). The lack of continuous sodding and sparse vegetation cover are peculiar for such areas. After the disturbance of grass and soil cover secondary successions are especially slow and durable in these conditions.

The main types of the anthropogenic impact in the alpine zone of the Carpathians are grazing and trampling. The phytocoenoses on the slopes are very vulnerable to these factors. In terms of the species composition the most sensitive are the communities with considerable percentage of *Cetraria islandica*, which is easily damaged mechanically. *C. islandica* seems to protect from unfavorable impact of ecological factors and forms the microclimate at the surface layer. Due to mechanical damage this protective function decreases or stops. Thus, even during moderate grazing and trampling the primary vegetation is replaced by the long-lasting secondary derivative communities, whereas the strong impact causes mosaic patches of bared soil and consequent erosion. For example, primary phytocoenoses *Uliginetum cetrariosum* in the Carpathians are replaced with secondary communities with domination of *Festuca airoides* and *Juncus trifidus*. The considerable participation of *Cetraria islandica* is characteristic for many phytocoenoses in Arctic tundra. The secondary succession was observed for 20 years in the alpine zone of the Ukrainian Carpathians on small plots between 400 and 2500cm^2 on areas where tussock has been removed (Fig. 2). Vegetation was restored fastest in tall-herbaceous coenoses, during 4 to 8 years. Slowest restoration was on convex top mountain plots with strong winds and little snow. The increase of barren sites due to water erosion and shifts of deep snow took place on steep slopes with 4-6m depth of snow.

During WW1 more than 25km of trenches were dug out in the alpine zone in the Chornohora Mts, with vegetation destroyed for the width of 1-5m. The restoration of vegetation was observed there in 1982-2003. It was found out that 70 years after the beginning of secondary succession in trenches the vegetation was restored almost to its original level. Restoration was fastest for *Juncus trifidus, Festuca airoides* and *Calamagrostis villosa* on even areas and gentle slopes with well-developed soil.

But in some coenoses secondary successions still has not been completed. This concern mostly the sites located on convex mountain tops and in places with strong winds and little snow. Secondary successions also proceed on rocky slopes with deep and long-lasting snow cover (Mt. Shpyci, Rebra and Brebeneska). Contribution of Arctic-alpine species (*Salix, Poa, Saxifraga spp., Loiseleuria etc.*) is high in such communities.

Rocky coenoses are most appropriate for colonization by the species of pioneer strategy, as plots devoid of any vegetation appear regularly due to wind erosion and soil sloughing. However, new populations appear there only very rarely.

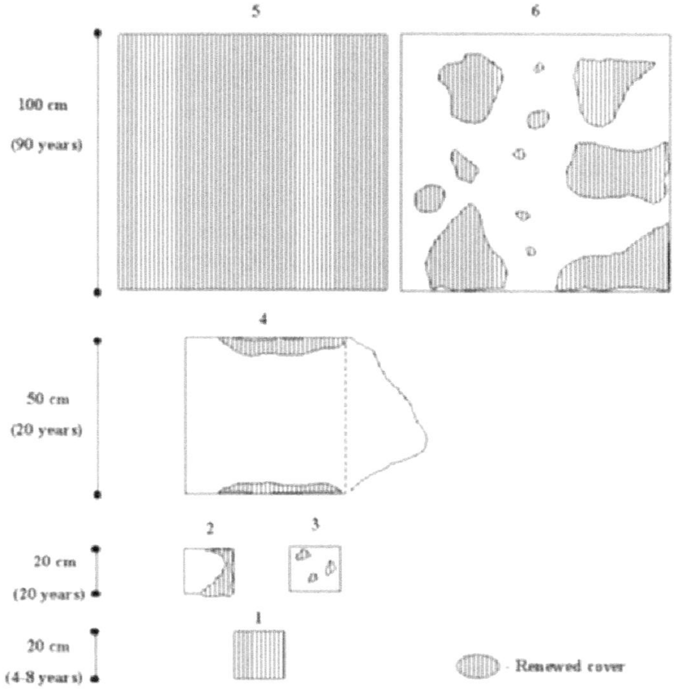

Figure 2. Typical variants of plant colonization in different alpine coenoses of the Carpathians:
1 Steep and gentle slopes with well-developed soil.
2,3 Rocky mountain tops with strong winds.
4 Steep slopes with much snow.
5 Gentle slopes with well-developed soil.
6 Rocky top areas, rocky slopes and areas with long-lasting snow.

Floristic composition of plots originally described 120 years ago by Zapałowicz (1880) was compared with their present condition, involving a 20x1km stretch on the Chornohora ridge. The comparison involved 12 species of plants. There is ample evidence to confirm the appearance of new populations of two species: *Ranunculus thora* (Pozhyzhevska Mt. and Dantser Mt.) and *Festuca carpatica* (Pozhyzhevska Mt.). New localities are 2-4km away from the nearest populations of these species.

High stability of floristic composition in communities on rocky plots can be attributed to the fact that petrophilous plants, with many being Arctic-alpine species, belong to stenotopic plants, with each species requiring a very specific set of environmental factors and colonizations of rocky habitats proves it. Most populations occupy only a small part of appropriate sites. For instance, such behavior is typical of *Erigeron alpinus, Saussurea alpina, Primula halleri* and several species known both to Jan Mayen and the Carpathians (*Poa glauca, Cerastium alpinum, C.cerastoides, Oxyria digyna, Veronica alpina*).

Vegetative and seed reproduction of species is interrelated. Undoubtedly, seed dispersal is crucial for distant propagation of diaspores. But as the seeds germinate and the seedlings survive in Arctic or alpine conditions, vegetative spread and reproduction takes the lead. Despite low rates of individual annual vegetative propagation amounting to a few centimeters, the total mobility of a population can be rather high.

Vegetative mobility of a population is defined as a multiplication of average annual mobility of mature individuals (modules) by their density (Table 2). This indicator can be used to project colonization rates of vegetatively mobile plants.

In Arctic tundras vegetative expansion and propagation has a crucial role in the successful invasion and holding the territory during the colonizing process. The characteristic feature of plant distribution in the altitudinal direction in high mountains and the latitudinal one in tundras is gradual loss from the floristic composition the species with the special "vegetative" stage. Many species under extreme ecological factors do not produce fruits or their fruiting is scanty and irregular. Among the Carpathian species of that kind are the small shrubs of the genus *Vaccinium* as well as *Homogyne alpina* and *Saussurea alpina.*

Table 2. Vegetative mobility of plants in the alpine communities of the Carpathians. A) Rate of vegetative mobility for inviduals (modules), cm·year^{-1}. B) Density of individuals (modules), cm·year^{-1}. C) Vegetative mobility of population, cm·year^{-1}·m^{2}.

Species name	Communities	A	B	C
Deschampsia caespitosa	Rhododendretum calamagrostiosum	0.5	30	15
Helictotrichon versicolor	Festucetum airoidi	1.0	30	30
Senecio carpaticus	Seslerietum cariceto-festucosum	4.0	11	44
Hieracium alpinum	Seslerietum cariceto-festucosum	0.5	110	55
Loiseleuria procumbens	Curvuletum cetrariosum	0.5	208	104
Juncus trifidus	Juncetum cetrario-myrtillosum	0.5	286	143
Calamagrostis villosa	Rhododendretum calamagrostiosum	3.0	107	321
Carex nigra	Festucetum airoidi	2.0	351	702
Vaccinium uliginosum	Uliginetum cetrariosum	2.0	484	968
Festuca airoides	Festucetum airoidi	1.0	1210	1210

4. SUMMARY

Out of 75 vascular species known to the Arctic island of Jan Mayen (Norway), 30 occur in the Carpathians, with 25 known to the Ukrainian Carpathians. For the common Arctic-alpine species, anemochores predominate in terms of their mode of dispersal.

The rates of plant colonization in the Arctic and alpine areas are very slow and tangible changes take place after tens and hundreds of years. Due to unfavorable environment and the negative impact of mosses on seed germination flowering plants show the slowest expansion rates at the pioneer stage. The survival ratio at the seedling stages of growth is very low. Colonization increases as vegetative propagation develops.

A model was developed to forecast the dynamics of further colonization whereby the vegetation mobility of a population can be determined by multiplying average annual mobility of mature individuals (modules) by their density. In the alpine zone of the Carpathians the vegetative propagation predominates over the seed reproduction on newly-colonized areas and during local expansion and recruitment.

The duration of secondary successions in the alpine zone of the Carpathians is over 20 years even on small plots of 400 to 10,000 cm^{-2}. On rocky convex tops, ravine slopes with scarce vegetation and areas with long-lasting snow cover the secondary successions (with *Saxifraga, Salix, Poa, Cerastium, Loiseleuria* spp.), takes up to 100 years. Restoration of vegetation coverage is the quickest for tall-herbaceous communities. For flat areas with a thick soil layer and domination of *Juncus trifidus, Festuca airoides, Calamagrostis villosa* secondary succession has taken about 70 years and vegetation has restored almost to its original state.

For their floristic composition, rocks, screes and mountain tops in the Carpathians are the closest to the Arctic tundra. Despite the sites with little competition, such areas show high stability of their floristic composition. A 120-year benchmark study on distribution of 12 species makes it possible to conclude that new populations of only two species (*Ranunculus thora* and *Festuca carpatica*) appeared.

The ways of dispersal of diaspores for large distances has not yet been properly studied and determined merely by the presence of species in certain locations. This is especially true of herbaceous plants. Potential research areas, therefore, may include endo- and exo-ornithochory and transfer of seeds by sea on pieces of wood. Additionally, it is important to develop models for dispersal of diaspores with various weight and different adaptations to anemochory by hurricane winds and ascending air flows.

REFERENCES

Birkenmajer K 1960. Raised marine features of the Hornsund area, Vestspitsbergen. *Studia Geologica Polonica* 5, 1-95.

Brochmann C, TM Gabrielsen, I Nordal, J Landvik, and R Elven 2003. Glacial survival or tabula rasa? The history of North Atlantic biota revisited. *Taxon 52*, 417-450.

Chernov JI and NV Matveyeva 1997. Arctic ecosystems in Russia. Polar and alpine tundra. Pp 361-507 in F.E.Wielgolaski (ed.) *Ecosystems of the World 3*. Elsevier Science.

Chervona knyha Ukrainy 1994. Tvarynnyi svit. Ukrainsky Encyclopedia, Kyjiiv, 456 (in Ukrainian).

Elvebakk A 1997. Tundra diversity and ecological characteristics of Svalbard. Polar and alpine tundra. Pp 347-359 in F.E.Wielgolaski (ed.) *Ecosystems of the W orld* 3. Elsevier Science.

Gabrielsen GW, B Brekke, IG Alsos, and JR Hansen 1997. Natur- og kulturmiljøet på Jan Mayen. *Norsk Polarinstitutt Meddelelser* 144, 1-127 (in Norwegian).

Inozemtsev AA 1987. Ptitsy i les. *Agropromizdat* 302 (in Russian).

Kuc M 1964. Deglaciation of Treskelen-Treskelodden in Hornsund, Vestspitsbergen, as shown by vegetation. *Studia Geologica Polonica* 11, 197-206.

Lid J 1941. Bryophytes of Jan Mayen. *Norges Svalbard- og Ishavsundersøkelser Meddelelser* 48, 1-13.

Lid J 1964. The flora of Jan Mayen. Norsk Polarinstitutt Skrifter 130, 1-108.

Matveyeva NV 1994. Floristic classification and ecology of tundra vegetation of the Taimyr Peninsula, Northern Siberia. *Journal of Vegetation Science* 5, 813-828.

Nechayev VA 2001. Ptitsy - potrebiteli i rasprostraniteli plodov i semian drevesnykh rasteniy v Primorskom kraie. *Byulleten MOIP, otdel biologicheskiy* 106 (2), 14-21 (in Russian).

Pirozhnikow E and A G rniak 1992. Changes in the characteristics of the soil and vegetation during the primary succession in the margined zone of the Werenskiold glacier, Spitsbergen. *Polish Polar Research* 13 (1), 19-29.

Rosenfeld SB and SV Volkov 2001. Kormovaia ecologia krasnozoboi kazarki na Zapadnom Taimyre. *Byulleten MOIP, otdel biologicheskiy* 103 (3), 52-57 (in Russian).

Tasienkevytch LO 1998. Pryrodna flora Karpat. Spysok vydiv sudynnyh Roslyn. *Derzhavnyi Pryrodoznavtchyi muzej NAN Ukrainy*, 610 (in Ukrainian).

van Tooren BF 1989. The ecological role of the bryophyte layer in Dutch chalk grasslands. *Stichting Pressa Trajectina* : 139.

van Tooren BF and HJ During 1988. Early succession of bryophyte communities on Dutch forest earth banks. *Lindbergia* , 14 : 40-46.

Zapalowicz H 1880. Ro linna szata g r Pokucko-Marmaroskich. *Sprawozdanie Komisyi Fizyjograficznej* , 24 : 1-390 (in Polish).

Zhyliaev GG 2001. Zhyttiezdatnist populaciy travianykh bagatorichnykiv. Avtoreferat dysertatsii doctora biolohichnyh nauk. *Dnipropetrowskyi Derzhavnyi Universytet* 34 (in Ukrainian).

Chapter 19 NATO Advanced Research Workshop

Biological Premises for Management of an Arctic Fox (*Alopex lagopus*) Population

The Demise and Future of the Hare-Footed Fox on Jan Mayen

Kirsti Rouvinen-Watt
Nova Scotia Agricultural College, Department of Plant and Animal Sciences, Truro, Canada
E-mail: krouvinen@nsac.ns.ca

Abstract: The arctic fox (*Alopex lagopus*), the only regularly occurring land mammal on Jan Mayen Island is now extinct. Fox colonization may have been transient and a reduced presence of sea ice now prevents further immigration. The small size of the island's fox population and former heavy selective trapping may be responsible for the demise of the fox. Reduced mating encounters and low availability of winter food may have reduced reproduction with environmental pollution having an additional adverse impact. Periodic presence of domestic dogs may have disturbed dens and introduced diseases. The re-establishment of the arctic fox population on Jan Mayen may be questioned, since the fox may not be part of permanent fauna. What makes the island ecosystem unique for scientific research is that it may be periodically void of land mammals.

Key words: *Alopex lagopus*, Arctic fox, environmental pollution, inbreeding, over-harvesting, population decline, re-establishment, re-introduction

1. INTRODUCTION

The arctic fox (*Alopex lagopus*; Greek for a "hare-footed fox") has been the only regularly occurring land mammal on Jan Mayen Island, but is now believed to be extinct (Barr 1998, 2003). This paper discusses the biology of the arctic fox and possible reasons behind its demise on the island. Current threats to the survival of the species are examined in light of possibilities for potential re-establishment of the population on Jan Mayen.

S. Skreslet (ed.), Jan Mayen Island in Scientific Focus, 207-218

Figure 1. One of the last arctic foxes of the Jan Mayen Island population, fed by a Loran-C crew member in 1964. Photo: L Markrud (original at www.jan-mayen.no).

The arctic fox is one of the most highly adapted Arctic mammals possessing a long, dense winter coat and short legs, ears and muzzle. The species has a circumpolar distribution occupying the Arctic tundra of Europe, Asia and North America. It is also present on the islands throughout the Bering Sea and the North Atlantic and Arctic Oceans. The foxes also move extensively over much of the northern polar ice cap (Garrott and Eberhardt 1999).

There are two colour phases of the arctic fox, white and blue. The white (polar) colour morph (Fig. 1) is more continental in its distribution, whereas the blue type is more prevalent on islands and coastal habitats (Garrott and Eberhardt 1999). During the early 1900s, when large numbers of arctic foxes were trapped on Jan Mayen, the blue genotype represented 90% of the harvest (Richter 1946). During the first season 1906-07, it is estimated that 200-300 foxes were caught on Jan Mayen, with the largest known catch, in 1917-18, 262 foxes (242 blue, Barr 1998, 2003). Many of these foxes were trapped live (Fig. 2) and were taken to fur farms in Norway (Einarsson and Skrede 1989, Nes *et al.* 1989). The fox harvest was high until 1930s but more variable until the Second World War. In winter 1929, a track survey showed a population size of no more than 7-10 foxes. Twenty-four foxes of reportedly poor quality were trapped in 1958-59, and only less than 6 foxes per year during 1960-63 (Barr 2003). From 1965 fox trapping was no longer allowed, but at the same time the fox population had more or less disappeared from Jan Mayen (Barr 1998).

2. LIFE HISTORY AND FOOD HABITS

The reproduction season of the arctic fox begins in February-March with the forming of monogamous pairs and seeking of den sites (Garrott and

Eberhardt 1999). Most dens are established in May. In coastal areas, the dens are usually in rocky outcrops or under boulder piles. The breeding pair holds a home range during the season or sometime all year round, with the size and shape adjusted according to food availability (Garrott and Eberhardt 1999). In Iceland, where foxes forage primarily along beaches, home range sizes vary between 8.6 and 18.5 km^2 (Garrott and Eberhardt 1999). In some coastal areas, the foxes appear to restrict their movements to productive beach areas year-round (Garrott and Eberhardt 1999).

Arctic foxes can be sexually mature in their first year of life. They are monoestrus and mating starts in the end of February and lasts until mid April. After a gestation period of 51-52 days the pups are born in May-June (Einarsson and Skrede 1989, Prestrud 1992). Arctic fox populations can be divided into two ecotypes. "Lemming foxes" feed mainly on lemmings on the tundra. "Coastal foxes" feed mainly on eggs, birds, and carrion from the marine ecosystem (Braestrup 1941). The lemming foxes have a mean litter size of 7 (maximum 18), while the coastal foxes have a mean litter size of 5 (maximum 10) (Tannerfeldt 1997). The cubs are born blind with a dark natal fur (Einarsson and Skrede 1989) and have an estimated body mass of 50-65 g (Prestrud and Nilssen 1995).

The cubs first emerge from the den when they are about 3-4 weeks old and are weaned at 4-5 weeks of age (Prestrud, 1992). They begin to be more independent between 8-9 weeks of age, and are rarely seen at the den by mid August (Prestrud 1992; Frafjord 1992). On Svalbard, the survival of the cubs, from when they were born until when they leave the den, is as high as 75-80% (Prestrud 1992). The cubs reach adult size at about 4-5 months of age (Frafjord 1992, Prestrud and Nilssen 1992). Adult wild arctic foxes weigh about 2.5-5.0 kg (Garrott and Eberhardt 1999). On Svalbard, the heaviest body weights have been recorded in December (juveniles 3.42 kg, yearlings 3.46 kg, adults 3.39 kg) (Prestrud and Nielssen 1995). Survival among juveniles is low (25%), but increases as the foxes get older being up to 70% at 6 years of age (Prestrud 1992).

Figure 2. Trap for catching of live arctic fox, Jan Mayen Island 1963. Photo: S Skreslet.

The opportunistic feeding strategy of the arctic fox, being both a predator and a scavenger, results in wide seasonal and regional variation in diet composition (Garrott and Eberhardt 1999). The foxes eat almost anything that is digestible and feed on berries, seaweed, insects and larvae, birds and their eggs, fish, seal placentas, small mammals and carcasses of large mammals (Tannerfeldt 1997). Arctic foxes residing in coastal habitats rely heavily on the nesting colonies of seabirds with species nesting in most accessible areas being preyed on the most heavily. During food abundance, in the early spring and summer, food is being stored for use later in the summer and during winter, when food supplies are limited (Garrott and Eberhardt 1999).

Coastal habitats occupied by the arctic fox are associated with islands in the Bering Sea, North Atlantic Ocean, and along portions of the Greenland Coast (Garrott and Eberhardt 1999). Climates range from maritime (North Atlantic) to severe (Greenland coast). In the latter, ocean waters freeze during winter. Foxes in these areas remain close to the coastline, deriving sustenance from marine life in the inter-tidal zones along beaches as well as nesting seabirds on rocky outcrops and cliffs. Marine habitats, on the other hand, consist of the constantly shifting sea ice of the Arctic Ocean. These habitats are characterized by low winter productivity and a limited fauna and include primarily marine invertebrates, fish, seals, polar bears, and arctic foxes (Garrott and Eberhardt 1999).

3. CHALLENGES TO SURVIVAL

3.1 Food Availability, Den Sites and Disease

Breeding success and juvenile survival, both dependent on food availability, are considered the most important and variable factors in arctic fox population dynamics (Tannerfeldt 1997). In Fennoscandia, one of the largest threats to arctic fox populations is food shortage. Other challenges to survival include population size, availability of long-term den sites, competition by red fox (*Vulpes vulpes*), diseases introduced by humans, predators and environmental pollution (Tannerfeldt 1997). In inland habitats, summer food supply of the arctic fox is limited by the lack of small rodent population peaks (Angerbjörn et al. 2002). In the winter, the fox depends on the carrion left behind by the wolf (*Canis lupus*) and the wolverine (*Gulo gulo*). This food source is also scarce due to the present low population numbers of these large predators (Tannerfeldt 1997).

On Jan Mayen, summer food supply can be considered plentiful due to large seabird colonies (Gabrielssen *et al.* 1997b). The most abundant prey base is provided by the nesting colonies of the kittiwakes, fulmars, little auks, guillemots and puffins, and their eggs and chicks (Gabrielssen 2003). The habitat on Jan Mayen Island is likely similar to that of coastal Iceland and Svalbard with a stable and reliable source of summer food but low and unpredictable availability of winter food (Hersteinsson *et al.* 1989, Prestrud 1992, Fuglei *et al.* 2003). Very importantly, on Jan Mayen, the lack of large predators has likely caused most severe winter food shortage and can be considered a serious threat to survival of the arctic fox population on the island.

The arctic fox depends on the availability of long-term den sites for successful reproduction. Some dens have been used for centuries by generations of foxes (Tannerfeldt 1997). Human interference at the den sites by tourism, and hunting with dogs as well as direct competition from the red fox, due to their more northern distribution facilitated by climate change, are listed as reasons for abandoning previously active dens (Tannerfeldt 1997). In Fennoscandia, the red fox species is continuously spreading north and poses a significant threat to the survival of the arctic fox by direct predation on the species, by taking over den sites and spreading diseases such as mange (Angerbjörn et al. 2002). On Jan Mayen Island, historic records show frequent presence of domestic dogs (Barr 2003). For example during 1963-65, several Newfoundland dogs resided with the crew on the island (Torbergsen 2004). It is likely that the dogs, similar to the red fox, may have caused disturbance of arctic fox breeding dens and have directly preyed on the foxes. A more serious threat may have been presented to the foxes by the dogs in the form of diseases. Several canine diseases, such as distemper, viral hepatitis, canine parvovirus, coccidiosis and toxoplasmosis, are transmissible between dogs and foxes (Kahn 2003). An outbreak of a disease foreign to the foxes on Jan Mayen may have put the already greatly diminished fox population in further danger of extinction.

3.2 Population Origin, Size and Ice Conditions

Regarding the existence of the arctic fox on Jan Mayen, it is likely that the colonization of this remote island by the species has been a relatively recent and transient event. This is supported by Dutch whaling records, which specifically note the lack of arctic foxes on the island in the 1600s (Hacquebord 1998). The fox population could have been established as recently as the late 1800s, when severe sea ice conditions (Gabrielssen *et al.* 1997b) may have facilitated migration of foxes from Greenland, which is the closest larger land mass. The recent absence of sea ice (Gabrielssen *et al.*

1997b), on the other hand, has prevented this migration from continuing resulting in an isolated island population. Therefore, the reduced size of the arctic fox population on Jan Mayen Island may have been a threat to survival *per se* as small, fragmented and isolated populations are at risk of becoming more inbred and less demographically stable, further reducing population size and rapidly increasing the level of inbreeding (Frankham *et al.* 2003). It is noteworthy that the majority of recorded extinctions since the 1600s have been of island forms of species (Frankham *et al.* 2003). Moreover, in areas where the arctic fox population is very fragmented, encountering a partner during the breeding season is also difficult (Angerbjörn *et al.* 2002). It is likely that the arctic fox on Jan Mayen Island has been predisposed to extinction simply by being a small population on an isolated island. In these conditions populations are known to quickly loose their ability to adapt due to the augmented loss of genetic diversity and the accumulation of deleterious alleles (Frankham *et al.* 2003).

In the *Alopex lagopus* species one such allele is the dominant Shadow (Ss) colour gene, which is lethal in the homozygous form (SS) (Einarsson and Skrede 1989, Nes *et al.* 1989). Although first reported on fur farms in Norway during the 1940-50s (Nes *et al.* 1989, Lohi 2004) it is possible that the gene may have existed in the wild population. Wild fox furs harvested from Iceland and Greenland during the 1970s have been observed to bear resemblance to farmed Shadow fox furs (Lohi 2004). It is plausible that the naturally higher level of inbreeding in the Jan Mayen Island fox population may have resulted in accumulation of deleterious alleles. The intensive and selective trapping of the blue genotype (Barr 1998, 2003) may have further enhanced these deleterious effects by impairing reproduction of the remaining population thereby only hastening the demise of the Jan Mayen foxes.

3.3 Environmental Pollution in the Arctic

Today, the arctic fox faces more indirect human threats. Parallel to what is seen in other species feeding at the highest trophic levels (~4-5.5) of the Arctic marine food chain (AMAP 2002), such as the polar bear (*Ursus maritimus*) and the glaucous gull (*Larus hyperboreus*), bioaccumulation and biomagnification of pollutants in the ecosystem adversely impacts the arctic fox. Due to the key role of fat reserves in winter survival and reproduction in Arctic mammals, the compounds of most concern are those with high affinity to the adipose tissue (Arctic Council 2000).

Polychlorinated biphenyls (PCBs) are a group of very stable chemicals that have reached the Arctic via long-range transport, but there are a number of local sources, such as abandoned military sites. Most PCBs found in seabirds on Jan Mayen have been shown not to originate from the local deposit, but rather have arrived to the area by long-range transport (Gabrielsen *et al.* 1997a). Liver concentrations of PCBs in glaucous gulls on Jan Mayen, exceeding 150,000 $ng \cdot g^{-1}$ lipid weight, are higher than those on Svalbard or Novaja Zemlja but lower than on Franz Josef Land (Gabrielssen et al. 1997b, AMAP 2002). Primary dependence on the marine food web by the arctic fox on island habitats likely exposes it to very high levels of persistent organic pollutants. Iceland's coastal foxes are reported to have PCB levels higher than Svalbard polar bears (AMAP 2002). According to long term studies on the congener patters in arctic foxes in Svalbard (Wang-Andersen *et al.* 1993), the levels of PCBs remained high and essentially unchanged from the 1970s to 1980s.

Similar to many Arctic animals, the arctic fox goes through periods of fasting (Fuglei *et al.* 1999), leading to an increase in contaminant concentrations in the remaining fat, as well as in blood and in other organs, increasing the risk that sensitive systems, such as the brain and reproductive organs, will be exposed to toxic levels (AMAP 2002). The PCB levels reported in the arctic foxes from Svalbard and coastal Iceland raise concerns about subtle neurobehavioural effects as well as effects on reproduction, cub survival, and immune suppression (AMAP 2002). The arctic fox, along with the polar bear and the glaucous gull, may be exposed to levels above the lowest observed adverse effects on reproduction in birds and mammals (Brunström and Halldin 2000). It is apparent that if these effects on individual animals are sufficiently widespread, a population can become vulnerable, especially if exposed to other stresses such as new pathogens and limited access to prey.

Whereas PCBs may not have been a problem in the years of the rapid decline of the Jan Mayen arctic fox population, other similar compounds existed in the arctic environment. One such "legacy" persistent organic pollutant is dichlorodiphenyltrichloroethane, more commonly known as DDT, which was widely used as an insecticide especially during the Second World War. Although many countries banned the use of DDT during the 1970-80s it is still used for the control of malaria and other insect borne diseases (AMAP 2002). Today, persistent organic pollutants (DDTs, PCBs, chlordanes) are the most prominent contaminants found in ringed seals in the Arctic, with the highest levels observed in East Greenland and around Svalbard (AMAP 2002). The contaminant levels in polar bears in the same geographic region indicate that there may still be significant sources of DDT and chlordane in the Russian Arctic (AMAP 2002). Although the levels of DDTs and PCBs are predicted to decline substantially in the coming decades, both are identified as contaminants, which may pose a population-level threat to some Arctic wildlife species (AMAP 2002). It is likely that the historic Global use of DDT, and perhaps also local use by the military, may have contributed to the decline of the arctic fox on Jan Mayen Island.

Compounds such as DDT and PCBs are a gradually diminishing problem, however, other persistent organic pollutants and heavy metals are present or emerging in the Arctic. Mercury accumulates in the Arctic as a result of precipitation from the atmosphere and deposits on snow playing a role as an important sink in the Global mercury cycle (AMAP 2002). It poses a risk specifically to piscivorous and other top predatory species via the formation of methylmercury in the aquatic ecosystem (Wiener et al. 2003), Cadmium also accumulates in birds and mammals and is known to cause a wide variety of toxic (Baker *et al.* 2003, Minami *et al.* 2001) and hormonal effects (Toppari 2002, Johnson *et al.* 2003, Yang *et al.* 2002). New compounds of concern include for example brominated flame retardants (BFRs, Herzke *et al.* 2003). In addition, the presence of higher levels of radionuclides in the ecosystem may pose an additional risk to the health and reproduction of mammals in the Arctic (AMAP 2002). Since the sources of these contaminants are largely elsewhere and the toxins persist in the environment, environmental pollution in the Arctic, including the Jan Mayen Island, will remain a serious concern. An environmental assessment of the island regarding the historic and present levels of persistent organic pollutants, heavy metals and other toxic compounds would give much valuable data about the past and aid in the future predictions of the health of the Jan Mayen ecosystem.

4. CONSIDERATIONS FOR RE-ESTABLISHMENT

When discussing the re-establishment of the arctic fox species on Jan Mayen Island, firstly, it is important that the reasons for the extinction of the original population be investigated. It is also necessary to examine whether these same circumstances prevail today resulting in a similar outcome. Secondly, we must consider whether the re-establishment should be in the form of natural recovery of the species by immigrating foxes or should there be a higher degree of human intervention by active re-introduction. If active re-introduction is considered, the justification and goals for such program must be defined. Is the goal to re-establish a population of arctic foxes that most closely resembles what once existed on the island or is it just to have foxes on the island as an environmental enrichment?

When aiding natural recovery or considering re-introduction, captive breeding provides means for conserving a species or a population that is incapable of surviving on its own (Frankham *et al.* 2003). The source of animals for such program could be live-trapped wild arctic foxes from nearby islands or the farmed blue foxes from Norway. The latter may be genetically the most closely related descendants of the original Jan Mayen gene pool as some of the fox harvest was transported live to Norway to serve as breeding stock (Einarsson and Skrede 1989, Nes *et al.* 1989). It is important to note, however, that the current farmed fox differs greatly from the wild arctic fox having been under severe selection pressure for many generations. A good example of this genetic difference is the body weight of the farmed blue fox (6.7-7.1kg, Einarsson and Skrede 1989), which is now many times that of the wild arctic fox (2.5-5.0kg, Garrott and Eberhardt 1999).

Before re-establishment of the arctic fox species on Jan Mayen can be considered, it is important that the genetic lineage of the original fox population on the island be determined. In order to do this, genetic analyses need to be carried out on archived animal material of the Jan Mayen fox population, tissues collected of arctic foxes on nearby geographic regions (e.g. Greenland, Iceland, Svalbard) as well as those raised on farms in Norway. Using molecular genetic methods such as microsatellites, short nucleotide polymorphisms or mitochondrial DNA, questions about genetic relatedness, diversity, level of inbreeding and the fragmentation of the population can be answered (Dalén *et al.* 2002, Frankham *et al.* 2003).

Further scientific research should be carried out on Jan Mayen to observe any signs indicating possible presence of the arctic fox on the island today, historic evidence of active den sites, and the types and amounts of suitable habitats for the foxes. Based on the number of bird cliffs, the species and the size of the nesting colonies, estimates could be developed for potential prey base. Furthermore, it would be very valuable to examine the historic role of the sea ice surrounding Jan Mayen in the colonization of the island by foxes and what impact the future ice conditions are expected to have on the island's fauna. Based on this information, predictive computer simulations could be employed to determine the carrying capacity of the island and the minimum viable population size for long-term survival in the given environment (Hanski 1997). Moreover, since the concentrations of environmental contaminants in the food webs on Jan Mayen are expected to be very high (Gabrielsen *et al.* 1997a, AMAP 2002), some level of human intervention seems necessary. This could include supporting feeding programs with non-toxic food sources, particularly during winter. The provision of artificial den sites may also encourage a larger number of foxes to breed on the island, if this is deemed a constraint in the natural habitat.

Lastly and perhaps most importantly, based on the historic records (Hacquebord 1998), it may be questioned whether the arctic fox should be considered part of permanent fauna on Jan Mayen. The answer to this question is critical in discussing whether re-establishment of the foxes on the island is realistic and if human intervention by active re-introduction would even be justified. Perhaps what makes the current Jan Mayen ecosystem unique for scientific research is that the island does not have any land mammals, including the arctic fox.

5. CONCLUSIONS

The arctic fox has been the only regularly occurring land mammal on Jan Mayen Island but is now believed to be extinct. The original colonization of the island by the species may have been a transient event, whereas the recent absence of sea ice has prevented further immigration. The small size of the island population *per se* together with heavy and selective trapping, causing loss of genetic diversity and the accumulation of deleterious alleles, are likely primarily responsible for the demise of the arctic fox on Jan Mayen. Reduced chances of encountering a mate and low availability of winter food may have resulted in poor reproduction of the remaining foxes. The frequent presence of domestic dogs may also have disturbed fox breeding dens and introduced diseases. Environmental pollutants, such as DDT, in the ecosystem may have further hastened the decline of the fox population.

In order to consider the re-establishment of the species on the island, the origin, genetic diversity and inbreeding depression of the historic fox population should be assessed. Further research should be carried out on the current status of the arctic fox on Jan Mayen, historic den sites, suitable habitats and potential prey base. The role of the sea ice in the colonization of the island and the impact of future ice conditions on the island's fauna should also be investigated. The carrying capacity of the island and the minimum viable population size could then be theoretically determined. As environmental pollution adversely impacts top predators in the Arctic, human intervention in the form of assisted feeding programs with food sources free of toxins would be necessary in order to support successful reproduction of the arctic fox and survival of the young. Artificial den sites may also be provided to encourage more foxes to breed on the island.

Most importantly, it may be questioned whether the arctic fox should be considered part of permanent fauna on Jan Mayen. The answer to this question is critical in discussing whether re-establishment of the foxes on the island is realistic and if human intervention by active re-introduction would be justified. Perhaps what makes the Jan Mayen ecosystem unique for scientific research is the absence of land mammals.

REFERENCES

Angerbjörn A, M Tanderfeldt, and B Elmhagen 2002. Lågan flämtar för fjällrävens släkt. *Fauna och Flora* 97 (2), 13-16.

Arctic Council 2000. Persistent Organic Pollutants (POPs). Factsheet #1. Produced for the Arctic Council by the Arctic Monitoring and Assessment Programme (AMAP). Nørhaven A/S, Viborg, Denmark.

Arctic Monitoring and Assessment Programme (AMAP) 2002. Arctic Pollution 2002. Oslo, Norway.

Baker TK, HB VanVooren, WC Smith, and MA Carfagna 2003. Involvement of cadmium in the sexual dimorphism of cadmium induced hepatotoxicity. *Toxicology Letters* 137, 185-192.

Barr S 1998. The modern history of Jan Mayen. *Circumpolar Journal* 13(2) , 93-101.

Barr S 2003. *Jan Mayen. Norges Utpost i Vest. Øyas historie gjennom 1500 år.* Kolofon AS &AIT AS.

Braestrup FW 1941. A study on the arctic fox in Greenland. Meddelelser om Grønland 13, 1-101.

Brunström B and K Halldin 2000. Ecotoxicological risk assessment of environmental pollutants in the Arctic. *Toxicology Letters* 112-113, 111-118.

Dalén L, A Götherström, M Tannerfeldt, and A Angerbjörn 2002. Is the endangered Fennoscandian arctic fox (*Alopex lagopus*) population genetically isolated? *Biological Conservation* 105, 171-178.

Einarsson EJ and A Skrede 1989. *Avl og Fôring av Rev.* A/S Landbruksforlaget, Ås, Norway.

Frafjord K 1992. Denning behaviour and activity of arctic fox (*Alopex lagopus*) pups: Implications of food availability. *Polar Biology* 12, 707-712.

Frankham R, JD Ballou, and DA Briscoe 2003. *Introduction to Conservation Genetics.* Cambridge University Press, New York,

Fuglei E and NA Øritsland 1999. Seasonal trends in body mass, food intake and resting metabolic rate, and induction of metabolic depression in arctic foxes (*Alopex lagopus*) at Svalbard. *Journal of Comparative Physiology* B 169 (6), 361-369.

Fuglei E, NA Øristland and P Prestrud 2003. Local variation in arctic fox abundance in Svalbard, Norway. *Polar Biology* 26, 93-98.

Gabrielssen GW 2003. Personal communication. Norwegian Polar Institute, Tromsø, Norway.

Gabrielssen GW, IG Alsos, and B Brekke 1997a. Undersøkelser av PCB i jord, fisk og sjøfugl i området rundt avfallsfyllingen på Jan Mayen (Investigations on PCB in soil, fish and seabirds around the waste deposit on Jan Mayen). Norsk *Polarinstitutt Rapport* 104, 1-26 (in Norwegian).

Gabrielssen GW, B Brekke, IG Alsos, and JR Hansen 1997b. Natur- og kulturmiljøet på Jan Mayen - med en vurdering av verneverdier, kunnskapsbehov og forvaltning. *Norsk Polarinstitutt. Meddelelser* 144, 1-127 (in Norwegian).

Garrott RA and LE Eberhardt 1999. Arctic fox. Pp. 394-406 in M Novak, JA Baker, ME Obbard and B Malloch (eds.) *Wild Furbearer Management and Conservation in North America.* CD-Rom Version. Queens Printer, Ontario.

Hacquebord L 1998. The first wintering on Jan Mayen. *Circumpolar Journal* 13, 70-81.

Hanski IA 1997. Metapopulation Dynamics. From Concepts and Observations to Predictive Models. Pp. 69-91 in IA Hanski and ME Gilpin (ed.) *Metapopulation Biology. Ecology, Genetics and Evolution.* Academic Press, San Diego, CA.

Hersteinsson P, A Angerbjörn, K Frafjord, and A Kaikusalo 1989. The arctic fox in Fennoscandia and Iceland: Management problems. *Biological Conservation* 49, 67-81.

Herzke D, GW Gabrielsen, A Evenset, and IC Burkow 2003. Polychlorinated camphenes (toxaphenes), polybrominated diphenylethers and other halogenated organic pollutants in glaucous gulls (*Larus hyperboreus*) from Svalbard and Bjørnøya (Bear Island). *Environmental Pollution* 121, 293-300.

Johnson MD, N Kenney, A Stoica, L Hilakivi-Clarke, B Singh. G Chepko, R Clarke, PF Sholler, A Lirio, C Foss, R Reiter, B Trock, P Soonyoung, and MB Martin 2003. Cadmium mimics the *in vivo* effects of estrogen in the uterus and mammary gland. Nature Medicine. Published online. 12 July doi:10.1038/nm902.

Kahn CM (ed.) 2003. The Merck Veterinary Manual 2003. 8th edition. Online version. Merck & Co., Inc. Whitehouse Station, NJ. URL address http://merckvetmanual.com.

Lohi O 2004. Personal communication. Fur animal research leader emeritus. Danish Institute of Agricultural Sciences, Foulum, Denmark.

Minami A, ATakeda, D Nishibaba, S Takefuta, and N Oku 2001. Cadmium toxicity in synaptic neurotransmission in the brain. *Brain Research* 894, 336-339.

Nes N, EJ Einarsson, and O Lohi 1989. *Beautiful Fur Animals - and their colour genetics.* Scientifur, Glostrup.

Prestrud P 1992. Arctic foxes in Svalbard: Population ecology and rabies. Ph.D. thesis, University of Oslo, Norway.

Prestrud P and K Nilssen 1995. Growth, size and sexual dimorphism in arctic foxes. *Journal of Mammalogy* 76, 522-530.

Richter S 1946. Jan Mayen i krigsårene (Jan Mayen during World War II). *Meddelelser Norges Svalbards- og Ishavsundersøkelser* 66, 1-24.

Tannerfeldt M 1997. Population fluctuations and life history consequences in the arctic fox. Ph.D. thesis. Stockholm University, Sweden.

Toppari J 2002. Environmental endocrine disrupters and disorders of sexual differentiation. *Seminars in Reproductive Medicine* 20 (3), 305-312.

Torbergsen E 2004. Personal communication. Resident of Jan Mayen during 1963-65.

Wang-Andersen G, JU Skaare, P Prestrud, E Steinnes 1993. Levels and congener patterns of PCBs in arctic fox, *Alopex lagopus*, in Svalbard. *Environmental Pollution* 82 (3), 269-275.

Wiener JG, DP Krabbenhoft, GH Heinz, and AM Scheuhammer 2003. Ecotoxicology of Mercury. Pp. 409-463 in DJ Hoffman, BA Rattner, GA Burton Jr, and J Cairns Jr (eds) *Handbook of Ecotoxicology.* 2nd ed. Lewis Publishers. CRC Press LLC, Washington.

Yang HS, AP Cheung, DK Han, JC Sim, EJ Lee, and JH Kim 2002. Cadmium-induced toxicity on testicular tissue and spermatogenesis in rats and the protective effects of α-tocopherol. *Fertility & Sterility* 78 (3), Suppl. 1, S274-S275.

Chapter 20 NATO Advanced Research Workshop

Research Opportunities for Jan Mayen Using Stable Isotope Analyses

Long-Term Monitoring of Marine and Freshwater Foodwebs

Keith A Hobson
Canadian Wildlife Service, Saskatoon, Canada
E-mail: keith.hobson@ec.gc.ca

Abstract: Polar environments pose logistical challenges for fieldwork designed to elucidate dietary history of individual organisms, trophic relationships among individuals and species, and the ultimate role of species in the flux of energy, elements, and contaminants. Stable isotopic tools delineate dietary histories of organisms that extend well beyond the periods when individuals are physically available. By examining tissues with both long- and short-turnover rates, a long-term dietary picture emerges. Tissues such as blood plasma and cellular fractions, feathers, nail, hair, and even breath carbon dioxide may often provide opportunity for non-destructive sampling protocols. This paper presents an overview of stable isotopic applications to time-integrated trophic reconstructions, source of feeding, behaviour of contaminants in the field of isotopic tracing in polar food webs with specific reference to Jan Mayen.

Key words: carbon-13, nitrogen-15, stable isotopes, seabirds, trophic level, source of feeding, long-term monitoring, contaminants, Arctic char, cannibalism

1. INTRODUCTION

Research in polar environments involves numerous logistical challenges. Studies on polar marine and freshwater food webs are particularly difficult since the organisms of interest are typically only available during a relatively short, ice-free period. When available, there are also numerous challenges to deciphering the dietary history of individuals due to problems inherent in conventional approaches to dietary reconstructions based on

S. Skreslet (ed.), Jan Mayen Island in Scientific Focus, 219-227

direct observation, stomach contents analysis, or on the examination of faecal material or regurgitated pellets (Duffy and Jackson 1986, González-Solis *et al.* 1997). While such investigations are often critical to understand detailed taxonomic components of diets, they are usually short-term in scope and suffer biases in favour of materials that are readily observed or well preserved. The last decade or so has seen the advent of stable isotope techniques that can augment these conventional approaches to understanding food web structure and the role of individuals and species in the flux of energy and contaminants. Stable isotope techniques are particularly suitable to the study of polar marine, terrestrial and freshwater food webs since they can also provide time-integrated information on assimilated foods and so reveal dietary history during periods when organisms are not physically available for study. In this paper, I provide necessary background information on the use of stable isotope techniques to investigate polar food webs and provide several examples of how such approaches have been applied to trophic-level delineation, source of feeding, and tracing contaminant flow. Emphasis will be placed on the use of stable-carbon ($^{13}C/^{12}C$, $\delta^{13}C$) and nitrogen ($^{15}N/^{14}N$, $\delta^{15}N$) isotopes. Specifically, the direct application of these techniques to future research and monitoring on Jan Mayen will be emphasized.

2. WHAT ARE STABLE ISOTOPES?

Stable isotopes are naturally occurring stable forms of elements that differ in atomic mass, and this mass difference confers specific properties with respect to biogeochemical processes. Stable isotope abundance of any element is typically expressed as a ratio of the rarer, heavy form to that of the more common, lighter form. Stable isotopes of light elements of most interest to ecological applications are those of carbon ($^{13}C/^{12}C$), nitrogen ($^{15}N/^{14}N$), sulfur ($^{34}S/^{32}S$), hydrogen ($^{2}H/^{1}H$) and oxygen ($^{18}O/^{16}O$). Those of heavier elements such as strontium (^{87}Sr) and lead (^{210}Pb) are also particularly useful but require more involved analytical procedures. Stable isotope ratios of the light elements are measured using isotope-ratio mass spectrometry (IRMS) and are usually expressed in abundance relative to international standards in delta (δ) notation and reported as parts per thousand deviation from those standards. The arbitrary choice of these standards results in $\delta^{13}C$ samples typically being negative (i.e. the standard is usually enriched in ^{13}C compared to the unknown) and $\delta^{15}N$ samples being positive (atmospheric air is typically depleted in ^{15}N relative to the unknown).

Carbon occurs in two stable forms, as ^{13}C and ^{12}C. Carbon isotopic compositions of animals reflect closely those of the diet (DeNiro and Epstein 1978). For this reason, stable-carbon isotope analysis is an ideal tool for tracing origins of nutrients in food webs since carbon may enter the base of food webs with characteristic isotopic signatures due to a variety of biogeochemical processes and these change little throughout the food web. The two main processes of interest here are those resulting in differences in $\delta^{13}C$ values of terrestrial vs. marine food webs and in differences between inshore, or benthically linked, food webs and offshore, or pelagic, food webs.

Carbon in the atmosphere is depleted in ^{13}C relative to dissolved carbonate in the oceans by about $7^{o}/_{oo}$ (Craig 1953). Effectively, heavier carbon dioxide enters into bicarbonate ion exchange at the ocean/atmosphere interface more readily than does lighter carbon dioxide. In general then, marine food webs tend to be enriched in ^{13}C compared with terrestrial C-3 food webs (e.g. Hobson 1990). For a variety of reasons, inshore marine food webs tend to be more enriched in ^{13}C than offshore or pelagic food webs (Hobson and Welch 1992) allowing the spatial analysis of foraging behaviour of a variety of organisms (e.g. Hobson *et al.* 2002b).

There are two stable forms of nitrogen, ^{14}N and ^{15}N. In terrestrial systems, nitrogen enters food webs through symbiotic fixation or through direct conversion of atmospheric nitrogen within plants. These processes typically lead to different isotopic signatures of plants adopting these strategies. In marine systems, inorganic nitrogen occurs in several molecular forms and relatively little is known about the isotopic fractionation effects occurring between these nitrogen pools and phytoplankton (Owens 1987, but see Wada and Hattori 1978).

As was found for stable-carbon isotopes, the stable nitrogen isotopic composition of animals is related ultimately to the isotopic compositions of their diets (DeNiro and Epstein 1981). Through processes associated primarily with differential excretion of ^{14}N, animals incorporate dietary ^{15}N into their tissues preferentially, an effect due to discrimination against the lighter isotope during protein amination and deamination (Macko *et al.* 1982). A broad survey of field and laboratory data confirms that there is a significant linear relationship between $\delta^{15}N$ values of an organism and its diet (DeNiro and Epstein 1981, Owens 1987). This phenomenon forms the basis of using $\delta^{15}N$ measurements to infer trophic level. Broad landscape-level differences in $\delta^{15}N$ values of soils and the food webs they support have also been measured and this has consequences for tracking some migrant birds (Hobson 1999).

3. ESTABLISHING ISOTOPIC MARINE AND FRESHWATER FOOD WEB MODELS FOR JAN MAYEN

Before long-term monitoring of trophic relationships and contaminant flow can be conducted for the marine food web off Jan Mayen Island, it is necessary to establish critical baseline information on trophic structure using stable-nitrogen isotope analyses. This can be done by sampling key elements from each approximate trophic position starting from marine phytoplankton or particulate organic matter (POM) and covering plants and animals from the inshore to the pelagic offshore system. Obviously, coordination with ship-based sampling of the marine foodweb about Jan Mayen represents an excellent opportunity to sample most species of interest. Suitable sample sizes of 7-10 individuals per species would be appropriate as a first step to investigating variance within species. After suitable pre-treatment to remove carbonates and lipids, samples are analyzed using CFIRMS. This approach has been used successfully several times in various Arctic marine food webs (Hobson and Welch 1992, Hobson et al. 1995, Hobson *et al.* 2002b) and effectively creates an isotopic base map which can serve as a template for future food web and contaminant studies. A bimodal plot of $\delta^{15}N$ vs. $\delta^{13}C$ values moreover, will reveal two clear trajectories of benthic- and pelagic-based food webs (Hobson *et al.* 2002b).

3.1 Using Isotope Values of Seabirds as Environmental Indicators

The seabird community of Jan Mayen Island is described by Gabrielsen *et al.* (1997), but briefly consists of species ranging from primarily piscivores (*Uria lomvia*) to planktivores and scavengers (*Fulmarus glacialis*). Seabirds are extremely useful as monitors of the health of marine foodwebs (Montevecchi 1993) and, on Jan Mayen, they are accessible for study during the breeding season and occur in large numbers. The broad trophic range of seabirds here allows the monitoring of key components of the marine food web through the indirect measurement of stable isotope values in the blood, eggs, and feathers of a few key species. The simultaneous and non-destructive measurement of several tissue types will further allow a time-integrated trophic analysis of each species from the period of egg formation through post-breeding moult, a period corresponding to about 6 months. Protein pathways will be determined through the analysis of proteinaceous components such as egg yolk and albumen as well as blood cellular and plasma fractions. Lipid pathways can also be delineated through the stable carbon isotope analysis of egg yolk lipids and blood lipids.

In addition to the regular monitoring of adult seabird trophic relationships, the isotopic analysis of the blood of hatch-year (HY) birds allows the direct comparison of adult and chick diets (e.g. Hobson *et al.* 2002a). Previously, knowledge of seabird diet was biased toward the diet of chicks since such diet can be monitored easily by conventional means at colonies. However, the use of stable isotope techniques now has revealed that chicks are often fed at a higher trophic level than their parents, having profound consequences for the interpretation of the role of seabirds in the flux of carbon through marine food webs and the vulnerability of various age groups to contaminant exposure.

The routine annual or bi-annual monitoring of seabird egg, blood and feather samples of at least two species (e.g. *Uria lomvia* and *Fulmarus glacialis*) at Jan Mayen Island represents a straightforward and yet powerful approach to ecosystem-level monitoring. Combined with decadal measurement of contaminants in seabirds and their eggs will also allow more critical interpretation of the contaminant data since it will allow individual data to be normalized for trophic level, a step crucial for those contaminants that are known to bio-accumulate.

3.2 Investigating and Monitoring Contaminants in the Marine Foodweb off Jan Mayen

The combined measurement of contaminants and stable isotopes in food web components of marine systems represents a dynamic and extremely useful approach to environmental toxicology. This is linked to the fact that stable-nitrogen and carbon isotope analyses provide information on trophic level and source of feeding and so can provide critical insight into the source and behaviour of contaminants in marine food webs. This approach has been used successfully recently in investigations of the Northwater Polynya, off western Greenland in northern Baffin Bay. For example, the bioaccumulation of organochlorines in that foodweb was quantified using the slope of the relationship between contaminant concentrations and organism $\delta^{15}N$ value (Fisk *et al.* 2001). This approach has been further applied to heavy metal accumulation in food webs involving marine mammals (Atwell *et al.* 1998).

Jan Mayen is situated adjacent to two major oceanic current flows. The East Greenland Current (EGC) flows to the west of Jan Mayen from the Arctic Ocean and so represents possible contaminant trajectories from that basin. To the east of Jan Mayen, North Atlantic waters circulate and represent contaminant inputs from Europe. Jan Mayen is thus strategically located to assess the potential differences between these two contaminant fluxes. While potential sampling strategies include at-sea sampling of the marine foodweb, the use of seabirds as bioindicators of contaminants also represents a viable means of conducting such analyses (Furness 1993). Moreover, the above proposed seabird trophic and foraging studies could readily be harmonized with a contaminant-based investigation.

3.3 Cannibalism by and Source of Nutrients to Arctic Char on Jan Mayen

Jan Mayen has a freshwater, landlocked population of Arctic Char, (*Salvelinus alpinus*). Some of these char occur as large morphs that are known to be cannibals having apparently escaped constraints of gape and body size (Skreslet 1973). Hobson and Welch (1995) examined isotopically a similar population of char in Char Lake, NWT, Canada, and found that char population was structured into a three trophic-level system with cannibalism occurring at intermediate (10-36.5cm) and large (>40cm) fork length. The common, intermediate-sized char apparently also fed on larval char as well as invertebrates. Hobson and Welch (1995) speculated that because such polar freshwater food webs are typically of low primary productivity, they place energetic constraints on food web components thereby limiting trophic structure. Escaping such constraints are possible for only a relatively few individuals that achieve a physical size to effectively create a new species feeding niche. The situation on Jan Mayen will allow a rare opportunity to investigate similar phenomena and so contribute to a better understanding of freshwater food web energetics in polar regions.

The food web of Char Lake, Canada, was found to be driven by a detrital food web based on phytoplankton, benthic algae and benthic moss. Stable isotope analyses using primarily $\delta^{13}C$ values, successfully established the relative importance of these three inputs to various species of primary consumers. Again, the situation on Jan Mayen will allow a valuable comparative isotopic approach to the freshwater food web of the Nordlaguna. The interesting difference between the nutrient base of Char Lake and that of Nordlaguna is the periodic influx of marine water into Nordlaguna. The effect of such infusions on the freshwater food web can be investigated isotopically since $\delta^{13}C$ values of marine-derived carbon is enriched in ^{13}C compared to terrestrial (C3) or freshwater based carbon. This would represent an extremely rare opportunity to monitor the periodic pulsed nutrient inputs from a marine source into a freshwater system, one that is also augmented by nutrient transfer from seabirds via faeces deposition.

3.4 Terrestrial Food Web Monitoring Opportunities

Stable isotope analyses offer a valuable means of tracking terrestrial as well as marine food webs. In particular, this approach has successfully delineated the reliance of terrestrial small mammals on marine resources on offshore islands. Specifically, the technique has been useful for evaluating the reliance of introduced rats and other small mammals on seabirds and their eggs (Hobson *et al.* 1999, Drever *et al.* 2000). If foxes were to be reintroduced to Jan Mayen Island, then routine isotopic analysis of fox tissues (e.g. hair, blood), will allow a quantitative estimate of their use of seabirds.

4. ISOTOPIC RESEARCH AND MONITORING OPPORTUNITIES ON JAN MAYEN

One of the primary constraints facing polar researchers is the often insurmountable logistical challenge of conducting field work. This is especially true of Arctic marine research that is typically forced to use ship-based sampling procedures at considerable cost. In this paper, I have presented the basis of an analytical technique which has already proven to be of immense value in overcoming some of these fundamental constraints. If long-term research and monitoring were sanctioned for Jan Mayen, the stable isotope approach will similarly contribute substantially to reducing costs and logistics related to long-term ecosystem-level research. While marine-based sampling from a research vessel off Jan Mayen would be extremely attractive and go a long way to establishing the isotopic base map required for some analyses, this is by no means critical. For example, other researchers have successfully used prey items brought back to colonies by seabirds as a means of acquiring needed food web components. Moreover, the establishment of the food web isotope base map can be created largely opportunistically through cooperation with fisheries vessels and other research programs. Nonetheless, several other contributors to this volume have argued for ship-based research opportunities relative to long-term monitoring at Jan Mayen and the stable isotope component of food web analyses will only benefit from coordinated sampling of the marine food web off such vessels.

Colony-based isotopic sampling of seabirds using serial blood, feather, and egg assays represents a fairly easy and straightforward field component of seabird research and monitoring. Here, the major advantage to seabird researchers will be to have a base camp to work out of during the relatively short summer breeding period. Again, it should be emphasized that the use of stable isotope analysis of several tissue types will provide information on diet, trophic level, and source of feeding well beyond the physical field season of seabird researchers thereby reducing costs and logistical constraints. Moreover, the combined isotope and contaminant investigations of seabirds represent an excellent opportunity to gain information on the behaviour of contaminants related to foraging behaviour of birds and their source of feeding.

A similar argument regarding logistics can be applied to the exciting possibility of conducting isotopic investigations of the freshwater food web of Nordlaguna on Jan Mayen, research and monitoring requiring really only the provision of a base-camp. This aspect of research is particularly interesting in light of the dynamic nature of the barrier separating the ocean from the freshwater impoundment containing char.

REFERENCES

Atwell L, KA Hobson, and HE Welch. 1998. Biomagnification and bioaccumulation of mercury in an Arctic marine food web: Insights from stable-nitrogen isotope analysis. *Canadian Journal of Fisheries and Aquatic Sciences* 55, 1114-1121.

Craig H 1953. The geochemistry of the stable carbon isotopes. *Geochimica et Cosmochimica Acta* 3, 53-92.

DeNiro MJ, and S Epstein 1978. Influence of diet on the distribution of carbon isotopes in animals. *Geochimica et Cosmochimica Acta* 42 495-506.

DeNiro MJ, and S Epstein 1981. Influence of diet on the distribution of nitrogen isotopes in animals. *Geochimica et Cosmochimica Acta* 45, 341-351..

Drever MC, LK Blight, KA Hobson, and DF Bertram 2000. Predation on seabird eggs by Keen's mice (*Peromyscus keeni*): using stable isotopes to decipher the diet of a terrestrial omnivore on a remote offshore island. *Canadian Journal of Zoology* 78, 2010-2018

Duffy DC, and S Jackson 1986. Diet studies of seabirds: a review of methods. *Waterbirds* 9, 1-17.

Fisk AT, KA Hobson, and RJ Norstrom. 2001. Influence of chemical and biological factors on trophic transfer of persistent organic pollutants in the Northwater Polynya marine food web. *Environmental Science and Technology* 35, 732-738.

Furness RW 1993. Birds as monitors of pollutants, pp 86-143 in R.W. Furness and J.J.D. Greenwood (eds.) Birds as monitors of environmental change. Chapman and Hall, London.

Gabrielsen G, B Brekke, IG Alsos, and JR Hansen 1997. Natur- og kulturmiljøet på Jan Maen - med en vurdering av verneverdier, kunnskapsbehov og forvaltning. *Norsk Polarisnstitutt, Meddelelser* 144, 1-127 (Norwegian).

Gonzalez-Solis J, D Oro, V Pedrocchi, LI Jover, and X Ruiz 1997. Bias associated with diet samples in Audouin's gulls. *Condor* 99, 773-779.

Hobson KA 1990. Stable isotope analysis of Marbled Murrelets: evidence for freshwater feeding and determination of trophic level. *Condor* 92, 897-903.

Hobson KA 1999. Tracing origins and migration of wildlife using stable isotopes: A review. *Oecologia* 120, 314-326.

Hobson KA and HE Welch 1992. Determination of trophic relationships within a high Arctic marine food web using $\delta^{13}C$ and $\delta^{15}N$ analysis. *Marine Ecology Progress Series* 84, 9-18.

Hobson KA, and HE Welch 1995. Cannibalism and trophic structure in a high Arctic lake: insights from stable-isotope analysis. *Canadian Journal of Fisheries and Aquatic Sciences* 52, 1195-1201

Hobson KA, WG Ambrose Jr and PE Renaud 1995. Sources of primary production, benthic-pelagic coupling, and trophic relationships within the Northeast Water Polynya: insights from $\delta^{13}C$ and $\delta^{15}N$ analysis. *Marine Ecology Progress Series* 128, 1-10.

Hobson KA, M Drever and G Kaiser 1999. Norway rats as predators of burrow-nesting seabirds: insights from stable isotope analysis. *Journal of Wildlife Management* 63, 14-25.

Hobson KA, G Gilchrist, and K Falk 2002a. Isotopic investigations of seabirds of the North Water Polynya: contrasting trophic relationships between the eastern and western sectors. *Condor* 104, 1-11.

Hobson KA, AT Fisk, N Karnovsky, M Holst, JM Gagnon and M Fortier 2002b. A stable isotope (^{13}C, ^{15}N) model for the North Water Polynya foodweb: implications for evaluating trophodynamics and the flow of energy and contaminants. *Deep Sea Research Part II: Topical Studies in Oceanography* 49, 5131-5150.

Macko, SA, WY Lee, and PL Parker 1982. Nitrogen and carbon isotope fractionation by two species of marine amphipods: laboratory and field studies. *Journal of Experimental Marine Biology and Ecology* 63, 145-149.

Montevecchi WA 1993. Birds as indicators of change in marine prey stocks, pp 217-266 in R.W. Furness and J.J.D. Greenwood (eds.) Birds as monitors of environmental change. Chapman and Hall, London.

Owens NJP 1987. Natural variations in ^{15}N in the marine environment. *Advances in Marine Biology* 24, 389-451.

Skreslet S 1973. The ecosystem of the Arctic lake Nordlaguna, Jan Mayen Island III. Ecology of Arctic char, *Salvelinus alpinus* (L.). *Astarte* 6, 43-54.

Wada E, and A Hattori 1978. Nitrogen isotope effects in the assimilation of inorganic and nitrogenous compounds by marine diatoms. *Geomicrobiology Journal* 1, 85-101.

Chapter 21 NATO Advanced Research Workshop

The Jan Mayen Whaling Industry

Its Exploitation of the Greenland Right Whale and its Impact on the Marine Ecosystem

Louwrens Hacquebord
Arctic Centre, University of Groningen, The Netherlands
Email: l.hacquebord@let.rug.nl

Abstract: After a relatively late discovery of the island, Dutch whalers used Jan Mayen as a base for their whaling industry. They built stations on the west coast of the island where they rendered whale oil from the blubber of the Greenland right whales. Altogether the whalers stayed for twenty-two years on Jan Mayen. In this period, approximately 1000 Greenland right whales (*Balaena mysticetus*) were killed and processed on the island leaving much uneaten plankton behind in the sea. Plankton-eating birds must have taken advantage of the situation and it is likely that the number of seabird rookeries on Jan Mayen increased. The combination of written sources and field data makes Jan Mayen a very challenging island for research.

Key words: Jan Mayen, history, archaeology, whaling, plankton, ice drifts, flora and fauna

1. INTRODUCTION

The island of Jan Mayen (71°N, 8°30'W) is located in the middle of the Atlantic Ocean, far away from the so-called civilized world. This is probably the reason why the island was discovered only in the seventeenth century. After its discovery Dutch whalers used the island for more than two decades as a base for their whaling activities. Centuries later the island was used by Norwegian foxhunters and today it is used as a base for the purposes of navigation and weather forecasting. However, this article focuses on the whaling period of Jan Mayen in the seventeenth century.

S. Skreslet (ed.), Jan Mayen Island in Scientific Focus, 229-238

2. THE DISCOVERY OF AN ISLAND

Although there are many stories of an earlier discovery of Jan Mayen Island, no traces of human activity older than the seventeenth century have been found there so far. It is generally recognized that the Englishman John Clarke, sailing for a Dunkirk firm, discovered the island on June 28 1614. He named the island Isabella and hunted Greenland right whales (*Balaena mysticetus*) there, of which he killed eleven (GAA.NA. 842, 1631).

It is possible that this discovery was preceded by discoveries of the island by two other Englishmen, Henry Hudson and Thomas Marmaduke. According to Thomas Edge, one of the first whaling captains of the English whaling company, Hudson saw the island in 1608 and named it Hudson's Tutches and Marmaduke saw the island in 1612 and named it Trinity Island (Muller 1874). However, there is no cartographical or written proof of these discoveries (Schilder 1988).

In 1614, when John Clarke made his discovery, three Dutch ships arrived at the island too: one from Delft, one from Amsterdam and one from Enkhuizen. The ship from Delft had been commissioned privately by a group of merchants from Delft who were also shareholders in the *Noordse Compagnie,* as the Dutch whaling company was called. The captain of the Delft ship, Jan Jansz Kerckhoff, named the island Mauritius. The other ships had been commissioned by the *Noordse Compagnie* and fitted out by merchants from Amsterdam and Enkhuizen, who were also shareholders in that company. The captains of these two Dutch ships, Jan Jacobsz May and Jacob de Gouwenaer, named the island Mr. Joris Eylant after the Dutch cartographer Joris Carolus who was on board one of the ships.

The present name of the island came from Jan Jacobsz May of Schellinkhout who was captain of the ship the *Gouden Cath*, which had been equipped by the shareholders from Amsterdam. The famous Amsterdam mapmaker Willem Jansz Blaeu named the island after him in his 1620 edition of the map of Europe which was originally published by Cornelis Doedz in 1606. Besides an outline map of the North Atlantic Ocean, the famous Zeespiegel by Willem Jansz Blaeu, dating from1623, provided the first detailed map of Jan Mayen Island. Ever since that moment the island has been called Jan Mayen Island or Jan Mayen for short.

3. THE ISLAND IN DISPUTE

The ship from Delft was sailing by private commission of the Delft merchants Nicasius Kien, Anthonie Monier and Dirk Adriaensz Leversteyn, all shareholders in the *Noordse Compagnie*. Back in the Netherlands the captain properly reported the discovery to his employers who, whether or not by design, failed to report it to the States General, the Dutch administration. The other two ships were fitted out by Amsterdam and Enkhuizen shareholders of the *Noordse Compagnie*. The crews of these ships saw the ship of Kyen, Monier and Leversteyn and thought that it was sailing for the *Noordse Compagnie*. They assumed that the merchants from Delft would report their discovery to the States General and consequently would demand a charter for the area for the joint company. However, the Delft merchants kept their discovery secret and returned to the island the following year to hunt whales for their own profit. When this came out it caused many arguments with the other shareholders of the *Noordse Compagnie* (Muller 1874). In this way the history of Jan Mayen started with arguments and soon after judicial disputes, which is why in official Dutch documents the island was for a long time referred to as the *eylandt in questie* (the island in dispute) (ARA‘s Gravenhage: Resolutions of the States General 30 October 1616 etc.).

The discussions about economic rights took place between Kyen, Monier and Leversteyn and the other shareholders of the *Noordse Compagnie*. Although in 1616 the States General agreed to the proposal of Kyen cs. to let both parties hunt whales around Jan Mayen, the question was put to the Dutch High Court. The High Court decided on 13 April 1617 that, based on the General Charter (the "General Charter" defined that the discoverer of a new territory had the right to make voyages to the newly discovered land in 4 consecutive years, excluding all others) and because both parties had a ship near Jan Mayen in 1615 and only reported their discovery afterwards, both parties had equal rights for whaling around Jan Mayen for four years, to begin with the year 1618. They should no longer obstruct each other and in the case of a possible conflict, the States General would arbitrate (ARA‘s Gravenhage: Sentensien van de Hoge Raad, 13 April 1617). In this way Kyen cs., who named themselves the *Kleine Noordse Compagnie,* had whaling rights around the island in the period 1618-1622 both as independent entrepreneurs and as shareholders of the *Noordse Compagnie*. The *Kleine Noordse Compagnie* expanded rapidly. Beside the Leversteyn family, the Rotterdam merchant family of Mush also joined the firm.

4. JAN MAYEN WHALING INDUSTRY

In the first period of exploitation, whaling took place in the bays of the island. We are well informed about this period because of the logbook of one of the whaling captains, Heertjen Jansz of Enkhuizen (1616) (GAE, 439/1597).

Approximately 200 men were working and living on the island. Temporary try works with furnaces and copper cauldrons were constructed at six locations on the west side of the island to cook the blubber of the whale and reduce it to whale oil (Fig. 1). At the end of the season the furnaces were taken down and the copper cauldrons were taken on board ship.

To lodge the workers in the try works, tents of canvas were pitched which could also be taken down at the end of the season. The painting by Cornelis Claesz van Wieringen, 'A Dutch whaling station on Jan Mayen' (ca. 1620) and the description in the logbook of Heertjen Jansz (1616) gives a good picture of the situation in the first decade of the Jan Mayen whaling industry (Fig. 2). In general the hauls were good and whaling was very profitable in this period. On average, more than ten Dutch whaling ships sailed to Jan Mayen every year in this period.

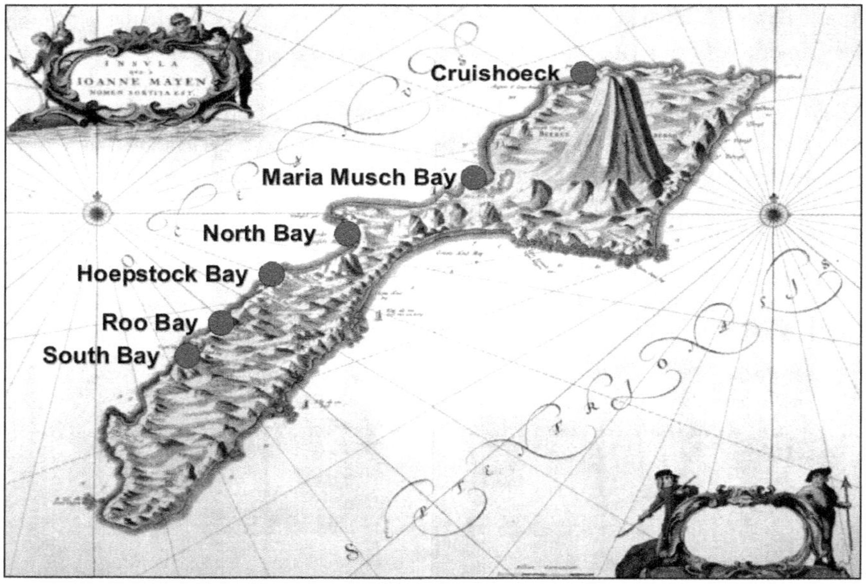

Figure 1. Historical map with whaling stations on Jan Mayen. Ioannes Blaeu, 1662 Atlas Maior Sive Cosmographia Blaviana. Library, University of Amsterdam The Netherlands.

Figure 2. Temporary station in the Northbay of Jan Mayen. The Kendall Whaling Institute, New Bedford Whaling Museum, New Bedford, Mass. USA.

The *Grooten Atlas ofte Werelt-beschrijving 1644-1665*, part I, page 19, by Joan Blaeu informs us that in 1624 ten wooden houses had been built in the South Bay to lodge the land men. Three houses were washed away in the winter but the people from Amsterdam built two new wooden houses the next summer. The same probably happened in the North Bay. From this it appears that the *Noordse Compagnie* switched to the construction of two semi-permanent land stations on the island, one in the North Bay and one in the South Bay. In 1983 and 1988 archaeological surveys showed bricks and other remains of buildings in both places (Fig. 3). The other places were abandoned. This fundamental change started a second period of exploitation on Jan Mayen. Much information from this period is given by the logbooks of the famous Dutch admiral Michiel Adraensz de Ruyter who, in his younger years, was a navigating officer on the whaler the *Groene Leeuw* and sailed to Jan Mayen in 1633 and 1635 (Honoré Naber 1930). The painting by Cornelis de Man (ca. 1639) gives a picture of a semi-permanent whaling station on Jan Mayen. The stations had big brick furnaces which were fitted with chimneys. The tents had been replaced by wooden storehouses and dwellings (Fig. 4). In this period 5 to 10 ships on average sailed to Jan Mayen every year.

In 1627 the plan to place artillery on the battery in the North Bay was mentioned for the first time (GAD, NA 1649, fol. s. nr. (4), 4 March 1627) and from *Wassenaer's Historisch Verhael* (part XIV. fol. 26) we know that in 1628 two fortresses with batteries were established to protect the station against the people from Dunkirk and the Basque country (Resolution of the States of Utrecht, 16 June 1628). Today, two of the guns, together with an anchor, are to be seen at the Norwegian weather station. The guns belonged to Jacob Meun, Frederik Gerritsz Wit, Jan Jansz Molenwaerff and Cornelis Claesz Veen de Jonghe, merchants from Enkhuizen (GAD. NA. 1649, fol. s nr. (4), 4 March 1627). Now they are part of the cultural heritage of Jan Mayen, as are the other objects found at Kvalrossbukta and kept at the meteorological station. In 1983 and 1988 more remains of Dutch whaling stations were found in the Maria Musbukta, the Kvalrossbukta and the Titeltbukta.

Despite the building of the fortresses, however, when in the late summer of 1632 two Basque whaling ships indeed landed on Jan Mayen, no one was there to defend the settlements against this raid. The Basques destroyed storehouses and installations and stole 600 casks of oil and 200,000 pounds of whale bone, causing a total damage of Dfl 100,000.

This action was the direct reason why the *Noordse Compagnie* chose to leave seven men behind at the island to protect the station in the winter of 1633-1634. It turned out not to be necessary because no Basques showed up that winter. However, the price was high because all the men died of hardship. They were found dead inside the houses when the ships returned at the beginning of June the following year. From their logbook much is known about the events during that winter. It seems that the weather was very bad and for that reason the men spent much time inside the houses, reading, writing and telling each other stories. They felt an earthquake, saw whales in the bay, killed only five polar bears and did not see any polar foxes during that winter. They ate the meat of the polar bears, became ill and died, probably of Trichinosis. The last page of the logbook was written at the end of April, approximately one month before the ships returned. The directors of the Noordse Compagnie decided not to let people winter on the island again.

5. PROFIT AND LOSS ACCOUNT

It is very hard to make a profit and loss account, because the figures are very incomplete. In the first exploitation period many good hauls were made around Jan Mayen. The logbook of Heertjen Jansz shows us that whaling certainly was profitable in the first years. Sometimes it was even exceptionally good, like the tremendous haul by Matthijs Jansz Hoepstock in the Hoepstock Bay. He caught 44 whales in 1619, producing 2300 casks of whale oil (GAR. NA.91, fol. 191, 13 September 1619 in Hacquebord 1998).

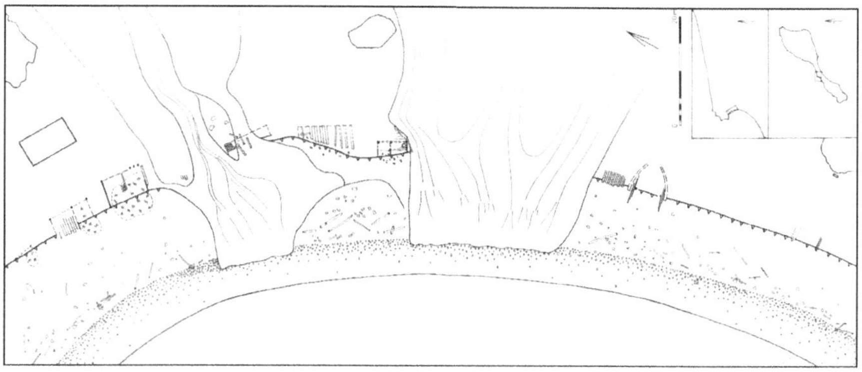

Figure 3. Cultural remains in the Northbay (Kvalrossbukta) on Jan Mayen in 1983. Arctic Centre University of Groningen, Groningen, The Netherlands.

In the second period, however, the yield was substantially lower. From the logbooks of Michiel Adriaensz de Ruyter (1633 and 1635) we know that the weather was very bad in this period, with much ice. Very often, when the whalers arrived at the island in June they could not land because of the ice. Then the ships kept sailing around the island, searching for a break in the ice. The catch was hardly profitable in those years. In 1632, owing to bad weather and ice, only eight whales were killed by nine company ships, giving a yield of Dfl. 23,360 (with a whale-oil price of Dfl. 60,00 per cask and a price for whalebone of Dfl. 8,00 per 100 pounds). The total costs of nine ships, including bounties for catching and trying was Dfl. 142,500, thus causing a loss of Dfl.119,140. Adding to this the loss of Dfl 100,000 which the company suffered because of the Basque raid on Jan Mayen in 1632, the total loss can be fixed at Dfl. 219,140 (Muller 1874, 263).

In 1633 the weather was bad again. Only 47 whales were caught by the whalers of the Dutch whaling company with a yield of Dfl. 146,640, minus the costs of eleven ships and the try works, Dfl. 175,000 in all, which makes a loss of Dfl 28,860. In 1635 42 whales were killed in the waters around Jan Mayen by ten ships, giving a yield of Dfl. 105,147. The total costs for ten ships including bounties were Dfl. 163,400, which meant a loss of Dfl. 58,253 that year. In the thirties the company lost Dfl 306,253 altogether in three years, the Basque raid included. Needless to say, the period of whaling at Jan Mayen was drawing to its close. Whaling around Jan Mayen did not pay anymore and the whalers moved their hunting grounds to the north into the bays of Spitsbergen.

Figure 4. Painting by Cornelis de Man, ca. 1639, Dutch whaling station in the Northbay of Jan Mayen. Rijksmuseum, Amsterdam, The Netherlands.

6. IMPACT OF WHALING

Based on incomplete historical source material, an estimation was able to be made of the minimum number of ships fitted out by the Dutch whaling companies (Fig. 5). Based on these figures and on some information on catches which has been preserved, it also turned out to be possible to make an estimate of the number of whales killed. For Jan Mayen alone this number can be estimated at 986. The figures show a downward trend in the whale catch and after 22 years whaling stopped around Jan Mayen.

The extermination rate here is comparable to the rates of extermination caused by the nineteenth century American northwest whaling trade in the Sea of Ochotsk and in the Chukchi and Beaufort Seas.The fact that the seas around Jan Mayen were dropping-areas will have been an important reason for this fast reduction of the Greenland right whale population (GAR NA. 101, fol. 77, 12 September 1619). As Weslavsky *et al.* (2000) have indicated this in its turn will probably have caused an increase in the populations of polar cod, seal and plankton-eating birds like the little auk (*Alle alle*). Many of the bird rookeries on the island probably date from the second half of the seventeenth century. If possible it would be most interesting to date the age of the rookeries on Jan Mayen.

During periods with ice-drifts around Jan Mayen many polar bears stayed on the island. In 1627 much ice was observed and 70 polar bears were killed during the whaling season. The fact that only 5 polar bears were killed during wintering on Jan Mayen (1633-1634) means that there were not always so many bears on the island.

7. CONCLUSION

Thanks to the restricting effect of the charter of the *Noordse Compagnie* whaling around Jan Mayen lasted for 22 years. At the end of the 1630s the whales were chased away from the waters around the island. The whalers moved to the edge of the pack-ice in the north, abandoning the whaling stations on the island. Based on written sources, approximately 1000 Greenland right whales (*Balaena mysticetus*) were killed and processed on Jan Mayen, leaving much uneaten plankton behind in the sea. Plankton-eating seabirds must have taken advantage of the extermination of the whales in the ecosystem, probably leading to an increase in the number of seabird rookeries on the island. Dating these bird rookeries might prove this assumption.

Written sources also show that during ice-drift periods many polar bears were on the island, even in the summer. Many of these bears were killed, indicating that the Jan Mayen whaling industry also had a substantial impact on the polar bear population in this region. It is an interesting fact that in the first half of the seventeenth century written sources did not mention any polar foxes on the island.

Finally, the combination of written sources and field data makes Jan Mayen a very interesting and challenging island for research.

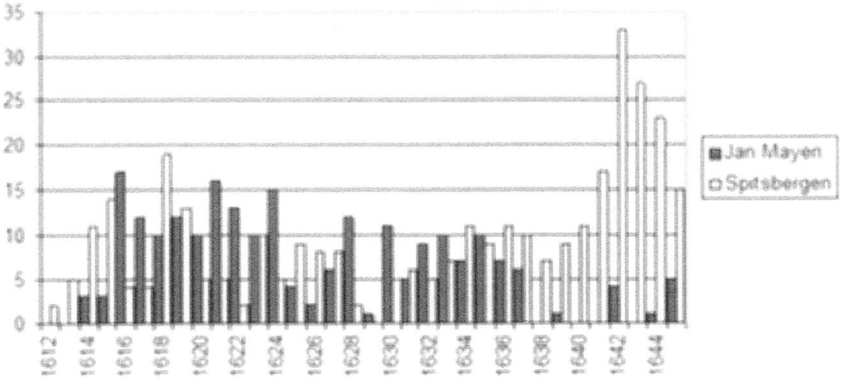

Figure 5. Number of ships to Jan Mayen 1612-1636. Arctic Centre, University of Groningen, The Netherlands.

REFERENCES

Hacquebord L 1984. Smeerenburg. Het verblijf van Nederlandse walvisvaarders op de westkust van Spitsbergen in de zeventiende eeuw (PhD-dissertation) Amsterdam/Groningen (in Dutch).
Hacquebord L 1998. Jan Mayen; its early history. *Circumpolar Journal* 13 (2), 44-69.
Honore Naber SP l', 1930. *Walvischvaarten, overwinteringen en jachtbedrijven in het Hooge Noorden 1633-1635*. Utrecht (in Dutch).
Muller SF 1874. *Geschiedenis der Noordsche Compagnie*. Utrecht (in Dutch).
Schilder G 1988. Spitsbergen in de spiegel van de kartografie. Een verkenning van de ontdekking en kartering. Pp 30-49 in L Hacquebord en WH Vroom (eds.). *Walvisvaart in de Gouden Eeuw. Opgravingen op Spitsbergen*. Amsterdam (in Dutch).
Weslawski JM, L Hacquebord, L Stempniewicz, and M Malinga 2000. Greenland whales and walruses in the Svalbard food web before and after exploitation. *Oceanologia* 42 (1), 37-56.

ARCHIVES VISITED IN THE NETHERLANDS

ARA: General State Archive, The Hague
GAR: Municipal Archive Rotterdam
GAD: Municipal Archive Delft
GAA: Municipal Archive Amsterdam
GAE: Municipal Archive Enkhuizen
RA Utrecht: State Archive Utrecht.

Chapter 22 NATO Advanced Research Workshop

Past and Present Infrastructure of Jan Mayen as Historical Documents

The Fate of Anthropogenic Remains

Susan Barr
Directorate for Cultural Heritage, Oslo, Norway
E-mail: susan.barr@ra.no

Abstract: Previous activity on Jan Mayen started with the 17th century whalers. Remains from this period and the 19th century scientific work are now only archaeologically interesting. The same applies to the buildings from the war period 1940-45. A few tiny trapping cabins and the post-war meteorological and navigation stations are the remaining standing buildings today.

Key words: History, archaeology, cultural heritage, buildings

1. INTRODUCTION

For such a small isolated island lying far out in the Norwegian Sea, Jan Mayen has seen a surprising amount of human activity throughout the past centuries. Possibly the history of human visits and activities there stretches back 1500 years, but then we are into the realm of speculations and interpretations of texts and ancient maps (Barr 2003). There are, however, various sources that tell us definitely of activity from the past 400 years, one such source being the actual monuments and sites that can be seen on the island today. With an eye to any future activities on the island, it may be interesting to note which buildings and structures are already to be found, as well as their current state and status.

S. Skreslet (ed.), Jan Mayen Island in Scientific Focus, 239-248

2. WHALING REMAINS

As described elsewhere in this publication, the early 17th century whaling activity was relatively huge and at its completion it left behind the remains of land stations in suitable bays around the coast of the island, particularly on the western side. Although these remains are protected by law from human interference, nature's ways are less easy to control. The lava basis of the island is unstable, to the extent that coastal erosion will soon have wiped out the last evidence of these stations. Luckily they have to some extent been documented by various means, including several archaeological excavations.

3. AN EARLY SCIENTIFIC BASE

During the First International Polar Year 1882-83 Jan Mayen was chosen as the base for a wintering group of 14 men from Austria-Hungary. The group spent 13 months carrying out mapping, observations and scientific work on the island according to the programme of the Polar Year, which included work particularly within meteorology, earth magnetism, aurora studies and astronomical phenomena, and hydrography. In addition the island was mapped in detail for the first time. The good results were obtained not least because the living and working conditions for the men were organised as optimally as it was possible to do within the limitations of geography and climate.

In its time the Austro-Hungarian station was roomy, comfortable and extremely practical for its purpose, which was quarters for both living and working for the 14 men under trying climatic and natural conditions. A great deal of thought and planning went into the construction of the prefabricated houses, and the expedition had no particular problems during the 13-month stay on the island. The houses had been built and tested before the expedition left home and painted in different colours inside and out to facilitate rebuilding them on Jan Mayen. They were well insulated and their form of an upturned boat made them impervious to the frequent storms and high winds during the year.

At the end of the expedition the station was left in excellent condition in order to serve any later expeditions. The buildings were later used sporadically by wintering trappers, but the harsh climate and lack of continual maintenance took their toll. The remains were more or less ravaged during the last war, when material from the ruined buildings was used to build anti-aircraft positions (Barr 1985, 2003).

4. WINTERING TRAPPERS AND THEIR CABINS

Through the first half of the 20th century the island's foxes were exploited by wintering trappers. The first expedition to attempt this spent the winter 1906-07 in the only building on the island, the Austrian Polar Year station. The next group in 1908-09 also lived here, but in addition built other small huts as extra shelter during trapping excursions on the island. One of these was a triangle-formed construction of two rows of short logs leaning against each other. This small tent-like structure beside Nordlaguna is the oldest building still standing (although only just) on the island (Barr 1985, 2003).

Figure 1. A trapper's cabin standing on the bar, *Bommen*, between Nordlaguna (left) and the sea. Built in 1926, the hut was destroyed in a storm in 1953. The wartime HF/DF house stood near to this hut. Photo from Barr (1985).

Figure 2. One of the small trappers' huts that still stand on Jan Mayen. *Camp Helene* was built in 1925 and measures inside 1.5 x 2m floor area and 1.35-1.58m in height. Photo: S Barr.

Some of the trappers' huts no longer exist (Fig. 1). Altogether there are now four huts still standing on Jan Mayen, including the tent-like structure. The three others are constructed of rough boards and logs (Fig. 2) that have been found amongst the huge piles of driftwood strewn in some of the bays (Barr 1985). On average the huts measure around 1.5 x 1.5 x 1.5 m, and attempts are made to keep them in reasonable repair. Apart from their status as cultural remains worthy of protection, they are also occasionally used as small shelters in connection with recreational foot tours by the current crews on the island.

5. THE PRE-WAR METEOROLOGICAL STATION

In 1921 the Norwegian state supported the establishment of a meteorological station on the eastern side of the island, just east of Eggøya. As an isolated landmass in the middle of the Norwegian Sea, in an area which is critical for weather forecasting in northern Europe, there were several countries that supported the idea of such a station. The radio and meteorological station was manned by three-four men for a year at a time, and came to consist of a large main building with an adjacent shed for the sheep and pigs that provided fresh meat throughout the year, a reserve station in case of fire or other damage, and various small outbuildings. High antenna masts which assured radio contact with the outside world completed the picture (Barr 1985).

Norway was occupied by German forces from 9th April 1940, but the station on Jan Mayen continued to function until September, when it was evacuated by the Norwegian Coast Guard ship "Fridtjof Nansen" and the buildings destroyed. When the island was reoccupied by Norwegian forces in March 1941, the original meteorological station, *Eldste Metten*, existed only as scattered remains. Its position was in any case too open for attack from sea or air forces, and only a small guard hut was erected in the same area to watch for approaching ships or planes (Barr 2003).

6. WAR-TIME CONSTRUCTION WORK

There was some considerable construction work during the war to support the small group of meteorological personell and the garrison that protected them. At the most there were 52 soldiers at one time on the island, and their main station was built in a small side valley to the broad valley Jøssingdalen that crosses the island from Sørlaguna to Nordlaguna. The four meteorological personnel occupied two smaller stations in Jøssingdalen, and in May 1944 a radio-sonde station was built nearby. In addition to this an emergency reserve cabin and various anti-aircraft and other positions were built, covering the middle part of the island. It was not possible to protect the north and south ends permanently, only by occasional patrols (Barr 2003).

In November 1943 an American high-frequency direction-finder (HF/DF) station with a personnel of eleven men was established beside Nordlaguna. Its purpose was to discover any German weather-reporting stations operating in the neighbourhood of Jan Mayen-East Greenland. The two-storey radio building stood on the bar between Nordlaguna and the sea (*Bommen*), while Nissen (Quonset) huts under the shelter of the slope to the north east of the lagune comprised the living quarters. These buildings, known as Atlantic City, were taken into use as the first Norwegian civilian meteorological station at the end of the war, while a completely new station was being built on the top of the adjacent slope (Barr 1985).

Of the buildings dating from this period only foundations and scattered remains are left, together with a couple of huts in an advanced state of disrepair. The lack of whole, standing buildings is partly because the Nissen hut type of construction did not withstand the weather once the buildings had passed out of use, and partly because war remains were not regarded as cultural heritage worth taking care of until relatively recently. The American two-storey HF/DF station house was destroyed by the sea in 1953.

7. THE POST-WAR METEOROLOGICAL STATION

The buildings that actually do stand today and could be considered for future use are all from the post-war period. In 1949 the meteorological crew moved into a newly-built station on the hill-top plain east of Nordlaguna. Known today as *Gamle Metten* (Fig. 3), the station consisted of, at the most, nine buildings, including the large main station containing the living quarters and radio room. A emergency reserve station, piggery, storehouse, and radio-sonde house were amongst the other buildings, and the whole complex was in full use with a crew of eight-ten until 1962 (Barr 1985, 2003).

Although there can be exremely windy conditions around the station, the setting is beautiful, with Beerenberg rearing majestically to the north east, the sea and Nordlaguna below and a view southwards along the west coast.

When the newest meteorological station was opened in 1962, *Gamle Metten* was abandoned more or less as it stood, with equipment and inventory almost intact. Unfortunately a considerable amount of stealing and damage was afflicted on the buildings, mostly by crews of occasional fishing vessels who took themselves ashore by Nordlaguna. Until the early 1980s the heritage value of the buildings was not appreciated. The newest building of the complex, *Fryden* built in 1957, was, however, kept in a good state of repair and has been in continual use as a recreational cabin for weekends away from the main station. The remaining buildings are in various states of repair.

In the 1980s repair and renovation work was done on the main buildings under the auspices of the cultural heritage authority, who clearly saw the need to preserve the remains of the station as an historical monument. In the past few years a repair-and-maintenance programme has been running which has laid a new roof and secured the windows of the main building, as well as clearing and airing the interior Although supported by the Meteorological Institute, the Military Logistics Division and the Directorate for Cultural Heritage, much of the work is done on a voluntary basis by the association *Ishavsforeningen Jan Mayen* (Arctic Society Jan Mayen) which is composed of present and former crew members from the current stations on the island. There are now three or four buildings on the site that can be adapted for future use.

Figure 3. The meteorological station 1949-62, *Gamle Metten*. The post in the foreground marks the spot where a crew member died after being blown off course by a winter storm. Photo: S Barr.

8. TWO MODERN STATIONS

In the period 1958-60 today's Loran navigation station was constructed on the east side of the island south of Sørlaguna. The station has an integrated E-shaped building containing the living quarters, recreational rooms, generators, communication room and the Loran operation room. A number of other buildings within the complex called *Olonkinbyen* (Olonkin City) cover various other functions such as carpentry shop, boat-building hall, gym, transport garages, and so on. Gennadij Olonkin was a Norwegian-Russian who joined Roald Amundsen's "Maud" expedition in 1918, 19 years old. He was radio-operator on the airship "Norge" on its flight from Rome to Svalbard in 1926, and he was manager of the meteorological station on Jan Mayen four separate years between 1928 and 1936 (Barr 2003).

Up to the 1990s a number of small cabins from the construction period were gathered around the station area. All except one of these have now been removed. The last one, known as *Kreml* (Kremlin), has been declared worthy of preservation and has recently been painted and had the roof repaired by the Arctic Society Jan Mayen, with support from the Directorate for Cultural Heritage.

The establishment of the Loran station at *Olonkinbyen* made it impractical to keep the meteorological station on the other side of the island. A new meteorological station was therefore built in 1962 a short distance to the east of *Olonkinbyen*. The two were not located together as the meteorological station needed more room around it for sending up radio-sonde balloons. Although the four-man crew lives at *Olonkinbyen*, the meteorological station is also equipped for living in, not least for periods when the weather makes it difficult to drive to and from *Olonkinbyen* (Barr 2003).

9. CULTURAL HERITAGE STATUS

The current cultural heritage regulations date on the whole from 1974, although some changes concerning administrative connections have been made since. According to these regulations all fixed and loose cultural remains pre-dating 1900 are automatically protected. This means that only the whaling remains and the Austro-Hungarian Polar Year station remains are protected by law. A new updated law is, however, planned which will bring the heritage protection regulations for Jan Mayen on a line with those for Svalbard. This will most probably mean that all structures and objects pre-dating 1946 will be automatically protected, thus including the hunting huts and war remains. The meteorological station 1949-62, *Gamle Metten*, will not be automatically protected, but may be declared worthy of protection anyway, allowing use within reasonable limits for preserving the historical qualities it still has (Anon. 2000).

If plans for the near future should include radical changes for the current building mass in and near *Olonkinbyen*, the cultural heritage authority would wish to be included in the discussion. This is after all an important Norwegian monument of the Cold War era, as well as a highly successful example of a fully-equipped polar station, unique in Norway for its dimensions.

As in most other matters, cost is also a parameter in cultural heritage work, and the variables Preservation, Cost, New Use, and Removal will have to be carefully considered.

REFERENCES

Anon. 2000. Miljøhandlingsplan for Jan Mayen år 2000 til 2005. Fylkesmannen i Nordland Rapport 2/2000, Bodø. 44 pp (in Norwegian).

Barr S 1985. Kulturminner på Jan Mayen / Historical remains on Jan Mayen. *Norsk Polarinstitutt Meddelelser* 108, 1- 67 (in Norwegian and English).

Barr S 2003. *Jan Mayen, Norges utpost i vest. Øyas historie gjennom 1500 år* (Jan Mayen, Norway's outpost to the west. The island's history through 1500 years). With a contribution by Louwrens Hacquebord. 2nd edition. Kolofon a/s, Oslo. 266 pp (in Norwegian).

Chapter 23 NATO Advanced Research Workshop

Research Operations in Remote Arctic Islands

The Model of the Polish Polar Station Hornsund in Spitsbergen

Piotr Głowacki
Polar and Marine Research Department, Institute of Geophysics, Warszawa, Poland.
E-mail: glowacki@igf.edu.pl

Abstract: Conducting research in the remote and inaccessible parts of the Globe often demands specific logistical actions. Long-term research programmes are thus usually carried out by permanent research teams working in particular stations. The working conditions under such circumstances frequently fall short of the standards set for the average research centers and laboratories. Still, the quality of the data and information to be produced has to meet the established quality norm. The Polish Polar Station Hornsund, that has been working in Spitsbergen since 1957 can serve as a model of the research and logistical solutions to be adopted for other stations in the Arctic as well as in other remote parts of the world.

Key words: Svalbard, Spitsbergen, Hornsund, polar station, Polish research, interdisciplinary science

1. INTRODUCTION

Although Poland is situated far from the Arctic and there are no Polish islands in polar regions, there is a long-established tradition of Polish scientists exploring the area. The activities date back to the year 1883 when Polish political prisoners of Siberia started to do research in the Arctic. Shortly afterwards Polish scientists participated in the Swedish–Russian Expedition to Spitsbergen (1891–1901).

Independent Polish expeditions to the Arctic started in 1932 and included exploring Bear Island (1932/1933), Spitsbergen (1934-1938) and Greenland (1937).

S. Skreslet (ed.), Jan Mayen Island in Scientific Focus, 249-259

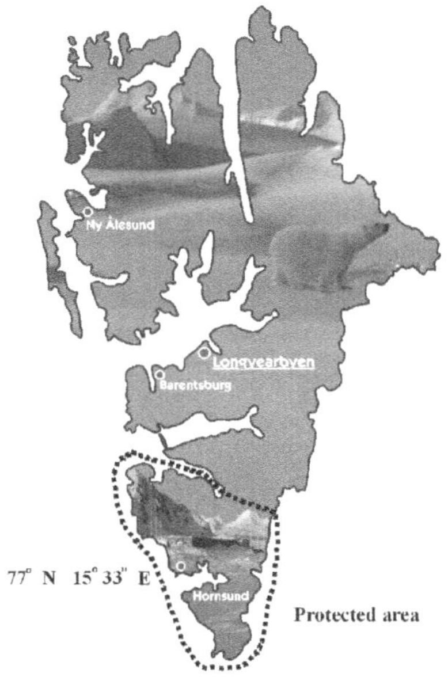

Figure 1. Location of the permanent Polish Polar Station Hornsund that is situated within the South Spitsbergen National Park (stippled area).

The independent Polish expeditions to the Arctic started in 1932 and included exploring Bear Island (1932/1933), Spitsbergen (1934-1938) and Greenland (1937).

In 1957 the Polish Polar Station Hornsund was established in the southern part of Spitsbergen. It has been in continuous work since 1978. As a site situated in South Spitsbergen National Park and therefore difficult to get to, it can be treated as an island on the island (Fig. 1).

2. POLISH RESEARCH IN HORNSUND

Building up continually working stations on a remote Arctic island is sensible as long as the following conditions are met:

- Pluridisciplinary research operations to ensure rational use of infrastructure.
- Observation programs are made.
- Programs are continual and long-term.
- Station is as a part of the international network and databases.
- Automatic stations and container field camps are fit enough for carrying out short-lasting experiments and measurements.

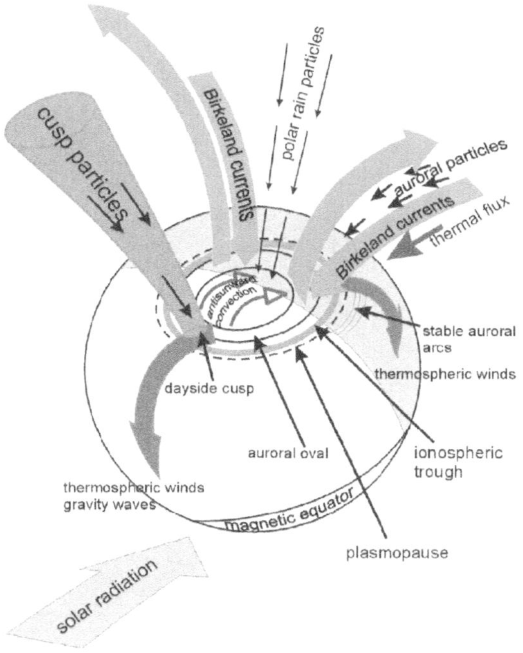

Figure 2. Solar-terrestrial relationships in the polar region (after Fremouw and Wittwer 1984).

To defend the costs of doing interdisciplinary research in the Arctic a particular area has to be interesting in a larger context. The Atlantic sector of the Arctic satisfies this condition in several ways:

- Being a region of the „polar cusp" effect (Fig. 2).
- Active oceanic rifts and changing structure of the lithosphere (Fig. 3).
- Occurrence of the permafrost phenomena (Fig. 4).
- Annual fluctuations in the strength of oceanic currents (Fig. 5).

These conditions have been addressed in the scientific program of the Polish Polar Station Hornsund in Spitsbergen. The spatial range of the observations made comprises 1000km above the ground and 10km into the Earth's crust. The main topic of the Polish research is „Continuous observations of geophysical fields at high geomagnetic and geographical latitudes". The Polish Polar Station Hornsund is situated in the Spitsbergen Archipelago, at the junction of the Asiatic and American Arctic. This location is very suitable for geophysical investigations and it is especially ideal for the study of lithospheric and atmospheric structures and processes as well as extraterrestrial space.

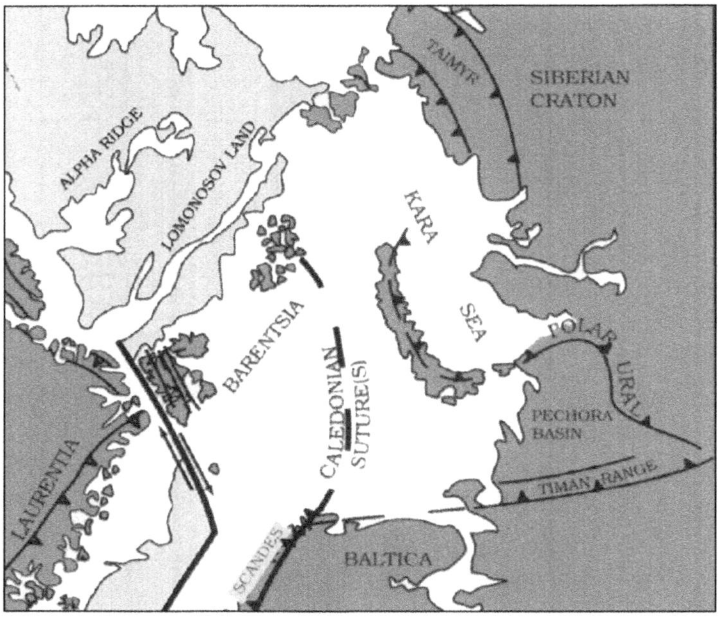

Figure 3. High Arctic terrains – early Tertiary relationships (Gee and Zeyen 1996).

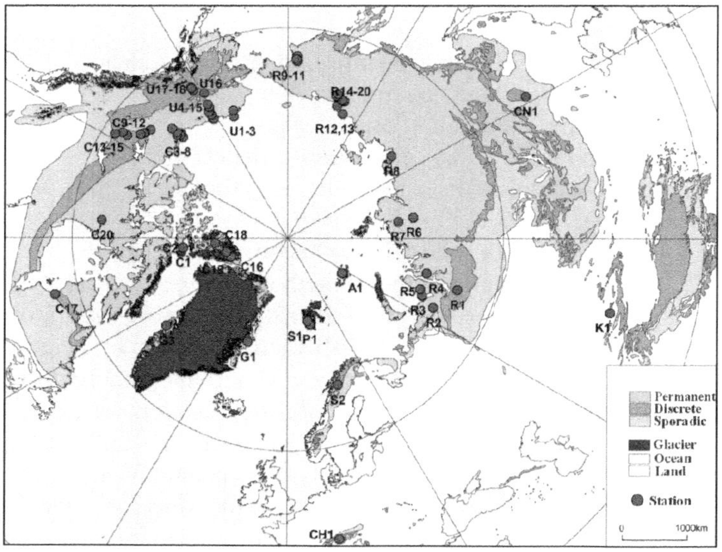

Figure 4. Stations within the permafrost extension of the Northern Hemisphere and CALM (Circumpolar Active Layer Monitoring) programs (after Brown *et al.* 1997).

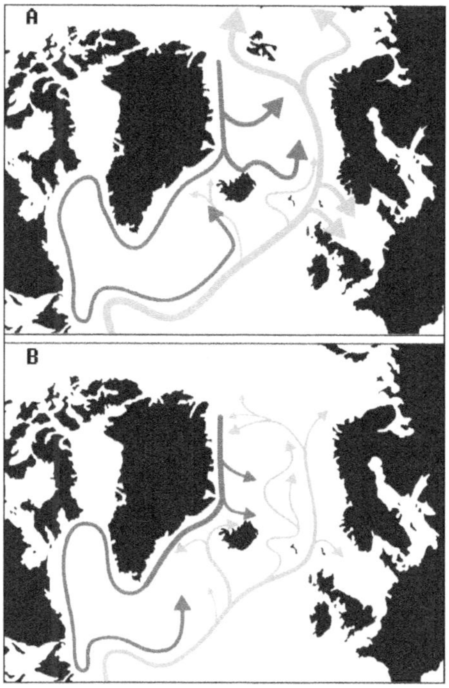

Figure 5. Circulation of Atlantic and Arctic water (source: Blindheim *et al.*, 2000). A: High North Atlantic Oscillation index. B: Low North Atlantic Oscillation index. Light grey: Atlantic water. Dark grey: Arctic water.

Glaciers near the Station are excellent for the study of interglacial physical processes. Seismological recording is the basis for the study of Arctic seismicity. In particular, it provides unique data on seismic events in South Spitsbergen and glacial seismic events. The location of Spitsbergen is optimal for studying physical processes of the Aurora Borealis and the polar cusp. Measurements of geomagnetic field components together with those of atmospheric electricity elements, ionospheric absorption, and auroral observations provide information on the processes that occur in the magnetosphere and ionosphere under the influence of solar wind. Another important problem studied there is the determination of factors that affect the solar radiation inflow to the Earth's surface in the Arctic. Many programs performed at the Station concern the physical parameters investigated in the framework of the International Program *Global Change*.

The current research comprises:

- Measurement of the structure of the Earth's crust by seismic methods.
- Registration of the changes in geophysical fields.
- Terrestrial cryosphere:
    - Permafrost.
        - Active layer thickness.
        - Physical and chemical processes.
    - Glaciology.
        - Glacier geometry changes.
        - Glacier dynamics (tidewater glaciers).
        - Hydro-thermal structure of glaciers.
    - Snow cover on glaciers and land.
        - Snow chemistry.
        - Snow thickness distribution.
- Climate changes and related processes.
- Marine and terrestrial ecology.

In 1980 the wooden Station Hornsund from 1957 was expanded by a separate part for summer expeditions. At present it is being modernised to meet the requirements of the 4th International Polar Year 2007/2008 (Figs 6 and 7).

3. MANPOWER

The team working in the station changes once a year at the beginning of the polar summer. On the whole in the summertime the team of technicians does all the needed repairs concerning the living space and the equipment. There are also Polish and foreign scientific groups that undertake specific research programmes in spring and summer.

Figure 6. Polish Polar Station Hornsund in the summer time.

Figure 7. Polish Polar Station Hornsund in the winter time.

From experience, the winter team has to comprise at least 7-8 people:

- Meteorologists (2 people when a synoptic station transmits data to the WMO network every 3 hours),
- Geophysicists (2 people when the programme comprises ionospheric to lithospheric investigations),
- Mechanical engineer (looking after the technical equipment),
- Electrical engineer (responsible for the power station and the feeding system),
- Electronics engineer (responsible for the communication and data transmition),
- Environmental observer (responsible for measurements and samples the terrestrial and aquatic environment)

There is no cook employed during the polar winter, so members of a team are on cooking duty in turns. On the other hand, during the short summer, when temporary experimental teams work at the station, a professional cook is indispensable.

4. TECHNICAL FACILITIES

Our experience suggests that a separation of the summer and winter teams is neither good nor actually possible. Moreover, one can neither switch off nor control the heating of the station's summer department during the polar night.

From 2004 the living facilities of the Polish Polar Station will be:

- Single bedrooms – 12
- Bigger bedrooms – 7 (30 places for seasonal groups)
- Science laboratories – 8

- Kitchen
- Dining room
- Drawing-room
- Bathrooms – 3
- W.C. – 4

Working and transportation facilities are:

- Power station (2 generators- each 70 kW)
- Garage and workshop
- Snow mobiles – 5
- Open boats – 2
- Zodiacs – 2
- Amphibians – 2
- Environmental pavilion – 1
- Geophysical pavilions – 2
- Boat house

All the work done in the area of the station requires the presence of people who secure others against the polar bears. The members of the team are rarely experienced hunters. Thus, a variety of weapons and deterrents are needed.

Transport of people, equipment and materials is the biggest problem. Vessels can anchor only far from the shore. Therefore, all transportation to and from the shore requires boats or pontoons. Furthermore, because of the lack of cranes, transportation of the big and heavy units demands prior preparation as it must be done by simple means and the help of all members of the team and the ship's crew. Amphibian vehicles that may traverse the shore even in the presence of floating growlers, are necessary for the handling of big fuel barrels.

The power supply of the Station has to have at least two working diesel aggregates that work alternatively to make possible periodic inspections without turning off the feeder. A separate building is needed for the aggregates to avoid noise and risks of fire.

The heat from the cooling system of the engine is used to heat the building of the power station, the workshop and the garage. The recuperation of the heat from the exhaust gas is used to melt the ice and the snow needed for the water supplies. Because of the permafrost the water from the nearby shallow lakes can be used only during the short polar summer.

The special sewage treatment plant has two reactors. One works during the long polar winter for a small number of people. Two others plants work during the time of the seasonal expeditions when the bathrooms and laboratories are far more used. Because of frost and snow cover the purified water cannot freely flow out on the ground and has to be pumped up under pressure.

Switching the generators and the problems with ground protective in permafrost demand feeding the instrumentation and measurement equipment from the supply warranted good quality. All the measurement equipment as well as all of the sensing devices have to be plugged into one main earthed conductor.

5. COMMUNICATION

Two satellite systems are used for communication (INMARSAT and IRIDIUM) while for safety a short - wave transmitter, which is occasionally used for getting in touch with other scientific stations and ships, is recommended.

Bearing in mind safety, the station should be within the range for the helicopter flights. Moreover, moving around in the area demands different means of transport in different seasons.

As research stations placed on the coast are a great attraction for cruise ships and sailing boats, it would be recommended to diminish or stream the tourist traffic away from scientific facilities where the presence of large groups of people interfere with the research and observations.

6. CONCLUSIONS

The presented problems and conditions experienced at the Polish Polar Station Hornsund could be applied to a scientific station located on Jan Mayen Island.

The geographical location of Jan Mayen challenges efforts to run a scientific research facility on a regular basis and at a steady state. According to experience from Hornsund, it is necessary to establish logistics that serve pluridisciplinary joint ventures that exploits the same general facilities, which ensures an optimal use of invested resources. Even better are truly interdisciplinary geophysical and environmental research programs that favour close scientific cooperation among the scientists A continuous monitoring program serves such goals. At Hornsund long-term geophysical observations of even simple measurements of pH in precipitation, air temperature and the position of glacier fronts (Fig. 8), provide information on environmental variability that is of value to a range of scientific disciplines.

The volcano Beerenberg that offers a change in altitude from the sea level to a height of 2277m seems to be ideal for studies of climate variability and air pollution. Sampling stations located at different levels in the height profile could split the elements of sea aerosols into long and short range transport of pollutants to and within the Atlantic sector of the Arctic. The exploitation of this opportunity on Jan Mayen could complete the network of transects in the European Arctic, including Northern Scandinavia, Bear Island and Spitsbergen.

Figure 8. Retreat of the Hans Glacier in Hornsund, Spitsbergen, from 1957 to 2003.

REFERENCES

Birkenmajer K, A Gaździcki, P Głowacki, M Górski, M Grad, A Guterch, J Jania, K Jażdżewski, K Kwarecki, R Ligowski, J Machowski, K Pękala, S Rakusa-Suszczewski, J Siciński, M Węsławski and M Zalewski 2002. Arctic and Antarctic Research Programme of Poland 2002-2010. Special issue (A Guterch ed.), Committee on Polar Research PAS, Warsaw 2002, 95 pp.

Blindheim J, V Borokov, B Hansen, SA Malmberg, WR Turrell and S Østerhus 2000, Upper layer cooling and freshering in the Norwegian Sea in relation to atmospheric forcing. *Deep-Sea Research* I, 47, 655-680

Brown J, OJJ Ferrians, JA Heginbottom and ES Melnikov 1997. International Permafrost Association Circum - Arctic Map of Permafrost and Ground Ice Conditions, *U.S. Geological Survey,* scale 1 : 10 000 000.

Fremouw EJ and LA Wittwer 1984. The Hilat satellite program: Introduction and objectives. 5 (2), 1-98 pp.

Głowacki P and WE Krawczyk 2002. Long range transport of pollutants – evidence from rainfall chemistry in Hornsund (Svalbard). Norsk Polarinstitutt Internrapport 10, 65-68.

Gee DG and HJ Zeyen (eds.)1996. EUROPROBE 1996 – Lithosphere Dynamics: Origin and Evolution of Continents. Published by the EUROPROBE Secretariate, Uppsala University, 138 pp.

Jania J and JO Hagen 1996, Mass balance of Arctic glaciers, *IASC Report* 5, 1-62.

Zalewski SM 1997. 40 Years of the Polish Polar Station Hornsund Spitsbergen. Pp 365-367 in Głowacki P (ed.) *Polish Polar Studies.* Institute of Geophysics of the Polish Academy of Sciences, Warszawa 1997.

Chapter 24 NATO Advanced Research Workshop

Jan Mayen Island and its Waters as Potential Targets for European Arctic Research

Graham Shimmield
Dunstaffnage Marine Research Laboratory, Oban, Scotland.
E-mail: graham.shimmield@sams.ac.uk

Abstract: Jan Mayen Island has featured in European scientific research since the first International Polar Year. Currently, Europe is funding a considerable number of national and international research programmes in the Arctic. Research on Jan Mayen Island and in surrounding waters, has the potential to contribute to long term monitoring and assessment of environmental change, as well as focussing on specific process studies. The current state of international and European funding is reviewed, together with some suggestions on the way forward for the island as an environmental observatory.

Key words: research, international programmes, Europe

1. INTRODUCTION

This short paper describes some of the current and proposed research programmes in the European Arctic. In this context, Jan Mayen Island has not featured significantly to date. However the island has been the subject of dedicated research campaigns predominantly in the International Geophysical Year (1957) and previous International Polar Years. Within this document, I review the international context of Arctic research looking closely at the co-ordination of research infrastructure which is crucial to delivering field work in the high north. By virtue of the island nature of Jan Mayen, I look more closely at some of the marine research programmes being conducted under European auspices. Turning to the future, for the

S. Skreslet (ed.), Jan Mayen Island in Scientific Focus, 261-267

International Polar Year 2007 potential objectives are identified, together with some closer examination of the way Jan Mayen Island may fit to these. Finally, and in conclusion, I comment on the next steps that could be taken to create a major environmental observatory on Jan Mayen Island for the future.

2. THE INTERNATIONAL CONTEXT

Research programmes in the Arctic are well co-ordinated through a variety of international programmes. In essence, these fall into three categories. Firstly, those dedicated to science programmes operated by individual nations or through collective funding as provided under the European Research Framework Programmes (currently Framework Programme 6). The second category involves research programmes dedicated towards monitoring and evaluation of environmental change and impacts on indigenous populations. The final category concerns research co-ordination for the use of scientific infrastructure including research bases, vessels and other environmental observing systems. Each of these are looked at in turn.

Scientific programmes fall under the auspices of the International Arctic Sciences Committee (IASC). The IASC reviews a wide range of research programmes and some of these have specific areas of interest such as the Arctic Ocean Sciences Board (www.aosb.org). Many programmes are reviewed by bodies such as AOSB and a short identification of some important ones are shown in Table 1.

Of course there are many international level programmes that have interest in Arctic science, such as the family of projects under the International Geosphere-Biosphere Programme (IGBP), or the World Meteorological Organisation (WMO).

Table 1. Selected science programs reviewed by AOSB (Arctic Ocean Sciences Board).

Acronym	Name of programme
JWACS	Joint Western Arctic Climate Study
CASES	Canadian Arctic Shelf Exchange Study
NABOS	Nansen-Amundsen Basins Observing System
SBE	International Shelf-Basin Exchange Working Group
IAPP	International Arctic Polynya Project
IAC	Arctic Coastal Dynamics
APARD/SIRRO	Arctic Paleo-River Discharge/Siberian River Run-Off

To examine the outcomes of monitoring and evaluation, several programmes collate the available scientific data. Principal among these are the Arctic Monitoring and Assessment Programme (AMAP) and the Arctic Climate Impact Assessment (ACIA). More recently some specific programmes such as the Arctic Climate Systems Study and Climate Cryosphere Programme (ACSYS/CliC) and the Study of Environmental Arctic Change (SEARCH) have now been established. An important characteristic of these international programmes for monitoring and assessment is their excellent use of the internet and up-to-date web-page and search engines to provide researchers with the latest environmental data.

The final category within the international context concerns research infrastructure co-ordination. In remote areas, such as the high Arctic or isolated islands within the Arctic, providing secure infrastructure co-ordination for scientific research is a primary objective. An outstanding success within the locus of European Arctic activity is the research village at Ny Ålesund in north west Svalbard. Co-ordination of these infrastructures is achieved through the Forum of Arctic Research Operators (FARO) and recent European co-ordination under the Human Potential and Research Infrastructures Programme such as ENVINET (European Network for Arctic-Alpine Environmental Research). The network provided an opportunity to link European high latitude and high altitude research stations across Europe (Fig. 1).

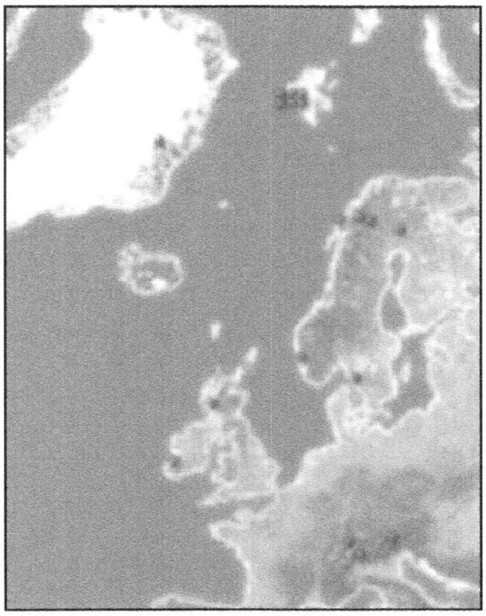

Figure 1. Location of ENVINET stations representing Arctic and high altitude observatories in Europe.

ENVINET succeeded in identifying three key areas to study in an integrated fashion. These were climate change, the impact of ultraviolet radiation and the long range transport of pollutants. The outcomes of ENVINET may be found on their web-page (envinet@npolar.no). More recently the Circum-Arctic Environmental Observatory Network emanating from North America (CEON) has started to gather pace. New opportunities are on the horizon concerning under sea telecommunications and observatories in a programme known as PLUTO. Logistic co-ordination is provided by the Arctic Logistics Information Access System (ALIAS).

3. EUROPEAN RESEARCH PROGRAMMES

As indicated in the introduction, Europe has funded research in the Arctic for over 100 years. Examination of the Commission's research site, CORDIS, indicates 263 listed programmes with 'Arctic' in the abstract. Clearly, reviewing these programmes is outside the scope of this brief report. Nevertheless, there are some current projects of high relevance to the development of Jan Mayen Island as an environmental observatory. These have been developed under Framework Programme 5 and some continue into the 6th Framework.

The European Science Foundation's Polar Board provides a high level strategic overview. Further details on the ESF Polar Board's activities and mission may be found on their website (www.esf.org). In 1998, the marine science and technology societies within Europe created a federation known as the European Federation of Marine Science and Technology Societies (EFMS; www.efms.org). The EFMS society brings together eight nations to determine matters of strategic importance to their marine member societies. EFMS represent academics and interested members of the public and provides an opportunity for dialogue on policy issues in a wider context. To date, the focus has been on the Mediterranean with the 2002 General Assembly taking place in Athens. Future areas of interest include the Baltic and possibly the Arctic.

At a national level there are a number of research programmes. At the Scottish Association for Marine Science (www.sams.ac.uk), we co-ordinate a number of European programmes in the Arctic as well as a major strategic programme funded by the UK's Natural Environment Research Council studying the northern seas from a standpoint of environmental change induced by both natural and man-made perturbation.

The European Science Foundation has published its EuroPolar Strategy which aims to introduce a Pan-European framework for operational capability in Polar Regions. This is linked to the 5 key thematic priorities in Framework Programme 6. These are:

- Global change in ecosystems.
- Impacts and mechanisms of greenhouse gas emissions on ozone depletion and carbon sinks.
- Operational forecasting and modelling – including Global climate change and observational systems.
- Biodiversity and ecosystems.
- Mechanisms of natural disasters.

Recently, the European Science Foundation has also examined its science policy in polar regions in respect of the Global Environmental Security programme. This was reported recently in the ESF Communi-cations Newsletter of the Summer of 2003, No 45.

4. RESEARCH INFRASTRUCTURES

Conducting scientific research in the Arctic requires the safe, efficient and economical deployment of resources to meet the scientific objectives of the research programmes. It has long been recognised that co-ordination of available facilities using a common approach may achieve significant results at lower cost. One such approach has been the establishment of the ENVINET Programme and particularly the Station Managers Forum for exchange of best practice. A significant question occurring within integrated research infrastructures is the degree to which individual process studies should be supported, in comparison to longer term monitoring and evaluation programmes. The latter carries significant responsibility for ensuring continued observation of the environment. Recently advances in technology have established the ability for scientists to work remotely from instruments within the field. A partnership with the robot laboratory at the University of Genova in Italy with the Scottish Association for Marine Science has demonstrated the capacity for underwater remote observation vehicles to be controlled via an internet link by scientists situated in a separate laboratory. In this case, studies of the underwater environment of the Kongsfjord in Svalbard were conducted by scientists sitting at the Dunstaffnage Marine Laboratory in Scotland. Other advances in marine technology look towards integrating seabed or moored instruments in the ocean, communicating to buoys on the sea surface, communicating to satellites and sending signals to appropriate ground stations with near-real time links to websites and other portals for information exchange. Such systems in the marine environment are still comparatively rare in comparison to meteorological stations on land, nevertheless there exists significant opportunities for utilising new technology in the Arctic environment. Existing facilities on Jan Mayen Island concerned with meteorological observations and navigation offer significant opportunity for controlling marine sensors in the close vicinity (within a 20 km radius of the island).

5. INTERNATIONAL POLAR YEAR 2007-08

2007 represents the 125th anniversary of the first International Polar Year which occurred in 1882. At the present time, various nations are co-ordinating their activity with a view to presenting plans for the 4th International Polar Year. There have been a number of suggestions to date, but in the context of Jan Mayen Island, there are discussions on the role that the Arctic Ocean has played in modulating Global climate especially in regard of the fluxes of heat, fresh water and carbon. Given the historical significance of Jan Mayen Island in former International Polar Years, it may be appropriate to consider a special task group for working up future plans and to report this to the co-ordinating committee for IPY4.

6. POTENTIAL OBJECTIVES AND STEPS FOR THE FUTURE

The following opportunities present themselves for studies on Jan Mayen Island relating to current European Programmes. These suggestions are very much my own and should not be considered as official recommendations:

1. To link and develop an observatory with common methodologies and protocols to other existing observing stations in the Arctic region (compared with the ENVINET Network). The emphasis for measurements should be on a continuous or semi-continuous nature using remote monitoring, including ocean moorings were appropriate. Data from the observatory should be made available through a common access interface such as a web-site.
2. To develop a programme of examining the impact of climate change at the regional scale.
3. To examine the time dependent history and magnitude of long-range transport of marine and atmospheric pollutants to the Arctic. (Jan Mayen Island lies on the return limb of both air mass and water masses exiting the Arctic region within the European sector).

4. To study the biodiversity, phylogeography and concept of refugia within the context of island communities evolving since the last ice age.
5. To study geological hazards at higher latitudes or altitudes using the opportunity provided by Mount Beerenberg.
6. To establish and calibrate proxies for palaeoenvironmental reconstruction from ice, soil, tephra and biogenic remains occurring on the Island.
7. To establish a protected area status for important and sensitive regions on the Island.

To take these objectives forward, it may be important to establish a task team for drafting scientific opportunities, threats and costs. In parallel, a liaison team should continue to outline the scientific objectives for Jan Mayen Island observatory to the International Arctic Science Committee, the Arctic Monitoring and Assessment Programme, the European Science Foundation Polar Board and the European Commission Directorate for Research. An additional activity may the creation of a focus group concentrating on objectives for the International Polar Year 2007. Finally, to establish the real long-term potential and to bring Jan Mayen Island into focus with the general public it will be important to set up a programme of public activity and presentations, potentially even fund-raising for some of the new research projects.

7. CONCLUSION

In summary, the Arctic continues, and is likely to remain, a major objective for research funded by the European Community. The geographic position of Jan Mayen Island and its relative isolation places it in an ideal situation to establish an environmental observatory in an extremely sensitive region of the northern hemisphere. Such an observatory can benefit from the latest technology available at relatively low cost. However, to establish the long term viability of a research programme it is necessary to recognise the current state-of-the-art in scientific objectives, and to ensure that the investment made is well-embedded in international scientific research objectives.

Chapter 25 NATO Advanced Research Workshop

The Scope for Norwegian Commitments Related to International Research on Jan Mayen Island

Odd G Skagestad
Norwegian Ministry of Foreign Affairs, Oslo, Norway
E-mail: ogs@mfa.no

Abstract: The main considerations that will govern the scope for Norwegian commitments related to international research operations on Jan Mayen Island are: 1) the physical characteristics of Jan Mayen Island, 2) its legal status in terms of international law as well as national legislation, 3) infrastructure and logistics, and 4) national policies concerning research or related activities. Any Norwegian support to proposed international research activities will require that the operation would be compatible with existing or prospective Norwegian national policies or strategies concerning research or related activities, which could entail the possibility of cooperation with key Norwegian research institutions.

Key words: Norwegian sovereignty, legal obstacles, access regime, environmental protection, regulations, Action Plan

1. INTRODUCTION

As a point of departure in describing the scope for Norwegian commitments related to international research operations on Jan Mayen Island, one will note that there are very few, if any, pre-ordained restrictions limiting the scope of operations that can be contemplated, subject to such variables as one's imagination, the financial resources at one's disposal, and one's priorities. However, there will always be a number of considerations that will narrow down the realistically available options and provide the

S. Skreslet (ed.), Jan Mayen Island in Scientific Focus, 269-278

more or less rigid modalities for the kind of activities that would seem worthwhile or possible to embark on. This, of course, goes for research operations as well.

2. GENERAL CONSIDERATIONS

For the present purpose it is necessary to identify and outline the main parameters that must be taken into consideration when considering any project dealing with international research operations on Jan Mayen. This would not only apply to prospective endeavors on the part of the Norwegian government, but also to any other activity by any other party, insofar as they would have to take account of relevant rules or guidelines established by Norwegian authorities.

The main parameters that together form the scope for Norwegian commitments (actual or possible) in our context may be roughly grouped in the following categories:

- Physical characteristics including limitations (location, geography, climate).
- Legal status, in terms of *viz.*:
    - international law.
    - national (domestic) legislation.
- Infrastructure and logistics (existing and prospective).
- National policies concerning research or related activities.

3. PHYSICAL CHARACTERISTICS

Any presentation or discussion relating to the scope for prospective operations of any kind would be incomplete without at least a quick look at the physical properties of the object of our attention. Jan Mayen is a small island of just 373 km^2 (slightly less than 146 square miles). It is located in a rather remote part of the North Atlantic Ocean, between 70 and 72°N, and between 8 and 9°W. To illustrate the remoteness: The island is situated 600 km NE of Iceland, 500 km E of Greenland, 1000 km W of mainland Norway, 1000 km SW of the Spitsbergen Island of the Svalbard archipelago.

Geologically and geo-tectonically, the island is a volcanic outcrop of the Mid-Atlantic Ridge. The island is rocky and mountainous, with the highest peak at 2277m (Mt. Beerenberg), which is a snow-covered and still active volcano. The climate is windy and rather cold, as the island lies in the border zone between the high Arctic and the sub-arctic transitional zone. The coast-line is rough, with no natural harbors. As a result of the climate, the size of the island and its distance from other land areas, the local flora and fauna do not display a particularly rich biodiversity, at least not with regard to terrestrial species (the surrounding marine areas is a different matter).

Jan Mayen may be considered a desolate mini-universe, a far-off place with a rather inhospitable natural environment. This, obviously, has also had an impact on the character and the extent of human activities.

The island has no indigenous human population and no long-term settlers. Since the early 17th century, the island was visited on and off by a diverse lot, including whalers, adventurers, explorers and even scientists (i.e. the Austro-Hungarian expedition during the 1st International Polar Year 1882-83). No permanent or continuous settlement existed until the establishment of the first meteorological station in 1921. The population which has always consisted of temporary residents reached an all-time high in the late 1960's with approximately 40 souls, mostly personnel from The Norwegian Defence Communication and Data Services Administration, an outfit which is now known as the Defence Logistics Organization. Since then, the number of inhabitants has shrunk to 18 persons, and a further downscaling is envisaged as a result of the forthcoming phasing-out of the Loran-C station.

The logistics and infrastructure associated with the Loran-C system has been both difficult and costly to maintain and upgrade, and has added to the problem that the island is a difficult and costly place to reach. However, it may be worthwhile noting the apparent paradox that those very physical characteristics that make Jan Mayen such a difficult place for research operations (the remote location, the forbidding climate etc.) seem to be the very same factors that make the island such an interesting and attractive place for carrying out scientific observations.

4. LEGAL STATUS IN TERMS OF INTERNATIONAL LAW

4.1 Norwegian Sovereignty

Any international activities with regard to Jan Mayen Island must take into account the status of the territory in terms of international law, or more specifically, take note of the trivial but nevertheless crucial fact that this is Norwegian territory.

Up until 1922 the legal status of Jan Mayen was *terra nullius* i.e. "no-man's land". In that year the island was annexed by the Norwegian Meteorological Institute on behalf of the Kingdom of Norway, and by Royal resolution (Order-in-Council) of 8 May 1929 the Norwegian government reconfirmed the legal status of the island by a formal decree of Norwegian sovereignty.

The further legal basis for the exercise of Norwegian jurisdiction was established by the Jan Mayen Act of 27 February 1930, which, *inter alia*, decided that Jan Mayen is an integral part of the Kingdom of Norway.

Thus, the legal status of Jan Mayen is basically the same as the Svalbard archipelago, and unlike the Norwegian possessions in the Antarctic, which are dependencies, and not integral parts of the Kingdom. Until 1995, the island was administered by the Governor of Svalbard. However, the Svalbard Treaty does not apply to Jan Mayen, which means that Norwegian jurisdiction can be exercised without any considerations concerning restrictions contained in that Treaty. Or, to put it in another way: The position of the Norwegian government with regard to international research operations or any other kind of operations on Jan Mayen, is not affected by any specific obligations or restrictions in terms of international agreements or any other instrument of international law.

The island's status as a part of the Kingdom means that any international conventions or agreements which Norway is a party to and is bound by, will also apply to the territory of Jan Mayen, unless Norway has taken a reservation in this respect in connection with the ratification.

It should be noted that the Norwegian sovereignty over Jan Mayen is unquestioned and unquestionable. The sovereignty is not dependent on any official presence or any particular kind of activity, and it has not been disputed by any other country.

4.2 The Continental Shelf and the Jan Mayen Fishery Zone

The 1982 U.N. Convention on Law of the Sea (UNCLOS) lays down the rules on what kind of international legal authority over sea, seabeds and underground that can be claimed on the basis of sovereignty over islands. The main rule is that islands provide the same legal basis as other coastal areas. The rights under the Convention apply to islands independently of such factors as area, population size and socio-economic considerations.

The decisive factor with regard to jurisdiction is whether the area satisfies the Convention's definition of "island". There is no doubt that this is the case for Jan Mayen. Thus, the Norwegian sovereignty over Jan Mayen gives Norway the right to the continental shelf around the island, and the right to establish maritime zones. Accordingly, and also pursuant to the Act of 17 December 1976 concerning Norway's Exclusive Economic Zone

(EEZ), the Jan Mayen Fishery Zone was established by government regulation of 23 May 1980.

4.3 Delimitations

The delimitations between *viz.* fishery zone/EEZ and continental shelf between Jan Mayen and Iceland were established by agreements of *viz.* 28 May 1980 and 22 October 1981 between Norway and Iceland. After the conclusion of the 1980 agreement Norway established a fisheries limit of 200 n. miles around Jan Mayen. The delimitation did not, however, follow the generally accepted method of using the median line. Instead, the boundary was drawn in a way that reduced the area in the border zone with Iceland, whereas Iceland kept a full EEZ of 200 n. miles. One argument in favour of this solution was that the land area of Iceland is larger than that of Jan Mayen. Another argument was that Iceland's population had a greater need for access to the fish stocks than Norwegians, or the Norwegians on Jan Mayen in particular. The validity of such arguments in terms of international law is doubtful, but Norway nevertheless opted for accommodating the Icelandic views.

The method for the delimitation in the sea between Jan Mayen and Greenland was settled by the International Court of Justice in the Hague in a verdict of 14 June 1993. The Court interpreted the 1958 Geneva Convention on the Continental Shelf in the light of common international law. The verdict confirmed that population size and other socio-economic considerations were not relevant to the delimitation issue. On the other hand it was taken into account that Greenland had a relatively longer coastline adjacent to the sea area in question, and that Greenland should be given a reasonable access to the fish resources in the disputed area. The verdict was subsequently used as the basis for the delimitation agreement of 18 December 1995 between Norway and Denmark concerning the area between Jan Mayen and Greenland.

4.4 Territorial Waters

The island of Jan Mayen has its own sea territory. By governmental decree of 25 February 1812, the territorial waters of Norway were ruled to comprise the waters within a distance of 4 n. miles from the coast. As a result of the annexation of Jan Mayen, this rule was also introduced here. The more precise definition of the territorial limits was established by Royal resolution (Order-in-Council) of 30 June 1955.

As from 1 January 2004, Norway has extended its territorial limits from 4 to 12 n. miles. This enlargement has also taken effect with regard to Jan Mayen.

5. LEGAL STATUS IN TERMS OF NATIONAL LEGISLATION

5.1 General and Basic Legislation

National (domestic) legislation defines how public matters are organized, and - to put it in simple terms - defines what you can and what you cannot do.

In terms of the Jan Mayen Act of 27 February 1930, § 2, Norwegian civil law, criminal law and administration of justice apply on the island. Other Norwegian laws apply only to the extent that this has been explicitly so decided by the government.

There is no privately owned land and no privately owned buildings, and the law forbids private acquisition of government property.

The Jan Mayen Act applies to the island's land territory and its territorial waters. Outside of the territorial limits, laws and regulations must have their legal basis in the aforementioned Act of 17 December 1976 concerning Norway's Economic Zone, or, they must be based on the presumption that Norwegian legislation apply in accordance with inter-legal rules (cfr. § 12 of the Norwegian Criminal Code).

With regard to the continental shelf, the legal basis for regulations is to be found in two acts from the 60's: *Viz.* the Act of 21 June 1963 and the Act of 29 November 1966. The latter concerns petroleum operations, the former concerns scientific exploration for and exploitation of other kinds of resources under the sea-bed. These acts have a general field of application, and do not relate to the seabed around Jan Mayen in particular.

As mentioned briefly above, the administration of Jan Mayen was previously the responsibility of the Governor of Svalbard. As this was felt to be an awkward and not very logical arrangement, the government decided on 25 August 1994 to transfer the administrative responsibilities to the County Governor of "Nordland fylke". The administration of justice is the responsibility of the Salten District Court, and police matters are the responsibility of the Police Commissioner of the city of Bodø.

The Head of the Loran-C station is the ranking local representative of civilian and military authorities, and is entrusted with the task of enforcing laws and regulations, and taking care of Norwegian interests.

If there were to be any mining operations, those would fall under the jurisdiction of the mine superintendant of Svalbard and Jan Mayen, whose office is presently located in Trondheim.

5.2 Specific Regulations Concerning the Access of Foreign Nationals to Jan Mayen

The present regulations (FOR 1962-06-01 nr 01) came into force in 1962 and contain 7 main articles with 15 paragraphs and numerous sub-paragraphs, spelling out detailed rules and restrictions concerning passport and visa requirements, entrance and exit control, residence permits, work permits, access denial and deportation rules. Of particular interest with regard to prospective international research operations are the paragraphs dealing with residence and work permits, which run like this:

§ 8. A foreign national arriving at Jan Mayen may be issued with a police permit to stay on the island for the maximum of one week provided that he will be able to leave the island by sea or air at the time of expiry of the permit. If a foreign national intends to stay for more than one week, he must obtain advance permission from the Ministry of Justice.

§ 9. The police shall keep a register of foreign nationals who have been issued with a permit to stay on the island.

§ 10. A foreign national who intends to seek any kind of work on Jan Mayen must obtain in advance a specific permit from the Ministry of Justice.

5.3 Specific Legislation Concerning Environment Protection

There is no unified body or single piece of comprehensive legislation covering the various aspects of environment protection on Jan Mayen. Most of the relevant rules are laid down in a number of separate governmental regulations which have their legal basis in legislation of a broader or more sweeping character, such as the Jan Mayen Act of 27 February 1930. Such rules include:

- Temporary regulation of 28 May 1971 concerning the natural environment on Svalbard and Jan Mayen.
- Regulation of 23 April 1976 concerning mining operations on Jan Mayen.
- Regulation of 11 August 1978 concerning the management of wildlife and freshwater fish on Svalbard and Jan Mayen.

- Regulation of 21 June 1974 concerning protection of objects of cultural heritage.

The question of revising these rules and amalgamating them into a single, consolidated regulation, was addressed and dealt with by the select committee which in 1998-98 prepared and drafted the proposed new Environment Protection Act concerning Svalbard, which was eventually adopted in 2002.

The outcome of this work as far as Jan Mayen is concerned, can be summarized as follows:

1. Jan Mayen is not included in or covered by the Svalbard Environment Act.
2. No further measures were taken to consolidate the various existing regulations concerning Jan Mayen.
3. The Jan Mayen Act of 27 February 1930 was amended to include the sentence "The King may issue regulations concerning environment protection on Jan Mayen".

Thus, the existing regulations were given a more up-dated legal basis. This does not make much of a difference in practical terms as far as the consequences for research operations are concerned. It should, however, be noted that when addressing these issues, the select committee re-iterated and reconfirmed the guiding principle which is considered paramount in Norwegian policies on the environment and nature management, namely that environment considerations shall be given preference whenever there may be a conflict with other interests or considerations.

The main point to be noted here is that any kind of activity on Jan Mayen - including research operations in general and prospective international research operations in particular - must take into consideration the stringent requirements that these regulations entail.

An environmental action plan for Jan Mayen for the years 2000 to 2005 has been published by the County Governor of Nordland fylke, in cooperation with the Norwegian Directorate for Nature Management, the Norwegian Polar Institute and the Directorate for Culture Heritage (Anon. 2000). It contains detailed and extensive information on the current environment regime as well as on prospective future rules and regulations.

6. NATIONAL POLICIES CONCERNING RESEARCH OR RELATED ACTIVITIES

When discussing the scope for Norwegian commitments related to international research operations on Jan Mayen Island, a key question that needs to be addressed is: How do such prospective research operations fit together with the broader national policies, strategies or plans?

To spell it out more specifically: Is there a *modicum* of compatibility? Is there a potential for useful synergies? Or perhaps even a potential for institutional co-operation, such as joint ventures with regard to projects? Or conversely, are there any contradictory objectives, any insurmountable obstacles? Such questions cannot be answered on a general level, but have to be considered on the basis of the character of the respective individual research operations that may be proposed. Meanwhile, it would be useful to familiarize oneself with the main plans and strategies that may be labeled "national policies" in this area.

It must also be added that, to the extent that one would envisage any "commitments" on the part of the Norwegian government, related to international research operations on Jan Mayen, it would not only be useful, but essentially necessary for any would-be operator to familiarize oneself with such plans and strategies. It probably goes without saying that any involvement by the Norwegian government in research operations would only take place in terms of, or on the basis of, declared government policy, as laid down in a relevant policy document.

Policy documents which deal with Jan Mayen in particular are scarce, but the Jan Mayen situation should be considered in the broader context of Norwegian research policies in general, and more specifically, in the context of Norwegian policies concerning research in the Arctic.

In December 1998 a working group headed by the Research Council of Norway presented a report in which it identified a number of subject areas where there were special opportunities for cooperation and synergy, and described future development projects. The subject areas concerned accord with the priorities established in the Research Council's strategic plan for Norwegian research in the Arctic (1995).

A brief description of the subject areas and projects were given in the Parliamentary (Storting) White Paper no. 9 (1999-2000) concerning Svalbard, chapter 8.5.4. A similar presentation was given in the Parliamentary White Paper no. 39 (1998-99) on national research policies, esp. chapter 5.2.2. Although the descriptions mentioned here deal exclusively with Svalbard, it stands to reason that similar considerations (with regard to past experience as well as current and future tasks) would be applicable to the Jan Mayen situation.

A more recent policy document which addresses these issues has been prepared by the Norwegian National Committee for Polar Research. This document, which is called "Strategic Plan for research in the Arctic 2003-2007", has apparently not yet been endorsed by the Norwegian Research Council, which means that the status of this Strategic Plan is still "unofficial". With this reservation, the draft Strategic Plan may give some useful indications concerning the key elements and main directions of Norwegian national policies on research matters in relation to Jan Mayen.

The obvious advice to any party entertaining plans for research operations related to Jan Mayen, would be to familiarize themselves with existing policies, plans and strategies. The equally obvious next step would be to establish contacts with the relevant responsible research institutions and agencies with a view to finding some common ground that would facilitate fruitful cooperation.

While noting that there is a number of potentially relevant institutions and agencies, one cannot overlook the fact that the Norwegian Polar Institute plays a central role in research activities in the Arctic. Thus, for any plans for research operations related to Jan Mayen to get off the ground, it would be a definite advantage to be able to solicit the cooperation of the Polar Institute (which of course would be contingent on the compatibility of such plans with the Institute's own priorities).

REFERENCES

Anon. 2000. Miljøhandlingsplan for Jan Mayen, år 2000-2005. Fylkesmannen i Nordland. Rapport 2/2000. 45pp (in Norwegian).

Chapter 26 NATO Advanced Research Workshop

Norwegian Protection of Arctic Islands

Restrictions and Consequences

Jan-Petter Huberth-Hansen
Norwegian Directorate for Nature Management, Trondheim, Norway
E-mail: jan-p.hubert-hansen@dirnat.no

Abstract: Norway has taken measures to protect Arctic and Antarctic islands. Recently two small islands near Svalbard were declared nature reserves. Approximately 65 % of the of the Svalbard archipelago is now protected. The Environmental Council for Jan Mayen has developed an Action Plan for 2000-2005 drafting protection measures that are under consideration by the Ministry of the Environment. Protective limitations affect all activities on the island, including research.

Key words: biodiversity, uniqueness, wilderness, conservation

EXTENDED ABSTRACT

Isolated, oceanic islands often have a biodiversity and a geological variety of unique character. Norway has several such islands – both in the Arctic and in the Antarctic – and has taken several measures to protect these landscapes and their flora and fauna.

Recently Norway has declared both Bjørnøya (Bear Island) between mainland Norway and Svalbard (2002), and Hopen Island southeast of Svalbard (2003) as nature reserves. Furthermore approximately 65 % of the terrestrial area of the Svalbard archipelago is protected. These designations both fulfil Norway's obligations according to international nature conservation conventions ratified, as well as national goals for conservation of flora, fauna and wilderness in the Arctic.

S. Skreslet (ed.), Jan Mayen Island in Scientific Focus, 279-280

The Environmental Council for Jan Mayen has in it's Action Plan for 2000-2005 focused the need for considering protection measures on the island (Anon. 2000). The Norwegian Directorate for Nature Management has worked out a memorandum for the Ministry of the Environment, documenting the conservation values on the island, and giving a clear recommendation to start a process aiming at nature protection on Jan Mayen.

If Jan Mayen is to become a platform for international monitoring and research it is of interest to identify and highlight restrictions and consequences for such activity – both limitations that exist today and those which may follow from a possible designation of a protected area on the island.

REFERENCES

Anon. 2000. Miljøhandlingsplan for Jan Mayen, år 2000-2005. Fylkesmannen i Nordland. Rapport 2/2000. 45pp (in Norwegian).

Chapter 27 NATO Advanced Research Workshop

Jan Mayen Infrastructure

Plans for the Future

Leif K Tviberg

Norwegian Ministry of Fisheries, Oslo, Norway

E-mail: leif.tviberg@fid.dep.no

Abstract: The present infrastructure on Jan Mayen Island is dedicated to the work of personnel who make meteorological observations, and who maintain the functions of a Loran-C station. Satellite navigation systems have now replaced the Loran-C system, and Norwegian authorities consider to withdraw from the Northwest European Loran-C agreement. If accepted by the Norwegian Parliament, the Loran-C station will be closed in 2006. The meteorological service will probably continue as a manned operation. An inter-ministerial working group have considered the future activities and infrastructure on the island. One of the conclusions from the working group is that a new main building should be built, either close to the current main station or near the meteorological station.

Key words: NELS, navigation, meteorology, investments, transportation, infrastructure

1. INTRODUCTION

The first official Norwegian presence on Jan Mayen was in 1921, when a meteorological and radio station was established. In 1960 a Loran-C station was established on the island, as a part of NATO's infrastructure programme. Both the Loran-C station and the infrastructure was run by Norwegian Defence Communications and Data Services Administration (NODECA), a part of the Norwegian defence organisation.

Till the end of 1994, the station was financed by the US Coast Guard. From 1995 the station is financed by Norwegian authorities, and the responsibility for budget and coordination of administrative issues is now transferred to the Ministry of Fisheries. This ministry has the responsibility

S. Skreslet (ed.), Jan Mayen Island in Scientific Focus, 281-284

for civilian navigation, and as the "largest user" also the responsibility for coordinating budget and activities on Jan Mayen.

Both in order to discuss the future needs for investments on the island and because of the uncertainty of the future of Loran-C, there was established a working group with civil servants from several ministries. The working group presented a report dated the 1st of June 2001 concerning future presence and investments needed on the island. Later the same group presented a report about the future need for personnel. There has also been made a report calculating the costs for different alternatives for a new main building.

2. CURRENT ACTIVITIES

The current activities on the island are:

- The Loran-C station (3 employees)
- The meteorological station (4 employees)
- Infrastructure functions (11 employees)
- Jan Mayen radio, which is run from Bodø. The personnel on Jan Mayen take care of maintenance.
- Four seismological stations with continuous data transmission to Norway. The data are used for research and as a part of the surveillance of the nuclear test ban treaty.
- The personnel on Jan Mayen do biological observations and have tasks concerning preserviation of the cultural heritage on the island.

3. THE FUTURE OF LORAN-C

The future of the Loran-C system is obviously a key issue concerning activities and infrastructure on Jan Mayen in the future. Loran-C is a radio navigation system that was established in Northern Europe around 1960, as a part of the NATO infrastructure.

In 1992, the Northwest European Loran-C agreement (NELS) was established. Participants of this agreement are Norway, Denmark, Germany, Ireland, France and The Netherlands.

As Global positioning systems (GPS) are now much more widely used than Loran-C, several of the members of the NELS agreement have signalled their wish to withdraw from the agreement. In the budget proposal for 2004, the Norwegian government have proposed that Norway should withdraw from the agreement. If this is accepted by the Parliament, the Norwegian

Loran-C stations will be run till the end of 2005, and closed the following year.

4. INFRASTRUCTURE

The main station that is often called "Olonkinbyen" (Olonkin City) was built in 1959-1960. It has living quarters, kitchen, offices, radio room, power plant, garage, stores and is a complete little community on its own. The Loran-C building with its radio mast is situated approximately 1 km to the southwest of the main station. The meteorological station is established 3km from the Olonkin City (Fig. 1) where it was built in 1962. In addition to facilities for the meteorological observations, the station can be used as backup living quarters in an emergency.

The air field that is close to the meteorological station is 1600m long and allows the C-130 Hercules air craft to land there. The landing strip is covered with lava sand and demands quite a lot of maintenance.

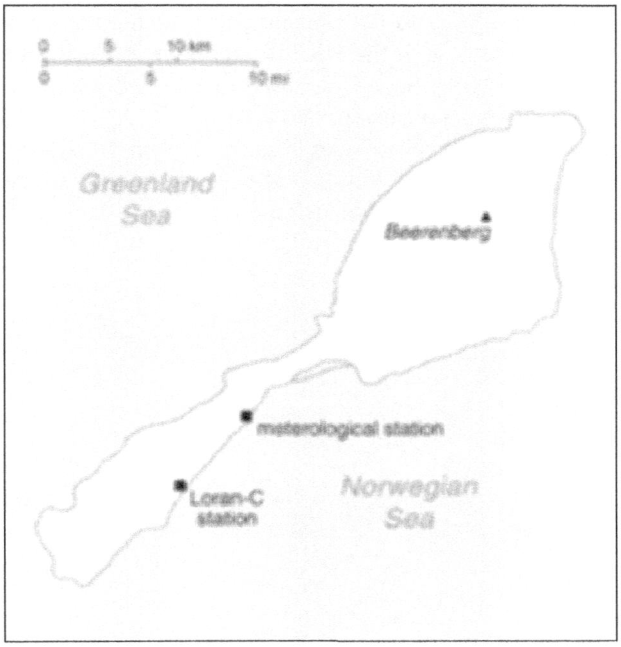

Figure 1. Jan Mayen Island with location of present governmental activities (Source: http://www.odci.gov/cia/publications/factbook).

Much of the physical infrastructure on Jan Mayen Island needs renewal or replacement, which is dealt with in the report of 2001. There is some serious damage on the main station, especially concerning the foundations of the main building. "Statsbygg" that is the agency for building and maintenence of governmental properties does not recommend repair of the buildings, and rather recommends building a new and smaller station. There are two alternatives for location, either at the Olonkin City or near the meteorological station. The costs are estimated to be 70-80 million NOK, depending on the location alternative.

An investment that is already under way in 2004 is the renewal of the system for unloading and storing fuel.

5. TRANSPORTATION

Jan Mayen Island has a very isolated location. There are at present eight flights per year with C-130 Hercules aircrafts from Bodø. Due to the weather conditions, the planes cannot always land, and some times they have to return to the mainland. This obviously gives extra costs.

Fuel and goods too large to be transported by plane, are transported by boat. Transportation ships visit the island twice a year, during the summer when weather conditions are best. Unloading of fuel and goods is a difficult and daring task, as there is no harbour at the island.

Jan Mayen cannot be reached with helicopter from Norway. Thus, it is important to be able to handle emergencies on the island, as there can be several days before external help can arrive.

Chapter 28 NATO Advanced Research Workshop

The Logistics of Past Scientific Research on Jan Mayen Island

How to Land, Exist and Work

Susan Barr
Directorate for Cultural Heritage, Oslo, Norway
E-mail: susan.barr@ra.no

Abstract: The logistics of activity on Jan Mayen are firstly governed by the challenges involved in getting to and from the island, and secondly by the natural conditions on the island. The climate and difficult weather conditions, combined with the lava sand and rock basis of the island can cause problems it is necessary to be aware of before engaging in field work and research there.

Key words: history, logistics

1. INTRODUCTION

Part of the charm and interest surrounding such an island as Jan Mayen is the fact that it is not easy to visit. Before September 1960 there was no facility for planes to land on Jan Mayen. This summer a landing strip was cleared in connection with the establishment of the Loran station, as "an extra broad and flat part of the road" to Kvalrossbukta (Barr 2003). There are not, and never have been, however, airport facilities for normal scheduled or private flights. Neither is there a harbour, quay or other landing constructions for ships and boats. Landing of personell or goods from ships is thus dependent on the weather and state of the sea. The volcanic origin of the island has formed a coast line with unexpected submerged or semi-submerged rocks off the sharp lava-rock coasts. Where there are long

S. Skreslet (ed.), Jan Mayen Island in Scientific Focus, 285-294

beaches, the slightest swell or waves pound the shore and make landing and re-launching wet and hazardous.

2. SEA TRANSPORT

The traditional way for expeditions to approach Jan Mayen is therefore by ship. Nowadays there is seldom fast ice around the island, and certainly not during the early summer. This has not always been the case. The Dutch archives from the 17th century whaling period tell of problems with both ice and bad weather during the summers, as well as for the sailing ships around the dangerous coasts. The International Polar Year expedition in 1882 met drift-ice 220 km south of Jan Mayen at the end of May and did not manage to get through to land before mid-July. On the other hand they could then use the ice-shelf still attached to the shoreline as a landing platform for all their building materials, equipment and provisions (Barr 2003).

There are several reports of other 19th century expeditions that met with problematic ice conditions in mid-summer. In July 1856 the French ship "Reine Hortense" and the British yacht "Foam" both sailed from Iceland to Jan Mayen, but it was only Lord Dufferin's expedition with "Foam" that managed to push through the ice and get ashore on a point near Beerenberg 13th July. In 1863 a Swedish expedition on a ship named "Jan Mayen" (!) could not reach the island because of ice, and likewise the French naval ship "Chateaurenault" was hindered in 1891. Changing ice conditions make it now seem unlikely that ice could cause any sort of hindrance to ship expeditions today, even outside the optimal summer period.

That expeditions have to give up landing attempts because of heavy seas and waves breaking on the beach happens also today, and is a problem that presumably will never be entirely solved on Jan Mayen. This happened to a German expedition with "Germania" in 1869, to the Duchess of Bedford with "Sapphire" and another English expedition with the yacht "Matador" in 1911, while later expeditions have not reported the problems at such lengths (Barr 2003).

3. THE QUESTION OF SHIPPING FACILITIES

The difficulties of landing and leaving Jan Mayen by ship have obviously led to ideas and plans through the years for a harbour or some form of mole or breakwater. Such plans were most advanced in the 1950s, while the construction of a landing strip for aircraft in 1960 took much of the impetus out of the matter.

In the 1950s there was rich herring fishing near to the island and the Director of Fisheries took up the question of a radio beacon on Jan Mayen. This was not pursued further at the time. A simple water-supply plant for replenishing the fishing fleet with fresh water from Nordlaguna was, however, constructed, and consisted of a small pumping engine, a long hosepipe and a buoy on the sea offshore. Apart from the fact that the offer was little used after the first year, the system also suffered from wave damage to the hose and its supports in 1952, while the buoy was taken by ice in April 1953. The system was started up again in 1954, but the pump and other equipment were washed away in a storm at the end of September (Barr 2003).

Nordlaguna was, however, still in focus as an aid to shipping. In July 1952, on the initiative of Member of Parliament Lars Ramdal, the Parliament sent to the Government a suggestion to make a harbour on Jan Mayen. The harbour's purpose was to serve the fishing and sealing fleets, rather than the island itself, but obviously both would benefit. The easiest way to create a harbour on the island was thought to be by opening a channel through "Bommen", the bar dividing Nordlaguna from the sea. That this would destroy the meteorological station's largest supply of fresh water was apparently not taken into consideration (Barr 2003).

Although drawings of the possible harbour (in fine, calm weather) seemed attractive, the plans were never realised. It would undoubtedly not have been easy either. The forces of wind, waves, ice and currents would have caused great problems, both during construction and later. From 1928 to 1953 Bommen was said to have shrunk to half its earlier width: from 200m to 100m. In November 1953 a cabin that was built on the middle of Bommen in 1926, together with a two-storey building nearby built in November 1943, were both washed away in a storm. A year later sealing ships visiting the island reported that the sea was now shallower outside Nordlaguna than before (Barr 2003).

The harbour plans were further debated in 1965 and 1979, but the conclusion was always that the natural conditions would make the construction and maintenance extremely difficult and the cost would be too high.

4. RECENT SEA TRANSPORT

The meteorological station that was established in 1921 was run up to 1962 with ship transport once a year. The old crew was then relieved, the new brought to the island, and all the necessary supplies and equipment for the coming year were left. The unloading process could be hazardous, even in relatively calm weather, if there were waves pounding on the beach. The heavy ship's boats could easily be swung broadside to the waves, filled with water and be almost impossible to turn and push out to sea again. Food, coal, live pigs, etc. all had to be handled in a way that kept them usable for the coming year (Barr 2003).

Despite the fact that an airstrip is now in use, it is still necessary to bring some goods and supplies to Jan Mayen by ship. Diesel fuel is brought by tankship which anchors off Kvalrossbukta and pumps the fuel ashore through a hose system. Obviously the weather cannot be too rough for this to function. It is planned to use the same method in the future off Båtvika, near to the main station Olonkinbyen. The cargo ship that brings bulky material and heavy goods and supplies once a year has a refined method of unloading, which occurs in Kvalrossbukta during optimal weather conditions. The ship is simply driven bow-first into the beach and anchored there. Bulldozers from the Jan Mayen station then pack sand beside the bow to form an unloading platform, and goods are unloaded directly on to a lorry on the platform. Simple, unusual and effective.

Jan Mayen has other unusual and effective solutions to various problems occasioned by its nature and climate. In use for many years from Olonkinbyen was a "cradle" for bringing ship's boats with passengers safely ashore (Fig. 1). The cradle was pushed into the sea off the beach by one of the station's heavy vehicles, the ship's boat could then be driven into this "dock" which was then quickly pulled on shore and the passengers could step dry-footed on to the beach. In the opposite direction the boats could be pushed into the sea safely despite breaking waves on the beach. In relatively calm weather large rafts can be pushed or pulled out to the ships and goods and people be unloaded on to these.

5. LOGISTICS ON LAND

The Austro-Hungarian Polar Year expedition 1882-83 was the first to spend enough time on Jan Mayen to really be able to experience the challenges and problems that travel on the island could present. Although the expedition's year-long scientific work was a great success, the leader – Emil von Wohlgemuth – commented in his report particularly on the difficult weather conditions that made the task relatively complicated. The quickly changing conditions often led to the men making the outward journey with sledges on good snow, and having to return dragging the sledges over the bare ground. The sea ice was too unstable and irregular to be used as a quick route round the coast and there were almost always open channels off the steep cliffs.

Figure 1. The "cradle" that has been used to bring boatloads of people safely ashore. Photo: S Barr.

The problem that the Austrians met was that although Jan Mayen on the one hand had Arctic, wintery conditions that gave special requirements for housing, food, clothing and logistics, it also had such a variation in weather conditions that any good, stable ice and snow regime could not be relied on. Other particular problems were presented by the abundance of iron traces in the ground – giving compass inaccuracy – and the frequent fog that also hindered the mapping work (Barr 2003).

In recent times it can be added that the topography gives blind spots for radio communication between various parts of the island (Fig. 2).

Other earlier reports tell of various conditions that made work on the island near impossible at times. The meteorological crew in 1922 could report that "it is pure hell here all the time". An apparently typical situation occurred in November:

> "In the night 22-23 November the wind increased to strong storm force with thick whirling snow. The anemometer broke and it was extremely difficult for us to repair it in the morning. An absolute hell on earth. We had to creep around and had more than enough to hold on tight when the worst wind gusts came. Inside the house we had the feeling of driving in a

car, not a comfortable feeling, but when the wind during the morning increased to gale force, we were in truth afraid that everything would be blown away. The window panes cracked and at 12 midday the entire antenna rig fell down on to the roof with a terrible crash. Again we had a hard job in the violent snow storm to get the rig down from the roof so that it would not damage the roof for us, and after half an hour's work we had the rig 'on dry land' – but we were wet. On the 24th there was still a strong gale with whirling snow so it was impossible to start the work to get the rig up again. Today we understood that it was no use waiting for good weather. We had to have a go, despite strong gale gusts and whirling snow ... Now everything is back in order again and we are fine and in good humour, but the Institute must provide better weather or we will row home." (Barr 2003).

The extreme wind conditions were again illustrated in January 1957 when the c. 80 knot wind exceeded the 73 knot top of the anemometer scale. It was suggested that a piece of paper could be added to the graph paper in order to record the tops of wind strength (hurricane force is 63 knots) (Barr 2003).

Figure 2. The volcanic landscape of the south end of Jan Mayen. Photo: S Barr.

6. A SAFETY NET FROM 1921 FOR FIELD WORKERS

The establishment of the meteorological activity on Jan Mayen in 1921, gave a safety net for visiting trapping and scientific expeditions through providing a relatively well-equipped government station with telegraphic connection to the mainland. The station itself was only relieved and re-provisioned once a year, and supplies for two years were kept in case of unavoidable delay for the relief ship. This logistic challenge was considerably lightened by cooperation with the expedition and relief ships that the Norwegian Polar Institute and its predecessor NSIU sent annually up to 1959 to the Norwegian trapping and meteorological stations in eastern Greenland. The loading and unloading operations on to the beaches remained, however, a hazardous affair dependent on the sea and weather conditions, and not least also on the strength and guts of the crews on land and ship (Barr 2003).

7. EARTHQUAKES AND OTHER VOLCANIC ACTIVITY

The first meteorological station 1921-40, “Eldste Metten”, lay in an area today known as "the moon landscape", from the black bleakness of the solidified lava rocks and sand. This area presented special problems, which included sharp-grained lava sand blowing into all chinks and crannies, for example into the instrument hut where it could damage the instruments or put them out of use. Friction and abrasion from the sand caused sand-blasting of the buildings and windows, leaving them opaque. Each year the leaving crew had to shovel and wheel barrowloads of sand away from the station area to leave it clear for the next year's team. It was cause for a bitter comment in the next report if this was not done before the new crew arrived. Sharp lava rocks cut and scraped boots and, in later years, do the same to vehicle tyres. Pockets of ice under the sand melt over time and the overlying sand crust can suddenly give way, leaving deep holes in unexpected places (Fig. 3).

Volcanic disturbances have caused evacuations during the last century, both of the entire crew and of single members who found the disturbances too uncomfortable to live with. Irregular small tremours were part of life on the island, but in October 1936 an unusually strong earthquake shook the whole area around the station, causing some damage to the buildings and opening cracks in the ground. It was felt that only repairs done to the buildings that summer had saved them from collapsing. Two of the three-man crew found the situation too nerveracking to continue. The crust of solidified lava the station stood on now "echoed hollowly when hit with a hammer as it had never done before". They asked to be relieved, and were exchanged at the end of November in a very-challenging ship operation in high seas. The three who were to spend the winter on the island received extra pay (Barr 2003).

A 100% evacuation was carried out in September 1970 when the volcano Beerenberg, which had been presumed extinct, awoke to life again. The entire personnel were evacuated by plane to the mainland, but returned over the next few days as the station and airstrip area were not affected. The lava that flowed into the sea on the north-eastern tip created 4 km^2 of new land. Since then fairly strong tremours have been felt at the station at various times, and a new, small eruption on the flanks of Beerenberg was recorded in 1985 (Barr 2003).

Figure 3. This hole suddenly opened in 2002 in the ground where the remains of the first meteorological station "Eldste Metten" stand. Photo: S Barr.

8. ACCIDENTS AND LOSS OF LIFE

A number of graves at Kvalrossbukta date from the whaling period in the 17th century, and witness the fact that whaling was a dangerous occupation, particularly off the shores of Jan Mayen. Graves and memorial plaques to victims of scurvy – winterers from both the whaling period and from the trappers of the early 20th century – exemplify this side of life on an isolated, Arctic island with little wildlife. This fate only occurred to winterers on the island before the meteorological station was established, as mentioned above (Barr 1985).

The winds that frequent the island are also strong enough to have caused fatal accidents in fairly recent years. The most serious concerned the University of London Beerenberg Expedition in 1961 which had its boat overturned by a katabatic wind from the mountain, with a loss of five lives. In December 1950 the leader of the meteorological station froze to death when he tried to reach the instruments situated 35 m from the station house and was blown off course by winds of over 80 knots (the wind gauge was blown to bits after this speed was registered). In October 1980 the 190 m high Loran C mast was blown down in a storm, and finally in 1987 a freak wave overturned a zodiac and caused the death of two of the three men onboard (Barr 2003).

9. CONCLUSION

Although this article has concentrated on the logistic difficulties and challenges posed by the particular climate and natural conditions on Jan Mayen, the impression would be very incomplete if it was to be left at that. Firstly, some of the worst accounts mentioned above occurred during the winter. Research activities at other times of the year, and also during the winter, can be a very pleasant experience. The weather is not always bad nor the sea always rough! In fact exactly the special, active and diverse natural conditions on Jan Mayen contribute to making it an exciting base for scientific work of various kinds.

REFERENCES

Barr S 1985. Kulturminner på Jan Mayen / Historical remains on Jan Mayen. *Norsk Polarinstitutt Meddelelser* 108, 1-67 (in Norwegian and English).

Barr S 2003. *Jan Mayen, Norges utpost i vest. Øyas historie gjennom 1500 år* (Jan Mayen, Norway's outpost to the west. The island's history through 1500 years). With a contribution by Louwrens Hacquebord. 2nd edition. Kolofon a/s, Oslo. 266 pp (in Norwegian).

Chapter 29 NATO Advanced Research Workshop

Logistics for Future Jan Mayen Island Research

Results from NATO Advanced Research Workshop

Stig Skreslet, Krzysztof Birkenmajer, Louwrens Hacquebord, Keith A Hobson, and Igor A Melnikov

S Skreslet, Faculty of Fisheries and Natural Sciences, Bodø Regional University, Norway
E-mail: stig.skreslet@hibo.no
K Birkenmajer, Institute of Geological Sciences, Krakow, Poland
L Hacquebord, Groningen Institute of Archaeology, University of Groningen, The Netherlands
KA Hobson, Prairie and Northern Wildlife Research Centre, Saskatoon, Canada
IA Melnikov, PP Shirshov Institute of Oceanology, Moscow, Russia

Abstract: A NATO Advanced Research Workshop explored the possibility of establishing an international station for multidisciplinary research on Jan Mayen, a remote Norwegian volcanic island situated in the Nordic Seas. Twenty two scientists from different disciplines shared opinions and were informed by six governmental representatives from Norway. Following the discussion of scientific options, the participants agreed that the infrastructure that may be evacuated when a Loran-C station expires, may serve research of local, regional, hemispheric and Global importance. The buildings should be maintained during the International Polar Year 2007-08 and in the years to follow. Recommendations were made with regard to scientific priorities and logistics for support from Norway and for operations on the island itself.

Key words: volcanology, climate change, thermohaline circulation, marine ecosystem, island ecology, biogeography, living resources, cultural heritage

1. INTRODUCTION

A NATO Advanced Research Workshop was organised in Oslo, Norway, 11-15 November 2003 to perform a feasibility study on whether existing infrastructure on Jan Mayen Island may be used as an international research station for multi-disciplinary purposes. One of the incentives was that the

S. Skreslet (ed.), Jan Mayen Island in Scientific Focus, 295-339

Loran-C navigation system is outdated and may be closed down, leaving a number of buildings unoccupied (Fig. 1). This situation created public and governmental concern (Anon. 2003, Tviberg 2004) and an interest in scientific research as a relevant future activity on the island was expressed.

To identify scientific problems that could be addressed by research activities on Jan Mayen and its oceanic surroundings, the participants were carefully selected according to their scientific training and experience (Table 1). Some were highly experienced from previous polar fieldwork and able to address particular Arctic problems that could be solved by adding observations from the Jan Mayen area, or that were unique in an Arctic context. Others were selected for their particular expertise, and were asked to identify more universal scientific problems that could improve the understanding within their scientific field by making observation on or at the island.

Figure 1. Jan Mayen Island with some place-names and the roads (curved lines) that interconnect maintained infrastructure. Nordlaguna and Sørlaguna are the two largest lakes on the island.

Table 1. Key participants in the NATO Advanced Research Workshop on a Joint International Scientific Observation Facility on Jan Mayen Island, Oslo 11-15 November 2003. *) Countries that are now full members of NATO. **) Co-Directors.

NATO Country Participants		Partner Country Participants	
Canada	Hobson, Keith A	Estonia*	Vana, Marko
	Rouvinen-Watt, Kirsti	Lithuania*	Stankunavicius, Gintautas
Germany	Hirche, Hans-Jürgen	Poland*	Birkenmajer, Krzysztof**
Netherlands	Hacquebord, Louwrens		Glowacki, Piotr
Norway	Barr, Susan		Piechura, Jan
	Gabrielsen, Geir W		Węslawski, Jan M
	Gulliksen, Bjørn	Russia	Byshev, Vladimir I
	Hagen, Jon OM		Klyashtorin, Leonid B
	Skreslet, Stig**		Melnikov, Igor A
United Kingdom	Shimmield, Graham B		Romanov, Vladimir F
	Walne, Anthony W	Ukraine	Kyyak, Volodymyr G

During one of the sessions, the invited scientists were provided with back-ground information from Norwegian governmental bodies involved in national administration and management (Table 2).

In their talks and in a number of working groups and plenary discussions the invited specialists identified scientific problems and observation methods potentially applicable to Jan Mayen research. They identified features unique to Jan Mayen and referred to its isolated position as an asset in large-scale local, regional, hemispheric or Global research. The present account is an edited version of group reports and notes made during the NATO ARW discussions.

Table 2. Governmental representatives providing information to the NATO Advanced Research Workshop on a Joint International Scientific Observation Facility on Jan Mayen Island, Oslo 11-15 November 2003.

Name	Affiliation
Bjørge, Arne	Institute of Marine Research, Bergen
Hov, Øystein	Norwegian Meteorological Institute, Oslo
Hubert-Hansen, Jan P	Norwegian Directorate for Management of Nature, Trondheim
Orheim, Olav	Norwegian Polar Institute, Tromsø
Skagestad, Odd G	Norwegian Ministry of Foreign Affairs, Oslo
Tviberg, Leif K	Norwegian Ministry of Fisheries, Oslo

2. GENERAL INCENTIVES

Jan Mayen Island attracts scientists from different fields of science (Skreslet 2004a). They often search for the unknown that may be revealed as unique in such remote places. This is a legitimate scientific driving force involving the need to gain new knowledge and understanding of complex systems.

The geographical isolation from all continents makes Jan Mayen unique. It represents a barrier for immigration of species, not only for terrestrial plants and invertebrate animals, but also for sessile marine invertebrates that only spread to new places by planktonic larvae. Human presence and other effects of anthropogenic activities may have opened the door for increased immigration of alien species, but the inaccessibility of Jan Mayen's shores and the risks of landing have restricted even that to a minimum. That may now be changing, because of the greater mobility of tourists travelling by cruise ships and seagoing private yachts.

There is an increased scientific awareness of decreasing biodiversity and the need for protection of nature, as well as our cultural heritage. It goes beyond the interest of individual scientists and is a growing political concern for nations that manage the heritage of the commons. The need to document a particular object or interest calls for the application of particular methods that have been developed by different scientific disciplines. However, the uniqueness of Jan Mayen is not just a conglomerate of results from random human impact, but also an effect of natural processes that are particular to the island and due to its geographical position.

The island is coupled to atmospheric circulation that makes the North American continent rather than Eurasia a source of direct atmospheric influence (cf. Skreslet 2004a, Vana 2004). On the other hand, it has a unique geographical position in relation to oceanic water movements from two directions (Piechura 2004). It is situated at the terminal ends of both the outflow of polar water from the Arctic Ocean and the northbound flow of Atlantic water. Jan Mayen is also positioned where Greenland Sea Deep Water is generated (Fig. 2). Thus, the island sits on geological structures at the proximal end of the World Ocean's Thermo-Haline Circulation (Skreslet 2004a) that is one of the processes that has to be studied to generate a universal understanding of Global climate, both on short- and long-temporal scales.

Each particular feature that is related to Jan Mayen Island, whether it is local, regional or Global, calls for disciplinary science. Disciplines may document information that seems important only to the specialist, but they also contribute to the understanding of coupled processes that are important both to the ecologist and the public at large.

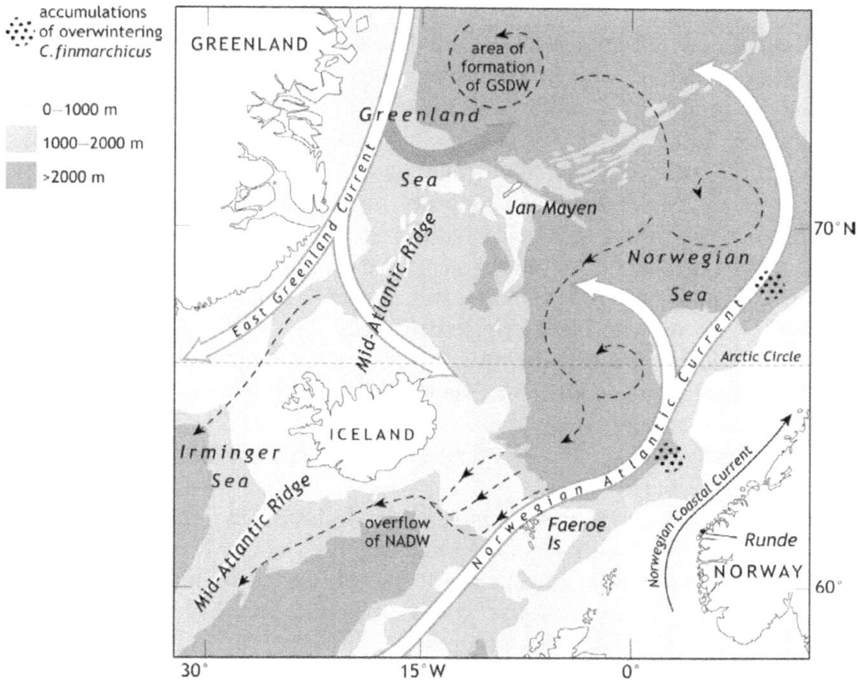

Figure 2. Jan Mayen Island in relation to bottom topography and currents in the Nordic Seas Basin. Unbroken arrows: surface currents. Dashed arrows: flows of Greenland Sea and North Atlantic Deep-Water. Two important wintering habitats of *Calanus finmarchicus* are indicated (Golmen and Skreslet 2004).

At present rather few scientific disciplines have been applied on Jan Mayen Island and in its shelf waters. Most have been allowed access for only brief periods. The knowledge of Jan Mayen biota and signs of past and present influence is therefore very incomplete (Gabrielsen *et al.* 1997).

3. SOME DIMENSIONS OF JAN MAYEN SCIENCE

3.1 The Geological Perspective

Beerenberg that rises 2277m above sea level is the dominating geological structure on Jan Mayen. It is the northernmost active subareal volcano on

Earth and the only active volcano on Norwegian territory. The study of lithospheric structures and processes on Jan Mayen above and below sea-level are important to understand linear features in the North Atlantic (Birkenmajer 2004).

The whole island consists of volcanic rocks erupted from fracture zones that separate the North American and Eurasian plates. The three latest eruptions occurred from 1970 to 1985. The geographical position of Jan Mayen and its distance from other volcanic structures on the Mid-Atlantic Ridge makes inter-comparisons interesting in many fields of volcanology:

- Petrology and geochemistry of rocks.
- Radiometry and dating of rocks in relation to stratigraphy and succession of volcanic events.
- Geothermal conditions.
- Exhalations, if they exist (location, chemistry, intensity, duration).

NORSAR operates a seismograph on Jan Mayen and receives seismic data for verification with regard to the Comprehensive Nuclear-Test-Ban Treaty (CTBT). However, the data are also used to locate distant earthquakes and record local seismic activity. Thus, it provides interesting background information for special research programs.

The glaciers of Beerenberg are a dominating feature of the Nord-Jan landscape (Hagen 2004). A complete understanding of the geomorphology will require:

- Studies of glacial deposits and the succession of their forms, supported by radiometric dating.
- Positioning measurements of advances and retreats of glaciers.

The permafrost on Jan Mayen probably varies in distribution and thickness due to different dynamic forcing factors like glaciological effects on soils, effects of regional and local climate, and volcanic activity.

Human activities on Jan Mayen have not been seriously endangered by volcanic activity or earthquakes (Barr 2004b). However, to be prepared for potential hazards (Birkenmajer 2004), some studies should be put on the agenda for future research:

- Evaluation of historical data in a geological context.
- Identify locations that may represent danger due to fault lines, avalanches and lahars.
- Calculation of risks related to future earthquakes, eruptions, and fumes.
- Planning of emergency actions based on predictions.

3.2 Global Geophysical Connections

Models that simulate atmospheric circulation in the Arctic are calibrated against meteorological data in crucial geographical points of interaction. Jan Mayen is positioned in an area where Arctic air meets air that blows in from the North Atlantic. The dynamic processes that results in this area of interaction reflects those in the North Polar region as well as forcing factors located at lower latitudes. Meteorological observations from Jan Mayen (Hov 2004) are therefore important for the understanding of geophysical processes on the northern hemisphere. Historical time series that document the variability of such processes are of importance to the calibration of numerical models developed for scientific simulation studies or for prognostic service. The validation and continuous refinement of such models needs access to standard meteorological data and special test parameters from Jan Mayen would be highly appreciated by the international community of meteorological scientists (Romanov 2004).

There are few meteorological stations in the Nordic Seas. Observations from Jan Mayen are therefore very important in the observation of atmospheric pressure gradients that define the weather in northern Europe (Stankunavicius 2004). The reliability of long-term weather forecasting will rely on analyses that take into account oceanographic forcing of atmospheric processes. Long-term oceanographic observations in the Barents Sea have proved to be of great interest in studies of atmospheric processes (Byshev *et al.* 2004). Similar studies in Jan Mayen waters would be of equal value.

Jan Mayen has a key position in relation to two water masses of Global importance (Piechura 2004). Its north-western side is exposed to the surface outflow of fresh water and sea ice from the Arctic Ocean, while its south-eastern side is exposed to inflow of Atlantic surface water. Thus, the two sides represent different regimes of air-sea interaction that influence the local and regional climate.

Seasonal and inter-annual changes in the advection of Arctic and Atlantic water cause the stratification and turbulence of the lower atmosphere to vary. Some changes are recorded by precipitation stored in the ice-cap of Beerenberg that rises 2277m. The base of the volcano is influenced by temperate and humid air. Radiation of heat from the top of this layer causes condensation of vapour, which results in cloud and fog formation that is a dominant feature of the island and its surroundings. The resulting decrease in solar radiation at lower altitudes causes abnormal ablation, i.e. reduced melting of glaciers at altitudes of 200-400m (Hagen 2004). On the other hand, the clouds reduce the radiation of heat from the sea surface and interfere with thermal conditions in the water column, which influences its stratification, turbulent processes in the mixed layer, and the formation of deep water. Thus, the geophysical processes involved in air-sea interactions around the island are important to understand, especially in relation to the Thermo-Haline Circulation (conveyor-belt) of the World Ocean.

The glaciers of Beerenberg contain a potential archive of data time series that directly represent or are proxies for geophysical processes of universal scientific interest. Information contained in ice at different altitudes may represent different atmospheric processes with particular spatial scales, and thus, different geographical origin. The lower locations are probably influenced by the regional oceanic regime more than the upper locations that may rather be affected by hemispheric tele-connection, e.g. with the North American continent.

The large amount of drift-logs that have accumulated on Jan Mayen Island represents another archive of information. Most logs have probably originated from Russian rivers, but a fraction has also been transported from rivers in Alaska and northern Canada. Identification of species, growth ring studies, radiocarbon dating, and analyses of oxygen isotope ratios in logs can be useful methods in palaeo-climate studies that, on Jan Mayen, involve advection of water between the Arctic Ocean and the Nordic Seas.

3.3 Biogeographic and Ecological Relations

The geophysical fronts that occur between masses of air and seawater in the Jan Mayen area also develop contrasts in biogeographic features (Gulliksen *et al.* 2004, Hirche 2004, Kyyak 2004). Their geographical position shifts with time and creates a changing physical environment that favours some species but not others.

Most marine species within the blend of Arctic, boreal and endemic organisms (cf. Skreslet 2004b) that live in the sub-arctic transitional zone could become elements of the marine flora and fauna of Jan Mayen. However, only migratory species, or the ones with life stages that can survive long-term transportation by advection of seawater volumes, being frozen into or being attached to sea ice (Melnikov 2004) or other floating objects, may disperse from the continents' coastal and shelf waters to the island.

Juveniles of endemic species in the sub-arctic transitional zone are supposed to be genetically adapted to an environment that fluctuates over a wide range, from typical Arctic to Atlantic conditions. Endemic sessile species like the Iceland scallop (*Chlamys islandica*) may therefore be expected to remain as a stable fraction of the island's benthic communities. However, if the intra-specific range of genotypic capabilities is subject to environmental selection, the DNA profile of endemic species may be expected to change with time and even between age-groups within populations. More typical Arctic and boreal species may rather be subject to periodic recruitment and presence, according to their vulnerability to physical changes in the sub-arctic environment.

Several migratory fish and marine mammals are present in Jan Mayen waters to feed between reproduction periods elsewhere within the Nordic Seas. Some are subject to commercial exploitation and their population ecology is fairly well monitored by ICES (International Council for the Exploration of the Sea) and managed by NEAFC (North-East Atlantic Fisheries Commission). There is far less information on other abundant species of fish and invertebrates that are not exploited, although they may represent major elements in the food chains of exploited stocks. They may be important prey to foraging seabirds and force the reproductive success of seabirds nesting on Jan Mayen. Research on such species may hold keys to the understanding of the occurrence of toxins observed in biological material from the island (Gabrielsen *et al.* 1997).

Terrestrial communities on Jan Mayen may have been established by immigration of species that have drifted by air, ice, drift-logs or other floating objects, or introduced by birds or human visitors, a process that applies to plants (Kyyak 2004) as well as animals. The communities are clearly influenced by the combined effects of geology and climate of Jan Mayen. The volcanic rocks are porous and subject to fog and rain with dissolved seawater constituents that cause chemical erosion. The water itself also causes erosion when it freezes and expands in pores and crevices in the rocks. Sand that results from mechanical erosion is itself an erosive force that grinds on rocky surfaces when it is swept over the landscape by winds. The environment is hostile and will only accept immigrants with particular tolerances, which to some extent accounts for the low biodiversity.

Water is not stored in the porous soil but may accumulate on top of the permafrost layer (Skreslet 2004b) where it may be available to plants. However, landscape forms with heavy carpets of moss (Gabrielsen *et al.* 1998) may absorb water and serve as moist and humid habitats for vascular plants and animals.

A few ponds and lakes occur on the island and may to some extent have been populated by invertebrates brought in by birds from continental habitats or from ponds on melting sea ice. However, the poverty of the Nordlaguna invertebrate fauna (Skreslet 2004b) may indicate either that such immigration is insignificant or that the oligohaline fluctuations of the lake water prevent survival of marine as well as freshwater species. In recent years, climatic change may have caused increased incidents of sea-water overflow to Nordlaguna with corresponding reductions of the land-locked population of arctic char (*Salvelinus alpinus*). This fish population is also under influence of toxic substances brought into the lake's food-web by seabirds (Gabrielsen and Strøm 2004). An understanding of the origin of toxins will require more knowledge on the genetic structures of seabirds in the Nordic Seas in general and the population ecology of seabirds nesting on Jan Mayen in particular. The tracing of toxin pathways through the food-web may be achieved by new scientific methods (Hobson 2004).

The communities of organisms on the island and in benthic communities of its littoral and sub-littoral zones (Weslawski 2004) may be subject to temporal variability. The abundance of each species may be periodical and related to the fluctuating physical environment, or due to long-term trends related to climate change or direct human presence. The uncertainty of the relative importance of factors may be an incentive for the management sector in Norway to establish routine ecological monitoring programs that accumulate time-series of biotic and abiotic data. The resulting time series may serve scientific purposes if the observations are frequent and systematic, and if the methods are subject to strict control and the objects are allowed to remain unchanged for long periods of time.

The scarcity of terrestrial mammals on Jan Mayen is an interesting scientific feature. Only the arctic fox (*Alopex lagopus*) has been observed in any abundance (Rouvinen-Watt 2004), but even this species may be a periodic member of the community. It appears to have been absent during the Dutch whaling in the 17th century (Hacquebord 2004). At present, there is no permanent population, although it was quite abundant during the first half of the 20th century. Intensive trapping may be the main reason for its subsequent extinction (Gabrielsen *et al.* 1997) but other factors may have been involved (Rouvinen-Watt 2004). These factors could be canine diseases from dogs kept in human settlements or non-degradable toxins accumulated in seabirds (Gabrielsen et al. 1997) used by foxes thereby reducing the chances of population growth. Decadal-scale reduction of sea ice cover may have reduced fox access to food resources of seal carrion (placentas from birth lairds, or carcasses from polar bear kills) left on the ice. Decades with reduced ice thickness and only brief seasonal presence of ice at the shores of the island would reduce new immigration and probably increase the loss of individuals that stray from the island onto the drift-ice.

The present status of arctic fox on Jan Mayen should be systematically studied to document any signs of new immigration. An annual census of tracks in systematically selected snow-covered areas should be performed to observe signs of immigration in periods when drift-ice reaches the shores. The present absence of arctic fox on Jan Mayen allows for comparative ecological studies on effects of arctic fox predation on bird populations in other parts of the Arctic. Such a study would also serve as a reference to studies of a new predation regime if arctic fox should again occur in substantial numbers.

According to historical sources, many polar bears occurred on the island in the winter and during summer periods when ice-drifts allowed access (Hacquebord 2004). Nowadays polar bears are rare at Jan Mayen, probably due to the decreased transportation of drift-ice to the island. However, today the settlement is situated on the south-eastern shore, while most previous settlements were on the north-western side that is more ferquently in contact with drift-ice. To avoid records of visiting polar bears and arctic fox from being biased by shifting observation points, automated stations could be established at strategic localities to record moving objects identified by infrared video equipment.

Jan Mayen is situated in an area where polar water and Atlantic water meet (Piechura 2004). Each water mass carries different species of plankton that sustain the region's food web (Hirche 2004) and either source may introduce toxins that found in local sea-birds and fish (Gabrielsen *et al.* 1997, Gabrielsen and Strøm 2004, Skreslet 2004b). The contaminants may follow intricate pathways being transported from their sources by plankton that enter local food-chains or are carried by migrating fish, birds or mammals from distant feeding habitats. To unravel how different organisms of the basin-scale ecosystem distribute contaminants will require sophisticated scientific methods. However, the use of stable isotopes in studies of food-webs has proved very useful to establish trophic relationships among organisms. The method may be applied on the Jan Mayen food web to study very interesting ecological relationships (Hobson 2004) and can contribute to the understanding of contaminant fluxes in the Arctic in general.

3.4 Opportunities from Data Time-Series

The long series of meteorological data acquired on Jan Mayen from 1921 is an advantage to investigations of climate change and variability in the Arctic. Their use in time series analyses is obvious, but they may also be used to calibrate information in ice core samples from Beerenberg glaciers. However, interesting constituents in ice cores like tele-commuted aerosols (Vana 2004) of marine and terrestrial origin (gasses, ions, minerals, organic matter) have not been subject to attention in the meteorological sampling routines.

Areas that have a heavy vegetation of moss may provide cores of permafrost that contain layers of organic matter. They could contain records of changes in the flora of the island (pollen, spores, radiocarbon age, oxygen isotope ratios as indication of local climate). The bio-indicators would supplement information from ice-cores sampled at higher altitudes.

Studies of recent and contemporary climate change and variability can be performed more effectively by establishing sampling and measurements that generate running time-series. Changes in the biology of terrestrial plants and invertebrates (reproduction, growth, and mortality) are good indicators of climatic influence on local biota. The geographical expansion or decrease in the extension of populations is an indication of general climatic change in the island's ecosystem. That also applies to the small populations of terrestrial birds, but their population success may be more dependent on the presence or absence of arctic fox.

Climate effects on local biota could also be studied in the ecosystem of Nordlaguna. There, careful annual sampling of gonads from a limited number of small mature arctic char, and monitoring of their physical environment could indicate annual variability in reproduction due to climatic factors. Back-calculation of body length increments from the width of ring structures in their otoliths would add information on environmentally induced growth. However, the lake ecosystem is influenced by the import of biomass provided by seabirds that forage on the elements of the marine food-web (Skreslet 2004b), which may conceal effects of climate.

The flora and fauna of the marine littoral and upper sub-littoral zone is influenced by wave action and grinding of floating sea ice (Weslawski 2004). Apart from such biota, the local benthic fauna and the physical environment do not reflect local climatic influence, and is rather being subject to basin-scale changes. However, this is exactly why the assemblages of marine species of Jan Mayen are so interesting. Planktonic juveniles of benthic animals may be members of basin-scale populations and recruited from assemblages that reproduce elsewhere around the periphery of the Nordic Seas (Gulliksen *et al.* 2004). Thus, the communities may change in composition, according to periodic changes in the large-scale advection of seawater of Arctic or Atlantic origin. Accordingly, long periods with a predominant import from either source would result in major changes. Investigations of the local growth and reproductive capacity of sessile marine organisms may reveal geographical changes in oceanic fronts around the island.

The growth and decline in seabird populations follow the availability of food. Thus, effects of climate on the food web seem to alter the abundance of prey (Klyashtorin and Lyubushin 2004) and possibly affect reproduction and mortality rates in bird populations. However, modern fishing exploits many of the major seabird prey populations that dominate the ecosystem. That makes the cause of any changes in seabird populations uncertain with regard to natural forcing factors. Fishing may reduce the abundance of planktivorous fish that provide prey for some birds, while the same reduction releases a predation pressure on plankton populations that is a food source for planktivorous birds (Skreslet 2004b). Other seabirds may profit from dead fish and fish remains that are thrown overboard from fishing vessels. Irrespective of such anthropogenic interference within the ecosystem, performance of a periodic census of nesting seabirds is a strong method for testing the ecological status of regional marine biota (Gabrielsen and Strøm 2004).

4. MANAGEMENT

4.1 Environmental Protection

Norwegian environmental authorities have adopted an action plan for protection of nature and the cultural heritage on Svalbard and Jan Mayen (Anon. 2000, Hubert-Hansen 2004). The knowledge of the environmental status of Jan Mayen is incomplete and not good enough to serve as a baseline for environmental management. The action plan was meant to upgrade the database, but so far difficult weather conditions and lack of logistics have prevented the plans from being carried out.

Although there are not enough local resources and means to prevent public access to the island, the Norwegian Ministry of Justice allows only authorized landings. The NATO ARW participants agree that scientific considerations are in accordance with strict regulation of public presence. Long-term studies of subtle ecological changes that are due to changes in the physical environment or relations between organisms may be totally disturbed by unregulated human impact. There is also a danger that instruments located in remote places may be destroyed or influenced by unauthorized human contact.

Because scientific considerations favor restrictions in public access, manned research programs may become an integrated part of authorized logistics that restrict human interaction with biota. A long-term scientific program will require that the scientific observations leave the least possible effects on the nature of Jan Mayen. Therefore, automated instruments and television cameras will be normal parts of the logistical set-up. Installment, inspection and maintenance in places away from the coastline will have to be made in rational ways by use of vehicles, snowmobiles, and helicopters, depending on landscape considerations. Both observations from automated stations and excursions by researchers may help in the monitoring and control of unauthorized activities.

In some locations it would even be necessary to establish an area of total exclusion of activities (including other scientific activities) to prevent contamination of sites for detailed geochemical studies of pollutants, or disturbance of long-term experiments.

4.2 Synergies of Science and the Cultural Heritage

Most human presence leaves unavoidable marks in the natural environment, sometimes creating a cultural landscape with pronounced examples of present or past infrastructure. Given time, many objects rise into the rank of becoming part of the cultural heritage. That is now the case with remains and buildings from periods of Dutch whaling, Norwegian arctic fox trapping, scientific expeditions, meteorological observations, warfare, military intelligence, and navigation service. Some are left to decay ("die in beauty"), because they are in a state beyond repair. Others may still serve human use, provided new activities are allowed to take responsibility for repairs and maintenance that do not degrade their historical values (Barr 2004a).

"Ishavsforeningen Jan Mayen" is an association of previous crew contingents. They perform voluntary maintenance of buildings under supervision of "Riksantikvaren" (Barr 2004a). Thus, they have maintained buildings located near Nordlaguna, on the lower slopes of Beerenberg (Anon. 2003). The infrastructure originally served the functions of nautical radio communication (Jan Mayen Radio) and meteorological field work. Some of the buildings are now fully repaired and may serve scientific purposes. They may be feasible for installment of automated instruments that monitor the local climate and physical variables in the sea, and in Nordlaguna. There is sufficient space to establish a multi-purpose field laboratory that could serve geological and biological investigations. The buildings may also serve as base-camp and emergency lodging for scientists or technicians who perform excursions with snowmobiles and helicopters onto the Beerenberg glaciers.

The present Loran-C facility that is situated at Sørlaguna represents a historical document from the Cold War era. Parts of it could be best maintained as such by serving new functions related to scientific activities. The buildings contain ample space for a variety of functions, but their quality may not meet modern health requirements and new standards of living.

Norwegian authorities will have to consider measures that meet all legal requirements related to management of nature, cultural heritage, and individual health. That may be decisive for the continued use and protection of existing infrastructure (Skagestad 2004).

5. JAN MAYEN RESEARCH LOGISTICS

5.1 Governmental Power and Responsibilities

"Fylkesmannen i Nordland" (County Governor of Nordland) who holds a governmental office in Bodø is the acting Governor of Jan Mayen. His office handles applications to land on the island and organises the monitoring of the Jan Mayen environment. He is responsible for the national administration of the island, but violation of Norwegian law by residents or visitors is handled by the Chief-of-Police in Bodø. Both authorities enforce their power by delegation to the Chief-of-Command on the island (Skagestad 2004).

The Norwegian Polar Institute is given the national mandate to organise polar research funded by Norway, which also includes Jan Mayen. The institute will therefore be involved in expeditions or campaigns with official Norwegian participation. It will also be involved in the planning and operation of logistics associated with a research station on the island.

5.2 Transportation to and from Jan Mayen Island

Most of the transportation to Jan Mayen goes from Bodø, a town in northern Norway, either by chartered ships or by airplanes. Civil air companies that operate long-range propeller crafts have some times been chartered but most of the flights are now made by C-130 Hercules operated by the Norwegian Air Force.

The standards of Bodø airport allow for international flights with airplanes up to the size of C-5A Galaxy that is the largest airplane operated by US Air Force. However, large or heavy cargo may also be transported by railroad to Bodø from any European country or transit harbour.

Bodø harbour has governmental priority and is the national port for shipping to the Norwegian Arctic, including Svalbard. The logistics contain interconnected terminal structures for the different means of transportation at land, sea and air, which allows for rational transfer between them. Thus, operators who transport cargo to Jan Mayen, preferably use this harbour for loading. For such reasons Bodø serves port functions for many national and foreign research vessels that receive cargo and exchange crew and scientists between cruises in the Nordic Seas.

Bodø Regional University operates the 29.4m (346t) research ship, R/V "Oscar Sund", that is well equipped for expeditions to and around Jan Mayen. It is classified for navigation in sea with <60% ice cover, and can carry cargo amounting to 150 tons (150m^3). The vessel's bunker capacities for fuel and water allows for continuous and unsupported operations for more than 30 days and it is certified for continuous navigation with a crew of four including a cook. There are cabins for six scientists, but the capacity is under expansion to 16. Its standard equipment for research operations comprises a split capstan winch for bottom and pelagic trawling to about 1000m depth, one ordinary winch with 2000m wire for light gear, and one winch with 1000m self-supporting cable for CTD, net-sondes and other towed instruments. The electronic instruments comprise a SBE25 SeaLogger and SBE32 carousel with 12 Niskin water samplers, Simrad EK500 Scientific echo sounder with integrator, and Scanmar acoustic trawl sensors. One wet lab and two dry labs provide ample space for work with samples. The university supports the Governor of Jan Mayen in scientific matters and may on request provide research assistance for him and other parties going to Jan Mayen.

Future coordinated international research operations on Jan Mayen would be facilitated by an organized logistical home-base that served and coordinated transportation. That could even apply to handling of visas to Norway and permits for visits to Jan Mayen. It should have at its disposal warehouse facilities that allow for collection, storage and loading of cargo transported to Jan Mayen, as well as for international distribution of cargo sent as shiploads from the island. It could also receive and distribute equipment for repairs and maintenance. As such, it would be rational to have that home-base established in Bodø.

5.3 Accommodation and Work Conditions

The demand for privacy may be met with private sleeping and bath rooms. The work should be organised according to a fixed schedule with daytime routines and if necessary, fixed change of watches. That will allow for fixed meals and proper work conditions for the cooking and mess personnel.

Some field work will be carried out over a couple of days in camps with varying standards of living and will require personal adjustment to work operations and weather conditions. Field researchers must therefore expect to prepare their own food and accept that sleeping conditions and hygiene is a matter of personal adaptation. However, researchers should at any time be able to require support that brings them back to security and comfort at the main station.

Health care is related to adequate nutrition, physical exercise and general well-being. Modern food requirements consider fresh vegetables and fruits as vital constituents in the daily intake of food. Regular transport of food supplies is therefore essential to operations even in remote research stations.

Weather conditions on Jan Mayen make outdoor exercise difficult and even risky during winter time. Rooms for indoor sports and fitness training allow the permanent crew to master general tasks as well as avoid accidental strain.

General well-being is related to the balance of privacy with social activities. Good living quarters should optimise this balance. However, proper leadership and careful selection of personnel with proper psychological profiles is equally important.

5.4 Local Transportation

Field operations will require transportation on land and at sea. Vehicles on wheels may only operate on prepared roads in the middle parts of the island. Transportation on the Beerenberg volcano can occur by skis, sledges and snow-scooters year round in high altitudes. In lower altitudes, and in the lower mountain range in the south, this is only possible when snow is present during winter. In periods with little snow and in summer, these areas can only be reached by foot. Environmental considerations (cf. Huberth-Hansen 2004) will not allow use of all-terrain vehicles (ATV). The assistance of a helicopter would be very feasible and valuable from any point of view.

Boats may navigate around the island when the sea is not ice-covered. However, the open sea and sudden katabatic winds from Beerenberg represent a hazard for navigation with open-decked boats. Solid amphibians and dinghies (zodiacs) may be much safer, but should not be used unless assisted by a ship or a helicopter in stand-by state.

5.5 Communication

Fog, blizzards and sandstorms frequently reduce the visibility on Jan Mayen and may sometimes require rapid retreat to the station. All field operations should therefore be assisted by positioning instruments (GPS), even in the neighbourhood of the research station. All personnel should carry portable satellite telephones and all vehicles should be equipped with radios for communication with the coastal radio station in Bodø. This precaution is not only necessary in cases of lost communication with the research station, but also if the research station should be evacuated in cases of emergency like fire, earthquakes or volcanic eruption.

5.6 Risky Operations

Some research will involve activities that represent risks. Both diving and mountaineering require skills and good equipment. The activities should be allowed but must be supported by firmly established regulations. Only certified and physically well trained personnel should do such work. Stand-by systems should be established to perform rescue operations if necessary. Equipment and trained personnel for first-aid treatment should be available and prepared to function until assistance from outside takes over the responsibility.

5.7 Health Care and Medical Help

Irrespective of how well research operations in remote places may be organised and equipped, accidents may require medical treatment. It is therefore essential to have adequately trained medical personnel permanently stationed on the island. Advanced surgery and follow-up hospital treatment should be available by transportation to Norway. Ships may be deployed from the Norwegian Coast Guard Base in Sortland, northern Norway. The transfer of patients from Jan Mayen to the ship may be performed at sea by the ship's helicopter. The helicopter may also transfer patients from the ship to the proper hospital, but at present the operation will require that patients stay on board the ship for some time. If the Coast Guard would permanently station a helicopter with crew on Jan Mayen, the ship could be used as a platform for the transfer of patients from one helicopter to another. That would reduce the patient's experience of seasickness and agony and cut the transportation time from a day or two to a number of hours.

5.8 Environmental Protection

Norwegian authorities will not allow any activity that disturbs the nature of Jan Mayen or objects of cultural heritage (Huberth-Hansen 2004). All sampling will have to be authorised.

All field equipment and refuse from field activities will have to be brought back to the research station for authorised disintegration or transportation back to Norway. However, automated stations that replace frequent traffic and wear on the landscape should be regarded as authorised parts of the cultural landscape and encouraged as being in accordance with the environmental policy.

5.9 Interaction with Tourism

Jan Mayen is remote and therefore an exotic place for tourists. At an increasing rate cruise-ships and large pleasure crafts have been observed to land people on-shore, even for excursions to the top of Beerenberg. Such activities may be detrimental to field research on the island. Sensitive equipment may be destroyed or record substances that cause bias in time series. Traffic may impact biota being studied. The general logistics of Norwegian presence on the island should therefore restrict such unauthorised activities. Authorisation should be strictly enforced and authorised landings by tourists should be channelled to a specified area where the scientific field work, as well as the general work at the station, is not disturbed.

6. A CALL FOR INTERDISCIPLINARY COOPERATION

Specific scientific campaigns to Jan Mayen Island should not expect to find special research equipment that is typical or unique to their disciplines. A permanent research station should rather be equipped to provide general support that is relevant to all operations, in order to reduce costs. Thus, the permanent station will provide subsistence, lodging, laboratories, and logistics for transportation, field operations and routine measurements. The scientific support system will therefore not require a permanent staff of fully qualified scientists within particular disciplines. The crew of researchers should rather consist of scientifically trained technical assistants up to the level of M.Sc. Their responsibilities would be to serve routine observations in field and laboratories, and assist visiting specialists that perform specific scientific work. That would be equivalent to the system adopted at the Polish research station in Hornsund, Svalbard (Glovacki 2004).

Both permanent and visiting scientific personnel should not waste time in being responsible for daily subsistence and maintenance routines. However, they should at any time be physically fit and prepared to take part in operations that require extra manpower. They will also have to accept responsibilities related to safety precautions and rescue operations.

6.1 Quality of Living Support

The permanent station should be well equipped with living quarters that allow for privacy. That includes private toilet, shower and closets. Common fitness rooms and sauna should be standard requirements, and a small swimming pool should be considered. Mess quarters must contain common living rooms and other social zones (TV room, cinema theater, bar, library, etc.). A well equipped kitchen manned with a qualified chef is mandatory for the well-being of research personnel.

Living quarters and laboratories in satellite stations will have simpler qualities and some field work will be based on camping in tents. In such circumstances scientists must perform or organize their own subsistence according to their personal requirements. However, all field campaigns must include common safety logistics with precautions that secure life and health of personnel who perform any kind of authorized scientific activity.

6.2 Standard Research Logistics

Research operations based on Jan Mayen will be field-based, but with a minimum effort of man-power. High-tech equipment installed in automated stations on land or attached to buoys and submerged rigs at sea will be paramount. Remotely operated or self-navigating vehicles for use on land and in the sea will probably reduce the demand for presence and transportation of scientific and technical personnel. However, some research tasks can only be solved by the presence of human skills, for observations, recovery of samples and data, and maintenance of equipment. These tasks require particular infrastructure and logistical procedures.

From a scientific point of view, laboratory facilities on Jan Mayen may not be very sophisticated, but they should be robust and match functional standards found in good research laboratories on the Norwegian mainland.

Wet laboratories should have taps with running freshwater and seawater. Temperature controlled cabinets for physiological work with terrestrial and aquatic organisms should be available. Dry labs for microscopy and analytical chemistry would be in demand from many disciplines of science, as well as workplaces for computer work and administration of projects.

Access to a well equipped workshop for electronic and mechanical construction and maintenance of scientific equipment cannot be overemphasized.

Maintenance of vehicles and other large units should be left with mechanical engineers who are part of the permanent crew of technicians. They should be qualified to maintain remotely operated monitoring stations on land and at sea. Two should have certificates for scientific diving to do maintenance, perform routine sampling and assist in specific diving operations. The technicians should also assist in transportation, including the piloting of helicopters and hovercrafts, and the driving of snowmobiles and trucks. They should also maintain the airstrip and be involved in procedures for landing and take-off of airplanes.

The operation of a research vessel will be mandatory for research on Jan Mayen. A well equipped ship about 30m long and with a gross tonnage of >300 tons, should be available at the island from May to September. It should only leave its station in the Jan Mayen area once every month for its home base in Norway to refuel, exchange crew and freight scientific cargo and research personnel. Underway, oceanographic observations should be made according to a standardized marine research program. To optimize ship time, observation schedules should use equipment that record data while the ship runs at full speed. Ship-mounted acoustic instruments, like the acoustic Doppler current profiler (ADCP), and echo integrators that record plankton and fish, and towed instruments like yo-yo instruments equipped with electronic, acoustical and optical sensors, and the SAHFOS Continuous Plankton Recorder (Walne and Reid 2004) are recommended.

The technical operation of the research vessel will not be a responsibility of the manager of the Jan Mayen research station, because the Law of the Sea leaves the ultimate command with the captain. However, the research manager will be responsible for planning of the schedule for normal scientific operations that the ship will serve in Jan Mayen waters.

6.3 Transportation Logistics

Large parts of Jan Mayen are under environmental protection, which will not allow free use of vehicles (Hansen 2004). However, there are a few main roads or tracks that are parts of the cultural landscape. Transportation with trucks and cars will therefore be available to some destination points. One is the road from the Loran-C station on the eastern side of the island to Kvalrossbukta on the western side. The other runs along the island to the old meteorological station at Nordlaguna. The maintenance of these two roads is important to minimise the impact on the nature, and also to secure rational scientific operations. All-terrain vehicles (ATV) should be limited to beaches where waves and wind rapidly wipe out any sign of traffic.

The use of snowmobiles and snow-scooters will usually not be possible on terrain in low altitudes, because the snow is readily swept away by wind. They may be used on the Beerenberg glaciers all year round. Thus, the buildings at Nordlaguna could be used as base-camp for such excursions. However, air-lift operations will be in demand from many disciplines.

A helicopter should be permanently stationed on the island. For the support of most scientific operations, a small craft may be sufficient. It would be used to deploy and maintain automated field stations at various places, exchange members of field parties, carry supplies, and perform rescue operations. However, it should be emphasised that helicopter flights may be prevented by fog, blizzards, sand-storms, and not least, strong catabatic winds from Beerenberg.

A larger helicopter with radar and other instruments for blind flying, and advanced equipment for rescue operations would greatly improve the safety precautions. It would allow more field-work to be done by use of ski-doos or hovercrafts in the drift-ice during winter, and during summer excursions with inflated crafts along the shores, without the assistance of a ship. Taking into consideration that Norway holds sovereignty over the Jan Mayen fisheries zone, the authorities may consider the possibility of deploying a Norwegian Coast Guard helicopter on the island. Its main purpose would be to control traffic in territorial waters, prevent unauthorised landings, support commercial fishing expeditions, and assist scientific operations.

A Coast Guard helicopter on Jan Mayen would be able to meet Coast Guard ships mid-way to Norway and transfer personnel who need hospital care or other service that will not be available on Jan Mayen.

Scientific personnel that install new equipment or train technical crew in new observation procedures would prefer short stays on the island. A frequency of two flights per month with C-130 Hercules for exchange of personnel and to renew supplies would be desirable, but may not be very realistic, due to the weather conditions. However, to increase the frequency of flights to Jan Mayen, the present airstrip should be upgraded to an airfield equipped with instrument landing support for aircrafts.

6.4 Particular Equipment Needs

Diving might become an important research activity at Jan Mayen. Some will occur from ships, but most operations may occur from land or from small boats in shallow water, as a permanent year-round activity. Sampling of organisms from marine benthic communities (Gulliksen *et al.* 2004) or in association with drift-ice (Melnikov 2004) may become routine, as well as inspection and maintenance of automated or remotely operated sensors. To be prepared for uncontrolled ascents from even shallow depths, diving operations from both ships and land should be supported by a portable decompression unit. It should be stored and operated in a heated room at the research base, but with an option to be removed for transportation by helicopter, airplane or ship in case of the need for prolonged medical treatment at a hospital.

There is a range of other support systems that should be available at a research base on Jan Mayen such as 1) satellite receiving station, 2) computer servers with Internet access, 3) data network with connecting points for laptops in sleeping rooms, living rooms and labs, 4) freezing and refrigerated rooms for storing of ice cores, botanical and zoological samples, and of archaeological material. Light microscopes with digital camera and a set of balances with different sensitivity and weighing range should be available at any time.

Standard electronic instruments should only be generally available if they are operated at year-round frequencies that ensure routine maintenance. That applies to 1) electrical aerosol spectrometer (EAS), 2) meteorological field instruments, 3) an instrument to measure black carbon content, 4) oxygen analyzer, 5) solar radiation measurer rations, 6) portable CTD (Conductivity, Temperature, Depth recorder, e.g. Sea-Bird Electronics SBE11 Sea-Logger), 7) an auto-analyzer or other instruments for chemical analyses of e.g. plant nutrients and pigments,.

Some items of field equipment should be available for routine work as well as more specialized work, like 1) CRREL SIPRE ice corer, 2) portable power stations, 3) diving equipment including air compressor, 4) walkie-talkies, 5) GPS instruments, etc.

6.5 Sites Requiring Special Attention (Hot Spots)

The top and the crater of the Beerenberg volcano are of particular interest for making meteorological and glaciological observations, and sampling of long-range aerosols and air plankton. Glacier sites at lower altitudes are interesting for the monitoring of their mass balance and contained oceanic aerosols (Hagen 2004).

The newest lava grounds above (Birkenmajer 2004) and below sea level (Gulliksen *et al.* 2004) are still under colonization and of particular interest to biogeographers who study the distribution of terrestrial and marine species. These places should be frequently visited to gain knowledge on geological and geomorphologic processes, as well as on the biota that establish themselves on new ground and sea-bottom.

Seabird colonies and sheltered littoral sites should be frequented to observe biogeographical changes related to oceanographic variability, climate change and toxic anthropogenic influence (Gabrielsen and Strøm 2004, Weslawski 2004).

For the study of terrestrial biota 40 representative areas of 20x50m should be selected and given particular protection. The biota should be different kinds ranging from barren ground to areas with thick vegetation layers (cf. Kyyak 2004).

Kvalrossbukta and Titeltbukta are of particular importance for archaeological studies (Hacquebord 2004). These places are also good localities for studies of near-shore marine biota.

Nordlaguna that contains a population of land-locked char is feasible for the study of trophic interaction between oceanic and oligohaline systems (Skreslet 2004b).

6.6 Support Observations

Several research operations will bring particular instruments but may request standard support observations from other disciplines. These may be meteorological observations at different heights, seawater temperature and salinity, tidal range, wave height, solar radiation, aerosol information (pH etc), and samples of particulates. Therefore, a standard monitoring program should be established to generate time series that may serve reference purposes, equivalent to routines at the Polish station on Svalbard (Glowacki 2004).

Research operations on Jan Mayen would benefit from information supported from new satellites (e.g. Cryosat and Icesat) for the regional mapping of cloud cover, sea ice extent, wave height, position of oceanographic fronts etc. Sites of deep water convection and variability could be monitored with certain satellite techniques (AVHRR, SAR, SeaWifs). The Aster satellite may allow collection of data in the visual wavelengths at smaller scales suitable for terrestrial cryogenic studies, ecology and geomorphology. Routine sampling of data that calibrate satellite monitoring in the Jan Mayen area may also serve to secure data in periods when fog and clouds prevent information from satellites.

6.7 Ecological Basin-Scale Network

Most of the natural immigration of species to Jan Mayen probably originates from shelves around the Nordic Seas Basin. Both terrestrial and marine organisms may be subject to advection of surface water towards the island (Fig. 2). The most important source regions may be the coasts of Norway and Greenland. The newly established research station at Zackenberg on North-East Greenland, and a planned research station at Runde in western Norway have strategic geographical positions that could provide reference data of particular interest for Jan Mayen research. However, coastal research performed at universities, institutes, laboratories and research stations by all nations around the Arctic Mediterranean is of relevance to Jan Mayen studies. Logistical links for exchange of data should therefore be established.

7. BASIC OBSERVATIONS FROM 2005 ONWARDS

The present meteorological observations on Jan Mayen will be continued with more or less the same number of personnel. Some of the support functions provided by the Loran-C personnel will have to be continued as well. That may result in some surplus work capacity that can serve additional functions, like serving equipment and protocols for routine scientific observations.

Since 1972, "Norsk Kartverk" (The Norwegian Mapping Authority) has performed periodical gravitational studies on Jan Mayen, and since 1994 established permanent monitoring of the relative sea level. The geodetic network of trigonometric points is now monitored by satellite technology and included in the national framework. At present, the organisation investigates the possibility of establishing a permanent geodetic observatory for measurement of the geoid, rift spreading and relative sea level change in this region (Harsson, personal communication). That will add to the permanent meteorological observations made on the island since 1921, and the local seismic monitoring established by the University of Bergen, Norway.

"Fylkesmannen i Nordland", the acting Governor of Jan Mayen Island, has enforced some environmental observations to be made by the crew stationed on the island. That will be continued and may even be expanded.

All permanent measurements will probably be incorporated in protocols enforced on Norwegian personnel attached to the meteorological station when the Loran-C system is terminated in 2005. Thus, there may be a redefinition of key skills of the future staff, and their role. Inclusion of advanced scientific equipment may require particular technical qualifications that are now not present. However, once a permanent crew of a sufficiently qualified permanent staff is present, additional science programmes can be established as expeditions or campaigns that top the multidisciplinary monitoring activity. Such expeditions should conform to high international standards.

8. INCENTIVES OF PARTICIPATING NATIONS

The participants of the NATO ARW were asked to identify national research strategies and interests that support the establishment of a permanent research base on Jan Mayen. They were also asked to identify how initiatives from the ARW could be forwarded to gain support from governmental authorities. The opinions are personal and do not necessarily express official national consent.

8.1 Canada

The past decades have seen considerable research and development in the field of eco-toxicology in polar environments, both marine and terrestrial. This work has resulted in expertise in stable-isotope, trace elements, molecular, and contaminant flux and mechanisms relative to polar marine food-webs, particularly those associated with the upper trophic levels. As such, Canadian participation in trophic-related studies associated with Jan Mayen is possible.

Remote sensing is also an area of expertise that can potentially be contributed from Canada. The Greenland current off eastern Canada is an area of oceanographic interest in Canada and this provides a natural link to Jan Mayen research.

8.2 European Union

At the European level, several opportunities for improved co-operation and co-ordination exist:

- Marine oceanographic expeditions. The UK, Germany, Poland and Norway will endeavour to publish their Arctic cruise programmes with the objective of identifying collaborative opportunities in waters around Jan Mayen Island.
- A web-site should be established allowing partner nations to contribute details of proposed work and identify links of interest. The website should also provide details and logistics concerning scientific support on Jan Mayen Island.
- The opportunity to obtain EU funding under a call dedicated to the 4th International Polar Year 2007-2008 (IPY 2007-2008) will be examined.
- The logistics and technology required to install and maintain ocean monitoring instruments will be evaluated.

8.3 Norway

The meteorological station on Jan Mayen will continue as a manned activity. The monitoring work delegated from Norwegian management authorities to the personnel will probably continue. It may be better structured and appear as a permanent activity that may be expanded. Thus, the staff of technicians may expand and include personnel with scientific training at the M.Sc. level.

The environmental action plan for Jan Mayen is overdue. The environmental authorities will probably support expeditions that may fulfill the program in the near future.

From a scientific point of view, governmental investments should not be allocated to dismantle the existing Loran-C infrastructure that may be feasible for international scientific use in the IPY 2007-2008. No building should be dismantled before its physical structures have been documented. The authorities may accept to postpone dismantling for some time to gain time for planning and funding of costly removal of debris from Jan Mayen. It would also allow more time for identification of new governmental responsibilities and construction of new infrastructure that adopts modern standards.

The Norwegian scientists who assist the Governor of Jan Mayen in scientific matters, and work with Jan Mayen projects at the University of Tromsø will cooperate to make scientific activities on Jan Mayen an official affair in Norway. They are encouraged to work for the establishment of a multidisciplinary expedition to Jan Mayen in advance of the IPY 2007-2008, preferably in collaboration with the international community. They will report to cooperative scientific organizations in Norway to gain wider support from the national research communities.

8.4 Poland

The Polish Academy of Sciences, especially the Institutes of Geophysics (IGFPAS) and the Institute of Oceanology (IOPAS), and the Polish Universities, would like to participate in an internationally coordinated research on Jan Mayen. Its surrounding waters are of particular interest to Polish scientists, for the recognition of climate change, and Earth and marine science. For such cooperation Polish facilities in the Arctic might be considered (Polish Polar Station Hornsund in Svalbard, and the research vessel "Oceania").

Polish research groups are ready to provide:

- Sampling and analytical equipment
- Field equipment
- Data analysis in Polish labs
- Time/personell costs covered for Polish participants
- Ship time of "Oceania" up to 6 days for joint marine program
- Data processing, scientific paper production

Furthermore the Polish scientists propose:

- Data exchange between the stations on Jan Mayen and Hornsund (especially data on natural environment and pollutants)

- Exchange visits for Norwegian scientists in Hornsund and r/v "Oceania" for a equivalent time in Jan Mayen facilities

8.5 Russia

As a member of the international World Ocean Circulation Experiment (WOCE), Russia has conducted time series observations during 15 years along N 60° in the Greenland Sea. Currently, physical, chemical, and geological data sets are available. In the context of recent Global climate change, Russia may be expected to continue oceanographic research in the North Atlantic based on national sources. However, Russian use of facilities on Jan Mayen would be very feasible, and should preferably occur in cooperation with NATO countries under the umbrella of international programmes like GLOBEC, CLIVAR etc.

8.6 Ukraine

Scientists in Ukraine have conducted Alpine vegetation research for 120 years. During the last 40 years, long-term ecological research of plant communities has been established on the population and community levels. It was recently established that a number of Arctic-alpine species are common to terrestrial ecosystems in the Carpathians and on Jan Mayen (Kyyak 2004). This feature can be used as a bio-indicator control of climate change in the Jan Mayen region and on the northern hemisphere.

9. RELEVANT INTERNATIONAL ORGANISATIONS

Actual scientific research is mostly funded from national sources. National initiatives may be formally coordinated with other nations, but are very often based on informal communication and exchange of practical information, material or data between individual scientists. However, much science is founded on international agreements based on formal exchange of information in international meetings and congresses.

Numerous scientific organisations are involved in Arctic research, both on national and international levels. The international research organisations are usually grouped in intergovernmental and nongovernmental organisations that cooperate in many ways that are not always evident.

9.1 Intergovernmental Organisations

Typical intergovernmental organisations have been established by agreement among nations. The nations are represented by delegates who have the mandate to express national interests and agree on terms negotiated during the meeting. There are a lot of organisations that relate to Arctic research, but most of them may be addressed through cooperation with a few large ones dealt with here.

9.1.1 Arctic Council

The Arctic Council was established as late as 1998 by eight nations having their own territories inside the Arctic. However, several non-Arctic states, intergovernmental, inter-parliamentary and nongovernmental organisations participate as observers. Several programs that are relevant to Jan Mayen science occur under the umbrella of the Arctic Council:

- AMAP (Arctic Monitoring and Assessment Programme).
- CAFF (Conservation of Arctic Flora and Fauna).
- PAME (Protection of the Arctic Marine Environment).
- ACIA (Arctic Climate Impact Assessment).

9.1.2 International Council for Exploration of the Sea

ICES give advice to the North-East Atlantic Fisheries Commission (NEAFC) and coordinate marine science that supports management of exploited living marine resources. Its activities are increasingly under development towards multi-species management and ecosystem analyses.

The ICES Cod and Climate Change Program has been assigned to a working group (ICES WGCCC) from 1992. It has been approved as one of the regional programs within GLOBEC and will continue its work until 2009 when GLOBEC will be officially ended. It will be associated with any engagement of ICES in the IPY 2007-2008.

9.1.3 North Atlantic Treaty Organisation

Since World War II, the Arctic has been a region of manoeuvre for armed forces that protect national interests. Some nations with territories inside the Arctic still maintain their status as neutral. Others have recently become members of NATO to participate in common defence actions (Table 1). However, the organisation has now accepted the Russian Federation and other former members of the Soviet Republic as Partner Countries that cooperate to decrease strategic tensions and promote national cooperation on several levels.

One of the instruments to increase hemispheric security is to stimulate contact and cooperation between scientists, organised by the NATO Scientific and Environmental Affairs Division. The priorities of the division change with the strategic decisions made within the alliance. Thus, international terrorism has at present been placed first on the agenda. However, climate change and other environmental threats that may promote conflicts of interest and destabilise relations between nations, are elements that fit into the next points of priority.

NATO will probably not directly support specific campaigns related to Jan Mayen research, but Collaborative Linkage Grants may possibly support planning processes where NATO and Partner Country members participate. Visiting Specialists may possibly qualify for funding when they train technical staff or fellow scientists in particular methods used in campaigns or monitoring.

The organisation may possibly support Advanced Study Institutes and Advanced Research Workshops that address Global environmental problems being reasons for doing research on the island. The geographic position of the island in relation to crucial meteorological and oceanographic processes that forces the World climate may warrant funding.

Any priority expressed by Partner Countries comes high on the list.

9.1.4 United Nations

UN organises scientific activities on behalf of member nations. Several of its organisations are involved in programs that apply to Jan Mayen:

- IOC (Intergovernmental Oceanographic Commission) is organised under the umbrella of United Nations. It assists governments in scientific matters related to marine environmental protection, fisheries and ecosystems, climate change, oceanic observing and monitoring, etc.
- WMO (World Meteorological Organisation) is established to coordinate, standardise and improve meteorological activities world wide. Thus, the meteorological station on Jan Mayen reports data within the WMO network and the activity occurs according to standards defined by the organisation.

9.1.5 European Science Foundation

ESF organises science within the European Union. In the 6th framework that is under development, one of seven programs concerns "sustainable development, Global change and ecosystems". Among several aspects ESF addresses biodiversity, climate research, coastal zone management, and marine research. One board in particular may become involved in the coordination of research related to the 6th framework and how it may relate to Jan Mayen research:

- EMB (ESF Marine Board) is not directly addressing Arctic research but the science plan (Boissonnas and d'Ozouville 2001) expresses interest in geophysical processes and living resources, biodiversity and anthropogenic effects that are unavoidably related to the Jan Mayen region. The science plan is comprehensive with respect to use of new technology for marine science.

9.2 Non-Governmental Organisations

Scientific NGOs are established by scientists or their organisations to support scientific incentives and have no direct obligations to governments. However, they may receive governmental funding and have considerable impact on the initiation and planning of governmental research established to develop the national economies and serve the management.

9.2.1 Arctic Ocean Sciences Board

AOSB is a non-governmental body that includes members and participants from research and governmental institutions in 17 countries. It addresses some missions that are particularly relevant to research on Jan Mayen:

- encouraging and supporting science-led international programs by offering planning, coordination and access to funding and logistics.
- establishing means of initiating and maintaining observational systems and the data produced

9.2.2 International Council of Scientific Unions

ICSU is the largest of the international scientific NGOs. It represents research in the university and institute sector. Among its committees are:

- IASC (International Arctic Science Committee), an organisation with 18 member countries. Arctic and non-Arctic countries are member based on their Arctic research programs. IASC has asked for ICSU membership recently.
- IUGG (International Union of Geodesy and Geophysics).

- SCOR (Scientific Committee for Oceanographic Research), an interdisciplinary committee that promotes international research programs, and is an advisory organ for UNESCO and IOC.

9.2.3 Other NGOs

Several international NGOs have been established by scientists who organises themselves on the basis of an individual membership:

- EFMS (European Federation for Marine Science and Technology Societies) is an organisation of national associations and societies that organise individual oceanographers in Europe. One of its roles is to take part in scientific decision-making that influences the establishment of framework programs for research in the European Union.
- EPB (European Polar Board).
- IUBS (International Union of Biological Science).
- IUCN (International Union for Conservation of Nature).

9.3 Relevant International Programs

Several intergovernmental and non-governmental organisations have common scientific problems on their agenda and therefore cooperate within international research programs. That is the case with some programs that address cases relevant to the Nordic Seas:

- ACIA (Arctic Climate Impact Assessment) is an initiative of the Arctic Council and IASC and executed together with CAFF and AMAP (Arctic Monitoring and Assessment Program). In 2004 ACIA will release its first comprehensive synthesis of the consequences of changes in climate and UV radiation in the Arctic. Follow-up programs may be expected and Jan Mayen could be a very particular and interesting point of reference.
- A station on Jan Mayen could also play an important role in CEON (Circumpolar Arctic Environmental Observatories Network) a joint project of IASC and FARO (Forum for Arctic Research Operators).
- CAFF (Conservation of Arctic Flora and Fauna) is developing a program to monitor circumpolar biodiversity. Jan Mayen may be a very feasible place to study the dynamic development of biota

subject to seasonal and inter-annual changes as a consequence of alternating Arctic and Atlantic influence.

- GLOBEC (Global Ocean Ecosystem Dynamics) was initiated by SCOR and IOC but is now part of IGBP (International Geosphere-Biosphere Programme) that is organised by ICSU.
  - ESSAS (Ecosystem Studies of Sub-Arctic Seas) is a new initiative that may become a hemispheric project within GLOBEC and will be of particular relevance to the Jan Mayen issue.
  - ICES/GLOBEC CCC (Cod and Climate Change) is one of nine programs where the International GLOBEC addresses how Global change may affect the abundance, diversity and productivity of marine populations.
  - IMBER (Integrated Marine Biogeochemistry and Ecosystem Research) project is presently being implemented by IGBP in cooperation with SCOR. Its goal is to understand how interactions between marine biogeochemical cycles and ecosystems respond to and force Global change. The Greenland Sea is a carbon sink and Jan Mayen is an appropriate place to monitor the hemispheric hydrological cycle, and study the fluxes of natural and anthropogenic chemicals. Zooplankton populations in the Nordic Seas have roles in the transfer of geochemicals in marine food-webs that are essential to production of human food in the North Atlantic. Thus, IGBP should be made aware of the potential for data acquisition and science in the Jan Mayen area, and its potential importance to IMBER.

10. RECOMMENDATIONS

The NATO ARW participants found that the establishment of an international station on Jan Mayen Island would involve scientific and logistical aspects, as well as political issues.

10.1 Priority Fields in Jan Mayen Science

Among the scientific problems that should be addressed on Jan Mayen Island are:

- General island biogeography and ecology (dispersion, colonisation).
- Airborne transport of pollutants.

- Integrated volcanological studies.
- Model study of colonisation of a remote Arctic island by diaspores that involves:
    - Their origin.
    - Transportation (wind, ocean currents, birds).
    - Effects of anthropogenic perturbation.
- Growth experiments with biological material from drift-wood and birds together with modelling of wind circulation.
- Formation, growth and transport of natural and anthropogenic aerosols and their role in cloud formation and climate change in the marine Arctic.
- Interdisciplinary study on the levels and impacts of environmental pollutants in the Jan Mayen ecosystem.
- Determination of a climate signal in the atmosphere/ocean system within the Jan Mayen region.
- To assess the carbon export from the Arctic Ocean by drift-ice in terms of ecological carbon sequestration in the Jan Mayen area.
- To intensify oceanographic research in adjacent marine areas (during the IPY 2007-2008 in particular), with special attention to:
    - Fresh-water balance.
    - Deep convection and deep water formation.
    - Mass and energy transport
    - Dynamics of the main currents, meso-scale features in particular.
- To better understand processes, their variability and consequences for the THC (Thermo-Haline Circulation) in the North Atlantic.
- History of the human occupation of the island and its impact on Jan Mayens environment.
- The role of Jan Mayen in the international polar years (IPYs).

10.2 Logistics

To develop interests for research on Jan Mayen the participants suggested:

- Expansion of the ongoing Norwegian meteorological observations.
- Provision of logistics & facilities for international expeditions:
    - Regular and reliable air transport of personnel, equipment and supplies. Upgrading of landing facilities.
    - Accommodation on-site.
    - Regular CTD and current measurements required within Jan Mayen region.

- Free access to long-term time-series on the temperature, atmospheric circulation and other climatic indices at Jan Mayen, Arctic, North Atlantic and the northern hemisphere.
- IPY 2007-2008 expedition with core group on Jan Mayen for longer time, and several groups for a shorter time. The approach should be multidisciplinary, but with regard to marine science restricted to inter-tidal work and reconnaissance for later, more extensive work. 12 people in 3 months for baseline surveys, sampling, and observations.
- Establishing permanent instrumentation, some oceanographic work starting now.
- The regional expedition activity in the Jan Mayen region should be accompanied with the current coordinated efforts in many countries with regard to diagnostic analyses of collected data aimed at modelling.
- Establishment of seasonal time series observations in transect "Jan Mayen - Greenland".
- High resolution coupled modelling atmosphere-ocean-ice as a tool for research.

10.3 Political Issues

The NATO ARW became well informed on national political and administrative issues related to Jan Mayen (Skagestad 2004) but did not have the qualifications to clearly identify political instruments that would facilitate the establishment of a scientific research base on the island. However, it is obvious that international research on the island needs to be organised within an intergovernmental scientific body. Being subject to trans-Atlantic natural processes and raising interests of nations on both sides of the ocean, the participants raised the question that NATO may serve a key role in this regard. However, nongovernmental organisations that represent the international scientific community should also become involved. The International Committee for Science (ICSU) plays a central role in this respect.

The participants expressed no opinions on how a Jan Mayen research base would fit into the Norwegian governmental structure. However, they strongly recommended that the Norwegian plans for a Jan Mayen environmental conservation management programme (Huberth-Hansen 2004) should be clarified as soon as possible and come to a reality well before 2007. The rationale is that the planning of the logistics for the 4th International Polar Year campaign in 2003-2004 will have to take into consideration adjustments that do not challenge environmental regulations.

10.4 Base-Line Census of Jan Mayen 2005 and/or 2006

The topography of Jan Mayen Island and the shallow parts of its shelf is well described and documented in maps. Otherwise, the existing general knowledge of Jan Mayen is weak and fragmentary and not very fit for deduction of new and interesting scientific hypotheses.

To provide a better base of information that may facilitate planning of more comprehensive scientific efforts, the NATO ARW pointed to the need for a base-line census that may reveal information being complementary to the knowledge that already exists (Gabrielsen *et al.* 1997).

The geomorphology of the landscape is known in some detail (Birkenmajer 2004), which is also the case with biota on the middle part of the island where human activities have been centered. Cliffs and rocky beaches make the northern and southern coastlines much less accessible. Therefore, the knowledge of their biota of terrestrial plants, invertebrates and birds are incompletely described. That is also the case with cultural remains.

Observations made from ships have been easier to make. Thus, the location of seabird colonies are known (Gabrielsen and Strøm 2004), but the abundance estimates of nesting pairs (Gabrielsen *et al.* 1997) are very inaccurate and need updating by inspection at closer range.

Several marine investigations have been made around the island, from ships and by diving (Gulliksen *et al.* 2004). The composition of the benthic fauna is therefore fairly well known, and fisheries resources of prawn (*Pandalus borealis*) and Iceland scallop (*Chlamys islandica*) on the island's shelf have been subject to investigations and management. However, marine biota in the inter-tidal zone and in the upper parts of the sub-littoral zone that can only be investigated by close visual inspection or by samplers operated from small boats remain to be investigated (Weslawski 2004). That is also the case with the geomorphology of beaches and inter-tidal rocks.

Most observations that have been made on Jan Mayen Island are old and may not adequately represent the present state. That is the case with several conditions related to management. The Norwegian authorities have planned to update knowledge by making new observations in a limited monitoring program (Anon. 2000). However, due to the lack of reliable logistics, these observations have not been made. If they are not made in 2004, it would be feasible to organize a more comprehensive expedition that combines the monitoring with a supplementary baseline census in 2005 or 2006. Such an expedition would serve valuable functions as an introduction to the IPY 2007-2008.

The NATO ARW recommended that a symposium could be organized in 2006 to summarize the information that exists in present data bases and the data that may emerge from new monitoring and a baseline census. The scope of the symposium should be to:

1. Identify time-series of parameters that have been or are presently being accumulated in data bases.
2. Suggest variables for which new time-series are needed.
3. Discuss standardization of measurements.

10.5 IPY Expedition or Campaign in 2007-08

An expedition to Jan Mayen Island in the 4th International Polar Year 2007-2008 (IPY 2007-2008), would allow studies of seasonal changes in many variables. Each would be interesting from a local and disciplinary point of view, but priority should be given to those that are more universally interesting, on regional or Global scales. They should also be used to initiate or modernize multi-year time-series.

During the IPY 2007-2008 routine observations in physical, chemical, and biological oceanography should be made year round, by automated observation facilities moored near-shore, and by short excursions with appropriate transportation technology, when the weather and safety precautions permit. However, the core activity would be interdisciplinary research vessel cruises made in three transects: From Lofoten in Norway to Jan Mayen, from Jan Mayen to Svalbard, and from Jan Mayen to Greenland (cf. Piechura 2004). The section from Jan Mayen to Greenland, i.e. across the East Greenland Current, should preferably be compared with observations from two parallel transects: From Svalbard to Greenland, and from Iceland to Greenland (cf. Melnikov 2004).

The IPY should plan to stimulate an intensification of the temporal and spatial dimensions of ongoing time-series to establish reference points in the periphery of the Nordic Seas. Thus, the program on Jan Mayen should be linked to other oceanographic work, at Zackenburg in Greenland, and field stations in Iceland, the Faeroes and Norway, including Svalbard. The Russian observations in the Kola section (Byshev *et al.* 2004) are obviously valuable as reference material.

In marine international programmes the freshwater balance, salt and heat transport, and deep water formation are linked to the ASOF programme. This follows the VEINS programme of which results have not been fully released. In the context of IPY 2007-2008, the marine activity should be linked to international scientific research objectives (Shimmield 2004). This would allow for access to costly data acquired from satellites and airborne observation platforms. During an IPY expedition to Jan Mayen, the work on the island in the fields of glaciology, geomorphology, geophysics, geochemistry, and ecology should be organized as an interdisciplinary campaign to optimise investment (Glowacki 2004). Process studies should be conducted in the context of routine observations (e.g. meteorology data). Jan Mayen should be proposed as a special site in the context of a wider Arctic observations network within the IPY 2007-2008 that will be aimed at the Arctic climate structure. Jan Mayen Island represents a key location in this structure, e.g. where oceanic and atmospheric synoptic activity is focused. It is an area of intense variability where observational data will make an important contribution to calibration and verification of existing models (Romanov 2004). Thus, the IPY 2007-2008 campaign should test observation procedures that may be included in a permanent research program established at an observatory on Jan Mayen.

The program of an IGY 2007-2008 campaign to Jan Mayen should have some elements on the priority list:

- Volcanological investigations (cf. Birkenmajer 2004).
    - Magma chemistry.
    - Temperature and composition of exhalations.
    - Study of tectonic elements, faults and fractures by aerial photography and terrestrial mapping.
    - Submarine mapping of the Beerenberg volcano's base by multi-beam sonar and video recording. Fractures in the Mid-Atlantic Ridge to the north possibly include hydrothermal vents with active geochemical processes going on. The depth range of such processes may be studied along the submarine slope that descends from the sea surface level.
- Cryosphere: permafrost, snow and ice.
    - Reconstruction of recent climate change (glacier mapping; ice coring) and trends in pollution levels. This will require one season to identify the best site with respect to temperature variation and glacier dynamics (cf. Hagen 2004).
    - Ongoing climate change should be monitored from one glacier by conducting mass balance observations, meltwater etc.
    - Distribution of permafrost across the island should be mapped. The complexity of the permafrost development will take into account the high heat flux from the island. This will allow the active layer to be identified. It is recognised that the active layer links to the NAO.
    - Elevation data (Beerenburg transect) for collection of snow samples will allow the estimation of marine boundary layer effects.

- Geomorphology (cf. Birkenmajer 2004).
  - Landform study of the Quaternary deposits.
  - Establishment of links to glaciology.
- Sediment/soil geochemistry reconstruction.
  - Collection of cores for dating of environmental change and sediment fluxes.
  - Palaeo-reconstruction of pollutants, fluxes etc. linked to the Arctic coastal basin dynamics projects.
- Airborne elements sampled from volumes of air, snow, freshwater and soil (cf. Kyyak 2004, Vana 2004):
  - Aerosols (seawater ions, contaminants).
  - Atmospheric plankton (seeds, pollen, juvenile spiders etc).
- Search for terrestrial or freshwater organisms that are immigrants in sea ice and drift-logs (cf. Kyyak 2004).
- Characteristics of the oceanic system (cf. Byshev *et al.* 2004, Gulliksen *et al.* 2004, Hov 2004, Melnikov 2004, Piechura 2004, Skreslet 2004b, Weslawski 2004):
  - Correlations of local air-sea interaction (meteorology, sea-surface exchange, thermal radiation, phase-shifts in water)
  - Seawater geophysics (seasonal stratigraphy, turbulence, deep-water formation).
  - Marine geochemical elements (cycling of plant nutrients, pollution in seawater and sea ice, import by advection of seawater).
  - Seasonal changes (phytoplankton blooms, reproduction periods in holoplankton and zoobenthos).
  - Allochthoneous invasions of plankton (phenotypic morphs, generation or population switches expressed by lipid profiles, alien larvae etc.).
  - Satellite location of drift-ice and oceanic fronts to identify seal reproduction and seabird foraging habitats, combined with *in situ* studies of trophic interaction with prey.
- Phenology of terrestrial plants, birds and selected marine taxa, including clutch initiation, life cycles and trophic position (cf. Gabrielsen and Strøm 2004, Gulliksen *et al.* 2004, Hobson 2004, Kyyak 2004).
- Growth, reproductive capacity etc. in organisms related to climate change (physiology of local morphs).

- Human dimension on Jan Mayen and the role of the island in the International Polar Years (Barr 2004a, 2004b, Hacquebord 2004).

10.6 Monitoring Program after IPY 2007-08

An inclusion of Jan Mayen Island in the IPY 2007-2008 may promote a more lasting inclusion in the framework of AMAP (Arctic Monitoring and Assessment Program), but this possibility is to be decided by Norwegian authorities (cf. Skagestad 2004). Steps should also be taken to identify the variety of current networks and ensure that measurements at Jan Mayen will be included. The NATO ARW identified several fields of disciplinary work.

10.6.1 Lithospheric Observations

Volcanic activity with periodic eruptions of lava from fissures on Jan Mayen is a consequence of regional tectonic processes at the northern end of the Mid-Atlantic Ridge. Observations that record such changes not only explain the local geology, but also contribute to the universal understanding of processes that work on larger scales. The geodetic, geological and seismological monitoring that already exists is valuable and should be continued, but new time-series should be established. The magnetic field is very complex in this region and information of Global interest may result from routine monitoring.

For the safety of personnel moving around on the island, the slope stability of mixed layers of snow, ice, lava, and mud should be monitored. Earthquakes, eruptions and toxic fumes represent hazards that may be expected to occur and should be monitored as a safety precaution as well as from a scientific point of view (Birkenmajer 2004).

10.6.2 Meteorology

The present meteorological observations on Jan Mayen (Hov 2004) are expected to continue for reasons of regional weather forecasting. The service should also include new routine measurements. Due to the cloudy conditions on Jan Mayen, optical techniques will not be so useful. However, collection of precipitation will serve monitoring of pollutants like pH, particulates, black carbon, metals, and aerosol size distribution (cf. Vana 2004). Such observations should be linked to the Arctic network that is based in Rovaniemi. It coordinates the atmospheric monitoring of long range pollutants transported to the Arctic (bulk aerosol deposition; ionic composition).

Scientific studies of the local climate in different places of the island should be included in the meteorological work. This is because the meteorological service may support scientific logistics that require weather forecasts in particular localities. It may also identify back trajectory projections that recognise local effects, and identify peak episodes in deposition of acid rain that have an impact on local ecosystems and environmental geochemistry.

Measurements in remote places of the island should be operated by automated stations that transmit data by ground-based links or by satellite. Some data may be sent to the manned base station on Jan Mayen, while other data may preferably be sent directly to users world-wide. On Jan Mayen such stations will be subject to strong winds, sand erosion and icing. They will therefore require repair, routine inspection and maintenance. Some stations may be accessed by trucks or snowmobiles, while others may require transportation by boat or helicopter.

10.6.3 Oceanography

Geophysical routine measurements in the sea should preferably be performed by sensors carried by moorings and offshore monitoring stations operated from the land base. However, ships are needed to establish and maintain automated stations, and their instruments should be controlled and calibrated by standardised ship-based methods. Monitoring of the air-sea interaction should be established as a scale of variation that represents the meso-scale variability of the atmosphere, for instance the Rossby radius of deformation in the marine boundary layer. This would require uninterrupted measurements of characteristics which record regional and local effects. These measurements must be linked into a wider observing network. Parameters will include temperature, salinity, etc from moorings over long periods of time.

Chemical and biological oceanography can increasingly be performed by automated sensors, and optical and acoustic instruments (Boissonnas and d'Ozouville 2001). They may use moored stations as platforms and communicate by cables or acoustic transfer of data. However, the promising development of self-navigating remotely operated vehicles (ROVs) opens the possibility for unmanned cruises with sensors to geographical positions and depths that are not accessible by manned platforms. Still, many highly relevant measurements rely on ship-borne samplers like water-bottles, traps, nets, trawls, dredges and corers. Some kinds of sampling require visual observation of organisms, like in diving-based studies of sympagic (under-ice) communities (Melnikov 2004), and observations and sampling of seabirds, seals and whales.

Ships are the best platforms for routine surveys of rather conservative features. However, hovercrafts may be useful when drift-ice enclose the island and prevents adequate navigation. Helicopters that should be kept for safety precautions are also feasible for spotting of non-conservative features, e.g. aggregations of foraging birds in oceanic fronts, lairds of juvenile seals on drift-ice etc.

11. ECONOMY

The NATO ARW participants were not prepared to estimate costs associated with maintenance or building of infrastructure, and running costs associated with scientific operations. However, they drew attention to the standards that they are acquainted with during previous Arctic field campaigns and would accept the infrastructure standards that are presently prevailing on Jan Mayen. They understood that the removal of buildings from Jan Mayen would require considerable governmental expenses and found that to be detrimental to Arctic science.

The participants also noted that several nations operate research vessels on a regular basis in the Nordic Seas. Thus, joint international scientific work on and around Jan Mayen is a matter of international agreements rather than being a matter of new and large expenses to Norway per se.

National and international funding agencies should be notified of the potential for doing science on and around Jan Mayen when the Loran-C activity expires. International organisations that fund scientific cooperation should be invited to consider the inclusion of a Jan Mayen research station in their future programs.

It was understood that Norway runs inadequate transportation logistics for scientific operations on Jan Mayen, and that extra costs will be required. However, the geographic position of the island may be so important for future international research on Global scientific issues that other nations may be prepared to contribute logistical funding, if the station may be given a trans-national rather than national status.

12. CONCLUSION

The participants concluded that an increased scientific research activity on Jan Mayen Island will be of significant importance to the future understanding of Arctic ecology and geophysical feed-back couplings in Global geophysical processes. They expressed willingness to share this responsibility with Norwegian colleagues.

The recommendations from the NATO ARW must be made available to the Norwegian inter-ministerial group that is planning the use of the Loran-C infrastructure that is left vacant on Jan Mayen Island after 2005 (Tviberg 2004). The communication with Norwegian authorities should be followed up on higher administrative levels, by formal contacts with the Norwegian Ministries of Science, Environment, Fisheries, Defence, and Foreign Affairs, the Norwegian Research Council, and the Governor of Jan Mayen.

The participants agreed to inform their respective institutes and motivate national non-governmental bodies to consider scientific efforts on Jan Mayen. National scientific organisations are expected to bring forward the issue in international organisations.

The NATO Scientific and Environmental Affairs Division is expected to consider the results from the Advanced Research Workshop in Oslo within its bodies. It is hoped that the organisation will make steps to facilitate the inclusion of a Jan Mayen research station in the network of collaborating international institutions and organizations in Europe, Asia and North America.

REFERENCES

Anon. 2000. Miljøhandlingsplan for Jan Mayen. År 2000 til 2005. Fylkemannen i Nordland, Report 2/2000. 45 pp. (in Norwegian).

Anon. 2003. Jan Mayen og fremtiden. Betenkning fra Ishavsforeningen Jan Mayen November 2003. 17 pp. (in Norwegian).

Barr S 2004a. Past and present infrastructure of Jan Mayen as historical documents. Pp. 239-247 in S Skreslet (ed.) *Jan Mayen Island in Scientific Focus*. Kluwer Publishers, Dordrecht.

Barr S 2004b. The logistics of past scientific research on Jan Mayen Island. Pp. 285-294 in S Skreslet (ed.) *Jan Mayen Island in Scientific Focus*. Kluwer Publishers, Dordrecht.

Birkenmajer K 2004. Geology of Jan Mayen Island and surroundings, Pp. 13-26 in S Skreslet (ed.) *Jan Mayen Island in Scientific Focus*. Kluwer Publishers, Dordrecht.

Boissonnas J and L d'Ozouville 2001 (eds.). An integrated Marine Science Plan for Europe. ESF Marine Board Working Document. 116 pp.

Byshev VI, VG Neiman, and JA Romanov 2004. On the origin of climatic signals in the Nordic Seas. Pp. 79-89 in S Skreslet (ed.) *Jan Mayen Island in Scientific Focus*. Kluwer Publishers, Dordrecht.

Gabrielsen G and H Strøm 2004. Seabird research and monitoring on Jan Mayen. Pp. 181-000 in S Skreslet (ed.) *Jan Mayen Island in Scientific Focus*. Kluwer Publishers, Dordrecht.

Gabrielsen GW, B Brekke, IG Alsos and JR Hansen 1997. Natur- og kulturmiljøet på Jan Mayen. Norsk Polarinstitutt, Meddelelser 144. 127 pp. (Norwegian).

Glowacki P 2004. Research operations in remote Arctic islands. Pp. 249-259 in S Skreslet (ed.) *Jan Mayen Island in Scientific Focus*. Kluwer Publishers, Dordrecht.

Golmen L and S Skreslet 2004. Two new Nordic research facilities ... of western Norway ... and on Jan Mayen? *Ocean Challenge* 13 (2), 18-19.

Gulliksen B, F Beuchel, T Brattegard, and R Palerud 2004. The marine sublittoral fauna of Jan Mayen Island. Pp. 159-171 in S Skreslet (ed.) *Jan Mayen Island in Scientific Focus*. Kluwer Publishers, Dordrecht.

Hacquebord L 2004. The Jan Mayen whaling industry. Pp. 229-238 in S Skreslet (ed.) *Jan Mayen Island in Scientific Focus*. Kluwer Publishers, Dordrecht.

Hagen JOM 2004. The potential of the Beerenberg glaciers for climate studies. Pp. 27-36 in S Skreslet (ed.) *Jan Mayen Island in Scientific Focus*. Kluwer Publishers, Dordrecht.

Hirche HJ 2004. Zooplankton habitats of the Greenland Sea. Pp. 123-133 in S Skreslet (ed.) *Jan Mayen Island in Scientific Focus*. Kluwer Publishers, Dordrecht.

Hobson KA 2004. Research opportunities for Jan Mayen using stable isotope analyses. Pp. 219-227 in S Skreslet (ed.) *Jan Mayen Island in Scientific Focus*. Kluwer Publishers, Dordrecht.

Hov Ø 2004. Meteorological observations on Jan Mayen. Pp. 37-40 in S Skreslet (ed.) *Jan Mayen Island in Scientific Focus*. Kluwer Publishers, Dordrecht.

Huberth-Hansen JP 2004. Norwegian protection of Arctic islands. Pp. 279-280 in S Skreslet (ed.) *Jan Mayen Island in Scientific Focus*. Kluwer Publishers, Dordrecht.

Klyashtorin LB and AA Lyubushin 2004. Cyclic variability of climate and fisheries resources. Pp 147-157 in S Skreslet (ed.) *Jan Mayen Island in Scientific Focus*. Kluwer Publishers, Dordrecht.

Kyyak V 1004. Dispersal of plants in remote Arctic-alpine habitats. Pp. 195-206 in S Skreslet (ed.) *Jan Mayen Island in Scientific Focus*. Kluwer Publishers, Dordrecht.

Melnikov IA 2004. Sea ice export as an indicator of changing Arctic Ocean ecology. Pp. 113-122 in S Skreslet (ed.) *Jan Mayen Island in Scientific Focus*. Kluwer Publishers, Dordrecht.

Piechura J 2004. The circulation of the Nordic Seas. Pp. 91-99 in S Skreslet (ed.) *Jan Mayen Island in Scientific Focus*. Kluwer Publishers, Dordrecht.

Romanov VF 2004. Northern hemispheric air circulation. Pp. 55-66 in S Skreslet (ed.) *Jan Mayen Island in Scientific Focus*. Kluwer Publishers, Dordrecht.

Rouvinen-Watt K 2004. Biological premises for management of an arctic fox population, Pp. 207-218 in S Skreslet (ed.) *Jan Mayen Island in Scientific Focus*. Kluwer Publishers, Dordrecht.

Shimmield GB 2004. Jan Mayen Island and its waters as potential targets for European Arctic research. Pp. 261-267 in S Skreslet (ed.) *Jan Mayen Island in Scientific Focus*. Kluwer Publishers, Dordrecht.

Skagestad, The Scope for Norwegian Commitments Related to International Research on Jan Mayen Island. Pp. 269-278 in S Skreslet (ed.) *Jan Mayen Island in Scientific Focus*. Kluwer Publishers, Dordrecht.

Skreslet S 2004a. Research operations on Jan Mayen Island. Pp. 1-11 in S Skreslet (ed.) *Jan Mayen Island in Scientific Focus*. Kluwer Publishers, Dordrecht.

Skreslet S 2004b. Jan Mayen Island ecology. Pp. 101-112 in S Skreslet (ed.) *Jan Mayen Island in Scientific Focus*. Kluwer Publishers, Dordrecht.

Stankunavicius G 2004. North Atlantic influence on Lithuanian climate. Pp. 41-54 in S Skreslet (ed.) *Jan Mayen Island in Scientific Focus*. Kluwer Publishers, Dordrecht.

Tviberg LK 2004. Jan Mayen infrastructure. Pp. 281-284 in S Skreslet (ed.) *Jan Mayen Island in Scientific Focus*. Kluwer Publishers, Dordrecht.

Vana M 2004. Natural and anthropogenic aerosols in Arctic areas. Pp. 67-77 in S Skreslet (ed.) *Jan Mayen Island in Scientific Focus*. Kluwer Publishers, Dordrecht.

Walne AW and PC Reid 2004. Continuous plankton recorder observations. Pp. 135-145 in S Skreslet (ed.) *Jan Mayen Island in Scientific Focus*. Kluwer Publishers, Dordrecht.

Weslawski JM 2004. The marine fauna of Arctic islands as bioindicators. Pp. 173-180 in S Skreslet (ed.) *Jan Mayen Island in Scientific Focus*. Kluwer Publishers, Dordrecht.

Chapter 30 NATO Advanced Research Workshop

Summary of NATO ARW

Blue-Print for Jan Mayen Island Research Development

Stig Skreslet and Krzysztof Birkenmajer

S Skreslet, Faculty of Fisheries and Natural Sciences, Bodø Regional University, Norway
E-mail: stig.skreslet@hibo.no
K Birkenmajer, Institute of Geological Sciences, Krakow, Poland

Abstract: Eleven NATO Country scientists and eleven Partner Country scientists met in Oslo 11-15 November 2003 to discuss the possibility of taking the Loran-C station based on Jan Mayen into use for international multidisciplinary research. They documented wide research interests expressed from five NATO Countries and five Partner Countries. When vacated in 2005 the infrastructure should be maintained to accommodate visiting scientists and a permanent group of research technicians. Adequate logistics would require special safety precautions, environmentally acceptable transportation on the island, and means of frequent transport between the island and a home base in Norway.

Key words: natural science, history, cultural heritage, aviation, nature conservation

1. INTRODUCTION

The NATO Advanced Research Workshop (ARW) that was organized in Oslo, Norway 11-15 November 2003 dealt with the possibility of establishing a "Joint International Scientific Observation Facility on Jan Mayen Island". The text of the present paper is excerpted from the general report sent by the two NATO ARW Co-Directors to the NATO Science Programme after the meeting.

S. Skreslet (ed.), Jan Mayen Island in Scientific Focus, 341-344

2. BRIEF BACKGROUND INFORMATION

Jan Mayen Island is a very remote and isolated island in the Nordic Seas. It is located on the Mid-Atlantic Ridge and is characterised by an active volcano that has generated the local geological structures. Its geographical position gives it an important role as an observation site in the Global network of meteorological stations, and several considerations warrant that more general research efforts should be made in Jan Mayen Island waters and on the island itself. However, the area is rather inaccessible to research vessels in winter, due to the formation and drift of sea ice. The large distance from European harbours also makes research efforts in summer rare and brief.

Norwegian authorities have considered terminating the Loran-C navigation service that is stationed on the island. The intention behind the present NATO ARW was to perform a feasibility study on whether vacant infrastructure on Jan Mayen Island may be taken into use as an international research station for multidisciplinary purposes. To identify scientific problems that could be addressed by research activities on Jan Mayen Island and in the oceanic region around the island, the participants were carefully selected according to their scientific training and experience. Some were highly experienced from previous fieldwork in polar regions and able to address particular Arctic problems that could be solved by adding observations from the Jan Mayen area, or that were unique in an Arctic context. Others were selected from their particular expertise, and were asked to identify more universal scientific problems that could improve the understanding within their scientific field by making observations on or at the island. In their talks the invited specialists identified problems and observation methods that were potentially applicable to Jan Mayen Island research.

3. RESULTS FROM THE PROCEEDINGS

Following group work and reports given in plenary sessions, the NATO ARW participants came up with some recommendations for the nearest few years ahead:

1. After the termination of the Loran-C navigational system in 2005, the infrastructure should be maintained for some years to allow for:
    a. Historical documentation of the facilities.
    b. Use of buildings for scientific purposes, allowing for minor modifications.

2. In the summers of 2005 or 2006, the existing infrastructure should be used to perform a broad census of the natural and anthropogenic features on the island and in its sublittoral zone.
3. In the 4th International Polar Year (IPY 2007-08), buildings and other infrastructure should serve year-round use by international cooperative science.

In a longer perspective, following an IPY 2007-2008 campaign or expedition to Jan Mayen the participants suggested some considerations for future actions:

1. Assess the need for modernization or reconstruction of facilities.
2. Upgrade the airfield and its instrumentation for more regular exchange of personnel and transportation of supplies.
3. Consider the use of unconventional aviation technology for transportation and observations, like air ships, air planes for vertical or short take-off and landing, and parachuting. Helicopters, hovercrafts and snow-mobiles have already proved their worth under Arctic conditions.
4. Conservation of the nature on Jan Mayen Island require logistics that remove refuse from scientific activities. Devices for compressing, packing and safe transportation to the Norwegian mainland, will be needed.
5. In Bodø, Norway, warehouse logistics will be needed for storing of supplies and gear for research groups making periodic visits to Jan Mayen Island, before transportation by air planes or ships.
6. To reduce unnecessary and frequent transportation of scientific gear to and from Jan Mayen Island, some basic infrastructure, general equipment, and instruments should be permanently based and serviced on the island.
7. To maintain and repair research station facilities, and automated or remotely operated instruments and robots, a permanent staff of about five research technicians should be established. Two teams should rotate on half-year terms. Considering five-year contracts to build up qualifications to a sufficient level, the turnover exchange would be one technician every year.
8. Considerable efforts should be put into safety precautions, taking into consideration the volcanic activity and especially, the strong winds and the low visibility due to fog and drifting snow and sand.
9. During summer, there will be a need for a medium-sized ship for research observations at sea, transportation of land parties, and as a security precaution for the stationed personnel. It may also serve local Coast Guard functions related to nature protection and unauthorized landings on the island. The ship should be able to operate oceanographic instruments and samplers, and perform

fishing, maintenance of drifters and moored installations. It should have sufficient fuel capacity to operate for about two months and have considerable capacity for carrying cargo to and from the island.

4. CONCLUDING REMARK

The group of scientists who took part in the NATO ARW unanimously expressed their interest in the possibility of doing research at a Norwegian station that could serve international science programs. It would especially allow the small nations with little polar research experience to take part in Arctic research at a location of great importance to North Atlantic resource management and Global Climate Change.

Index